Civil and Architectural Engineering

Civil and Architectural Engineering

Edited by **Seth Royal**

$\mathcal{CL}$ LANRYE
INTERNATIONAL

New Jersey

Published by Clanrye International,
55 Van Reypen Street,
Jersey City, NJ 07306, USA
www.clanryeinternational.com

Civil and Architectural Engineering
Edited by Seth Royal

International Standard Book Number: 978-1-63240-528-9 (Hardback)

Printed in the United States of America.

Contents

Preface

This book discusses the fundamental as well as modern approaches of civil and architectural engineering. It unfolds the innovative aspects of this field. Civil engineering is a field of engineering that deals with the construction of buildings, roads, hospitals, parks, bridges, dams and in-turn establishing and developing a new society. Architecture and civil engineering go hand in hand as both deal with planning and constructing buildings and other infrastructures. This book elucidates the concepts and innovative models around the prospective developments with respect to this field. It strives to provide a fair idea about this discipline and to help develop a better understanding of the latest advances within this field. It is appropriate for students seeking detailed information in this area as well as for experts, civil engineers and architects.

The researches compiled throughout the book are authentic and of high quality, combining several disciplines and from very diverse regions from around the world. Drawing on the contributions of many researchers from diverse countries, the book's objective is to provide the readers with the latest achievements in the area of research. This book will surely be a source of knowledge to all interested and researching the field.

In the end, I would like to express my deep sense of gratitude to all the authors for meeting the set deadlines in completing and submitting their research chapters. I would also like to thank the publisher for the support offered to us throughout the course of the book. Finally, I extend my sincere thanks to my family for being a constant source of inspiration and encouragement.

Editor

Finite Element Modeling and Nonlinear Analysis for Seismic Assessment of Off-Diagonal Steel Braced RC Frame

Keyvan Ramin[1],*, and Mitra Fereidoonfar[2]

Abstract: The geometric nonlinearity of off-diagonal bracing system (ODBS) could be a complementary system to covering and extending the nonlinearity of reinforced concrete material. Finite element modeling is performed for flexural frame, x-braced frame and the ODBS braced frame system at the initial phase. Then the different models are investigated along various analyses. According to the experimental results of flexural and x-braced frame, the verification is done. Analytical assessments are performed in according to three dimensional finite element modeling. Nonlinear static analysis is considered to obtain performance level and seismic behaviour, and then the response modification factors calculated from each model's pushover curve. In the next phase, the evaluation of cracks observed in the finite element models, especially for RC members of all three systems is performed. The finite element assessment is performed on engendered cracks in ODBS braced frame for various time steps. The nonlinear dynamic time history analysis accomplished in different stories models for three records of Elcentro, Naghan and Tabas earthquake accelerograms. Dynamic analysis is performed after scaling accelerogram on each type of flexural frame, x-braced frame and ODBS braced frame one by one. The base-point on RC frame is considered to investigate proportional displacement under each record. Hysteresis curves are assessed along continuing this study. The equivalent viscous damping for ODBS system is estimated in according to references. Results in each section show the ODBS system has an acceptable seismic behaviour and their conclusions have been converged when the ODBS system is utilized in reinforced concrete frame.

Keywords: FEM, seismic behaviour, pushover analysis, geometric nonlinearity, time history analysis, equivalent viscous damping, passive control, crack investigation, hysteresis curve.

1. Introduction

Since, the experimental researches are very expensive and time-consuming (Altun and Birdal 2012), the application of computer modeling methods as initial investigation and also in the next step, calibration of present computer models with a similar previous experimental research can be a certainty. In this study, the finite element model is calibrated in the first step for flexural frame and x-braced frame. Then the main model is simulated according to verified characteristics of the last step's model. The FE model is considered to different analysis and design.

The design of seismic resistant structures in seismic regions should satisfy two criteria. First, under frequent and low to moderate earthquakes, the structure should have sufficient strength and stiffness to control deflection and prevent any structural damage. Second, under rare and severe earthquakes the structures must have sufficient ductility to prevent collapse (Roeder and Popov 1977).

Reinforced concrete structures usually have dual behaviour against lateral loads, first the behaviour before cracking (pre-cracking) by limited resistant and the other behaviour after cracking (post-cracking) by increasing the ductility. Although the reinforced concrete behaviour of post cracking stages is complicated and extended by multi steps (elastic, yielding, elastic-perfectly-plastic (EPP), plastic and collapse), merely whatever is obvious in all steps after cracking is the high amount of ductility proportional to the other stage.

So, many kinds of lateral load bearings have been used in steel and Reinforced Concrete (RC) structures for recent years, which contain useful performances.

The important thing that should be mentioned here is if the additional system to RC frame being the occasion of imperfect energy absorption in reinforced concrete members, then some part of the structure will not contribute in energy dissipation and actually this system's application has no economic advantages. For example in a reinforced concrete x-braced frame, just upon the imposing lateral loads, the diagonal members of bracing system make a directional component with lateral loads and so they experience a high

[1]Structural & Mechanical Department, Advance Researches & Innovations, Aisan Disman Consulting Engineers Inc., 6718783559 Kermanshah, Iran.

*Corresponding Author; E-mail: keyvan_ramin@yahoo.com

[2]Department of Structural Engineering, Asians Disman Consulting Engineers, Kermanshah, Iran.

percent of those lateral load. Now if the lateral loads be increased, the axial plastic hinge will be formed in steel bracing members before the formation of flexural plastic hinge in RC members. By continuing the imposed lateral load on the x-braced RC frame, the limited nonlinear behaviour is started in term of large deformation in braced frame members. Based on the high capacity of energy absorption in diagonal steel bracing members proportional to reinforced concrete members, the RC frame will be collapsing in possible minimum time after occurrence the large displacement and collapse in diagonal steel members and it won't let reinforced concrete members to make plastic hinge in a short time domain. So this phenomenon will not let the reinforced concrete members being contributed in energy dissipation and on the other hand the plastic limit is abbreviated along this short time for the load transferring.

The off-diagonal bracing system induces new properties of the reinforced concrete frame. The different performances of the RC frame braced by off-diagonal bracing system (ODBS) are about the each member's opportunity and possibility for the formation of plastic hinges.

The specific geometry of the ODBS, one of the steel bracing members absorb the amount of energy until, cracking extend to RC members. Additional to large deformation in third member of ODBS, the cracking is increased in RC frame, even may be observed the frame's plastic hinges. While the other steel diagonal members, being oriented in parallel form, it can be a confident fuse to prevent declining the lateral resistance of the RC frame.

The preceding reasons demonstrate the ODBS braced frame has two stages behaviour for the elastic, elasto-plastic and plastic treatments. The extended amount of energy dissipation for ODBS system is not only because of the inherent nonlinear properties of materials but also the particular nonlinear geometry is effective on damping and energy dissipation and anyhow presents a particular system of passive energy dissipative.

The two basic requirements for seismic design are high stiffness at working load level and large ductility at severe over loadings. These requirements are difficult to be satisfied when the above conventional frames are used. On the contrary, Eccentrically Braced Frames offer an economical framing system satisfying both requirements.

In all types of this system, the vertical components of axial forces in the braces are held in equilibrium by shear and bending moments in short beams of lengths, which is the active links. Active links are designed to remain elastic at working loads and deform inelastic on over loading of structure, thereby dissipating large amount of energy. In this system the hazardous brace buckling can be entirely prevented since the link acts as fuse to limit the brace axial force. Also this frame has a much greater lateral resisting capacity than that of an MRF if the beam section used are the same (Mastrandrea and Piluso 2009a, b; Mastrandrea and Piluso 2009a, b). On the other hand in ODBS system the third member of bracing has similar treatment to active link but by another mechanism.

2. Aims and Objectives for this Research

The main aim of this research is related to investigation of ODBS system's behaviour. Corresponding to the title, in this assessment focused on seismic behaviour of ODBS braced RC frame under real registered earthquake records and also the spectral forces parallel to exact modeling by finite element method (FEM). FEM models were developed to simulate various RC frames with and without steel bracing systems of three full size frames for nonlinear response up to collapse, using the ANSYS program (ANSYS 2015). Then models verified for several analysis and investigation. As it is known to us, a considerable impact load induced at difference modes of vibration through earthquake and exerted on bracing components if the direction of forces change and components stretch under the influence of components' buckling caused by pressure (Ravi Kumar et al. 2007). This study expects that the amount of lateral forces being transmitted from earth to upper levels, subsequently the effect of impact will decrease as a result of using ODBS, the high energy absorption capacity, in lower floors of the structure. Time history analysis is done for high rise models by different properties. Also several comparisons assessed for applicable results and Conclusions from the current research efforts and recommendations for future studies are included.

3. Research Background

Basis on collapse prevention and life safety performance levels, a structure has to experience the large inelastic deformations in term of large capacity of energy dissipation during any excitation. Actually, the reason of structural stability of a system under any inelastic earthquake load is the condition of hysteretic loops, which means the stability of structural system depends on stability of hysteresis curves in each cycles. Such stable loops of a cyclic load or time history acceleration under an earthquake load can be a provision of sufficient ductility and large amount of energy dissipation for structural system's element (Khatib et al. 1988; Asgarian et al. 2010).

For high and medium rise buildings, structural steel has been used extensively due to its high strength and ductile properties. In general, bracing systems are divided into two general types: concentric and eccentric (Ghobarah and Abou Elfath 2001; Kim and Choi 2005; Moghaddam et al. 2005). Concentric braced systems are more desirable because of a relative high stiffness, along with their easy construction and economy aspects; hence these important criteria make this type more common than eccentrically braced frames (Davaran and Hoveidae 2009).

Eccentric braces need more construction accuracy thereby resulting in a decrease of construction speed and higher cost in spite of better stiffness performance and higher energy dissipation (Özhendekci and Özhendekci 2008; Bosco and Rossi 2009).

This system allows the architects to have more openings in panel areas (Moghaddam and Estekanchi 1995; Moghaddam and Estekanchi 1999). Moreover, because of the cyclic nature of seismic loads, these brace elements are designed symmetrically and so they should perform in two span to work symmetric.

The idea of steel bracing system application in reinforced concrete structures was first suggested for seismic strengthening of concrete buildings. From the viewpoint of both research and application, this idea has been very prevalent during past two decades because of its simplicity, implementation and its lower relative cost compare with shear wall. For example, Sugano and Fujimura performed a series of experiments on a model of one-story frame which had been strengthened through various methods. They examined x-bracing and k-bracing systems and compared them with samples strengthened by concrete and masonry in-filled walls. They aimed to determine the effect of each system on enhancement of in-plane strength and ductility of the samples (Sugano and Fujimura, 1980). Furthermore, Kavamata and Ohnuma demonstrated the possibility of the effective use of steel bracing systems in concrete buildings (Kawamata and Ohnuma 1981).

A model of two spans, two stories reinforced concrete frame in scale of 1:3 was chosen to represent the seismic weakness and behaviour. The strengthened frame was exposed to lateral and gravity loads and its displacements were allowed to increase by one fiftieth of the frame's Original height (inter-story drift). The strengthened inside frame by a ductile steel brace, demonstrated better behavior considerably than the preliminary reinforced concrete frame (Masri and Goel 1996) or applied from outside the frame (Bush et al. 1991).

In 1999, the direct internal use of steel bracing system in concrete frame was studied in laboratory. Experiments were carried out on five one span, one story frame samples with a scale of 1:2.5. Two of them had no bracing system but the other three samples strengthened by x-bracing system with different connector's component including bolt and nut, cover of RC column, and gusset plates placed in concrete. Results showed, depending upon various connectors' component, the bracing system considerably increases the equivalent stiffness of the frame and notably changes its behavior. When the bracing system's connector is implanted inside concrete, the performance of frame gets even better and further energy is absorbed. Generally, experiments demonstrated that the bracing system tolerates a major part of lateral load in reinforced concrete frame (Tasnimi and Masoumi 1999).

Dynamic behavior of the concrete buildings strengthened with concentric bracing systems has investigated by Abou Elfath and Ghobarah. A three story building was dynamically analyzed with various earthquake records and the effect of steel bracing system on building as well as the effect of bracing system distribution throughout frame's height was studied. The position of braces investigated the proportional seismic performance, inter-story drift, and damage index to show the effect of this type of bracing systems (Abou Elfath and Ghobarah 2000).

Maheri et al. (2003) first reviewed previous studies on strengthening by steel bracing systems and then investigated three models including a simple frame, a frame strengthened with x-bracing, and a frame strengthened with knee bracing system under lateral load until failure stage. They found that ductility of RC frame is considerably increases when using knee bracing system (Maheri and Akbari 2003).

In 1994, Moghaddam and Estekanchi modeled and tested Off-Centre bracing systems in steel frames for the first time. Later in 1999, they analyzed the seismic behavior of Off-Diagonal bracing system. They confirmed that this system's behavior resembles with seismic isolators and play a considerable role in reduction of seismic forces (Moghaddam and Estekanchi 1999, Moghaddam and Estekanchi 1994). All previous studies confirm the effectiveness of steel braces in rehabilitating and retrofitting of RC frame.

4. Finite Element Modeling

4.1 Models Characteristics

The flexural RC frame, the basic frame, was designed in according to ACI 318-02 concrete manual. In finite element models used scale of 2/5 same as the scale factor of experimental models and according to other geometric characteristics. The dimensions of 1.76 m for length, 1.38 m for height and 0.16×0.14 m of rectangular section dimensions is considered through the FE modeling of flexural frame. In both of the beam and column reinforcements, 4M10 are assigned for longitudinal reinforcement (Rebar's diameter of M10 is equal to 11.3 mm). In the plastic hinge regions (350 mm from each end), the transverse reinforcement of beams and columns used from 6 mm steel wires spaced at 35 mm and for other places far from plastic regions, the 6 mm steel wires spaced at 70 mm are used. The beam–column joint was transversely reinforced with two 6 mm wires corresponding to lateral shear force. Also in the x-braced RC frames, the two steel plates of $150 \times 150 \times 8$ mm were placed at each corner of the RC frames by cast in place method (Youssefa et al. 2007). Each plate was anchored to the RC frame using four 5/8-inch headed studs as shown in Fig. 1. Self-consolidated concrete with 28 days compressive strength of 40 MPa is used to concrete compressive stress of models. The yield stress of the steel reinforcement applied as 400 MPa. A brace by double-angle cross-section, consisting of two $25 \times 25 \times 3.2$ mm angles, by a cross-sectional area of 300 mm^2, is selected for the flexural frame F_1 and also a $C\ 30 \times 3.5$ mm channel with a cross-sectional area of around 500 mm^2 is selected for the frame F_X, as indicated in Fig. 1. The yielding capacity of 300 MPa is considered for the bracing members. The braced frame by off-diagonal bracing system (F_{ODBS}), the RC cross sections and semi-diagonal steel braces are such as characteristics of F_X model (x-braced frame) and their difference is only about third member of steel bracing system.

The beam–column joint of the moment frame was transversely reinforced with two 6 mm steel wires according to the special seismic provisions of the ACI

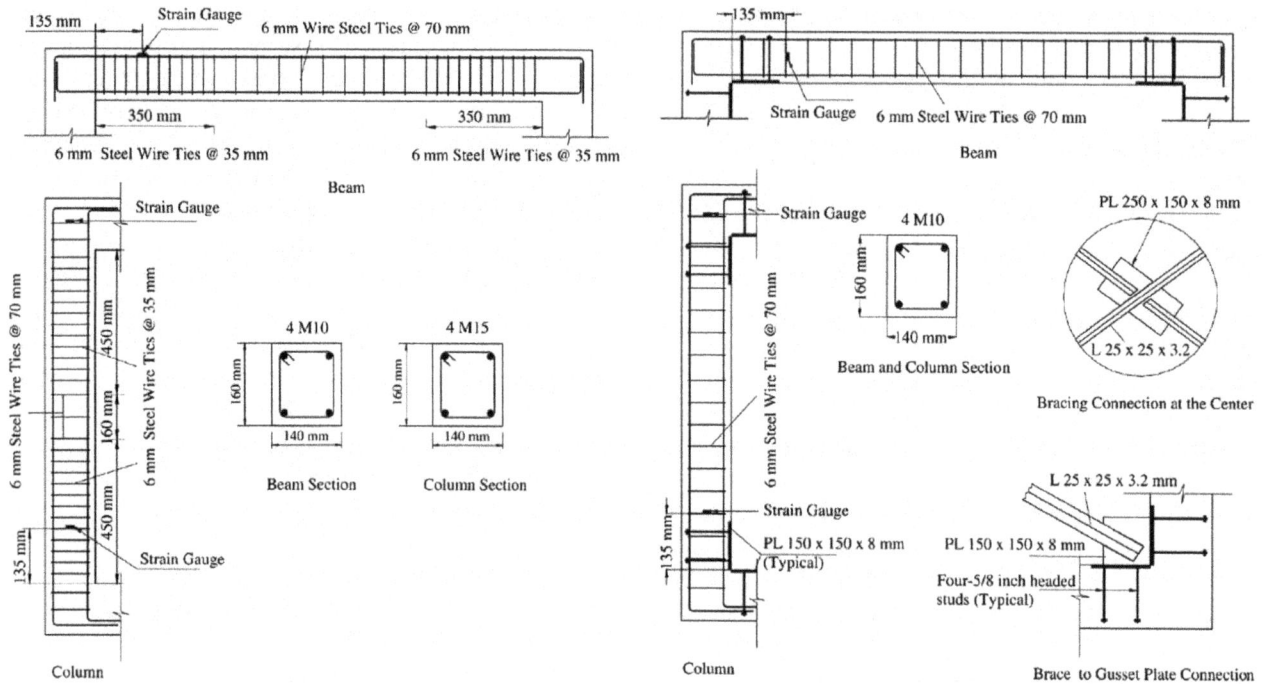

Fig. 1 Experimental characteristics of structural elements (Maheri et al. 2006; Youssefa et al. 2007), for F_1 model, flexural reinforced concrete frame (*left*) and for F_x model, x-braced reinforced concrete frame (*right*) to use for FE models and calibration process.

code (ACI Committee 318 2002). For the braced frame, the column stirrups were continued in the joint by one 6 mm wire in the joint area. Thus, the additional strength of gusset plates, during the braced frame, the beam–column joints inside the stiffeners are expected to eliminate the undesirable shear failure without the need for any special joint detailing. However, this needs further testing to reach final recommendations (Youssefa et al. 2007). Comparison results of Pushover analysis indicate that Off-Diagonal Bracing System can increase lateral flexibility by velocity control or on the other hand by increasing Damping characteristics. In X-Bracing system after yielding steel members, the strength of RC frame was decreased suddenly but in ODBS model, the steel and concrete materials have been contributed better than other models. The schematic finite element model of ODBS and its behaviour are brought in Fig. 2.

4.2 Material Nonlinearity

Concrete and steel are the two constituents of RC braced frame. Among them, concrete is much stronger in compression than in tension (tensile strength is of the order of one-tenth of compressive strength). While its tensile stress–strain relationship is almost linear, the stress–strain relationship in compression is nonlinear from the beginning. The Elastic-Perfectly-Plastic (EPP) model for steel, which is used in this work, assumes the stress to vary linearly with strain up to yield point and remain constant beyond that (Anam and Shoma 2011).

In this research the Willam and Warnke (1974), the yield and failure criteria, is considered for concrete model behaviour. Also, since the SAP2000 (2010) assumption applies the Drucker–Prager criteria for concrete material modeling and its behaviour, both of mentioned criteria's are considered in analysis of models. By this method, the analytical comparison of applied criteria's is done.

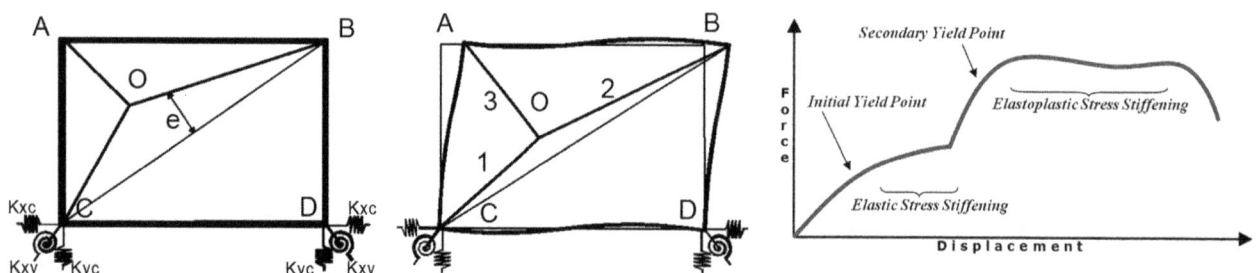

Fig. 2 (*Left*) Schematic models for off-diagonal bracing system; (*Right*) Schematic force–displacement diagram by two yield points and ductile behaviour to high energy dissipation.

In steel material modeling, the bilinear curve of behaviour is used. This model is included in two parts, linear and Elastic–Plastic behaviour. The elasticity modulus is $E_1 = 2 \times 10^6$ kg/cm^2 for linear part and $E_2 = 2 \times 10^4$ kg/cm^2 for the nonlinear part of the behaviour. These specifications are indicated in Fig. 3.

On the other hand, the uniaxial stress–strain relationship for confined concrete, known as the modified Kent and Park model, has been incorporated in the FE model constructed here. This model shows a good agreement with the experimental results (Kent and Park 1971; Scott et al. 1982) and offers a good balance between simplicity and accuracy (Taucer et al. 1991).

In non-linear dynamic analysis, the non-linear properties of the structure are considered as part of a time domain analysis. This approach is the most rigorous, and is required by some building codes for buildings of unusual configuration or of special importance. However, the calculated response can be very sensitive to the characteristics of the individual ground motion used as seismic input; therefore, several analyses are required using different ground motion records to achieve a reliable estimation of the probabilistic distribution of structural response.

Since the properties of the seismic response depend on the intensity, or severity, of the seismic shaking, a comprehensive assessment calls for numerous nonlinear dynamic analyses at various levels of intensity to represent different possible earthquake scenarios. This has led to the emergence of methods like the Incremental Dynamic Analysis (Bozorgnia and Bertero 2004).

4.3 Geometric Nonlinearity

For specification of nonlinear geometry of ODBS, concrete nonlinearity is added to material nonlinearity in this paper. Only steel nonlinearity for third member of ODBS is considered in this article's analysis. Steel, on the other hand, is linearly elastic up to a certain stress (called the proportional limit) after which it reaches yield point (f_y) where the stress remains almost constant despite changes in strain. Beyond the yield point, the stress increases again with strain (strain hardening) up to the maximum stress (ultimate strength, f_{ult}) when it decreases until failure at about a stress quite close to the yield strength.

Nonlinear static procedures use equivalent SDOF structural models and represent seismic ground motion with response spectra. Story drifts and component actions are related subsequently to the global demand parameter by the Pushover or capacity curves that are the basis of the nonlinear static procedures. Nonlinear dynamic analysis utilizes the combination of ground motion records with a detailed structural model, therefore is capable of producing results with relatively low uncertainty. In nonlinear dynamic analyses, the detailed structural model subjected to a ground-motion record produces estimates of component deformations for each degree of freedom in the model and the modal responses are combined using schemes such as the sum of squares square root.

Complete comparisons of the studied Retrofitted Frames in ANSYS program (version 15) with the Micro modeling structural element indicate that ODBS steel bracing RC frame has two yielding point that were related to main RC flexural frame and third steel member of ODBS. It's so useful for structures that are under Impact Loads and loads by high velocity specifically according to Fig. 2 of the previous page.

5. Finite Element Models

5.1 Finite Element Modeling Procedure

After doing the consecutive try and error, the possibility of FE models creation is provided for two models, flexural and X-bracing systems. Results verification is performed by more than 95 % of convergence accuracy based on two models. After this double precision, comparative models were found the best parameters for simulation processing. Some parameters same as materials definition, element type, elastic constants, nonlinear properties, contact surface and other structural characteristics, are contributed in ODBS modeling and processing. The finite element analysis is performed in term of various investigations to obtain the results after ODBS modeling. As indicated in Fig. 4d, other verification is based on cracking analysis for two comparative models in RC frame by comparing the experimental results; the created cracks of two models were being as same as each other. The 3-D finite element deformed shapes are given in Fig. 4a–c for each model.

The main Flexural RC frame is calibrated by results of experimental modeling of the same flexural frame and X-

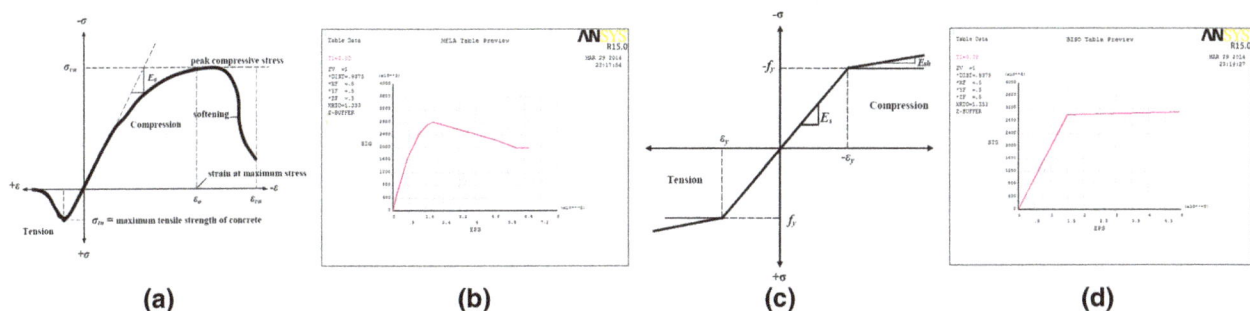

(a)   (b)   (c)   (d)

Fig. 3 Materials model properties, **a** parametric model for concrete; **b** concrete model definition in present research; **c** parametric model for steel; **d** steel model definition in present research.

(a)

(b)

(c)

(d)

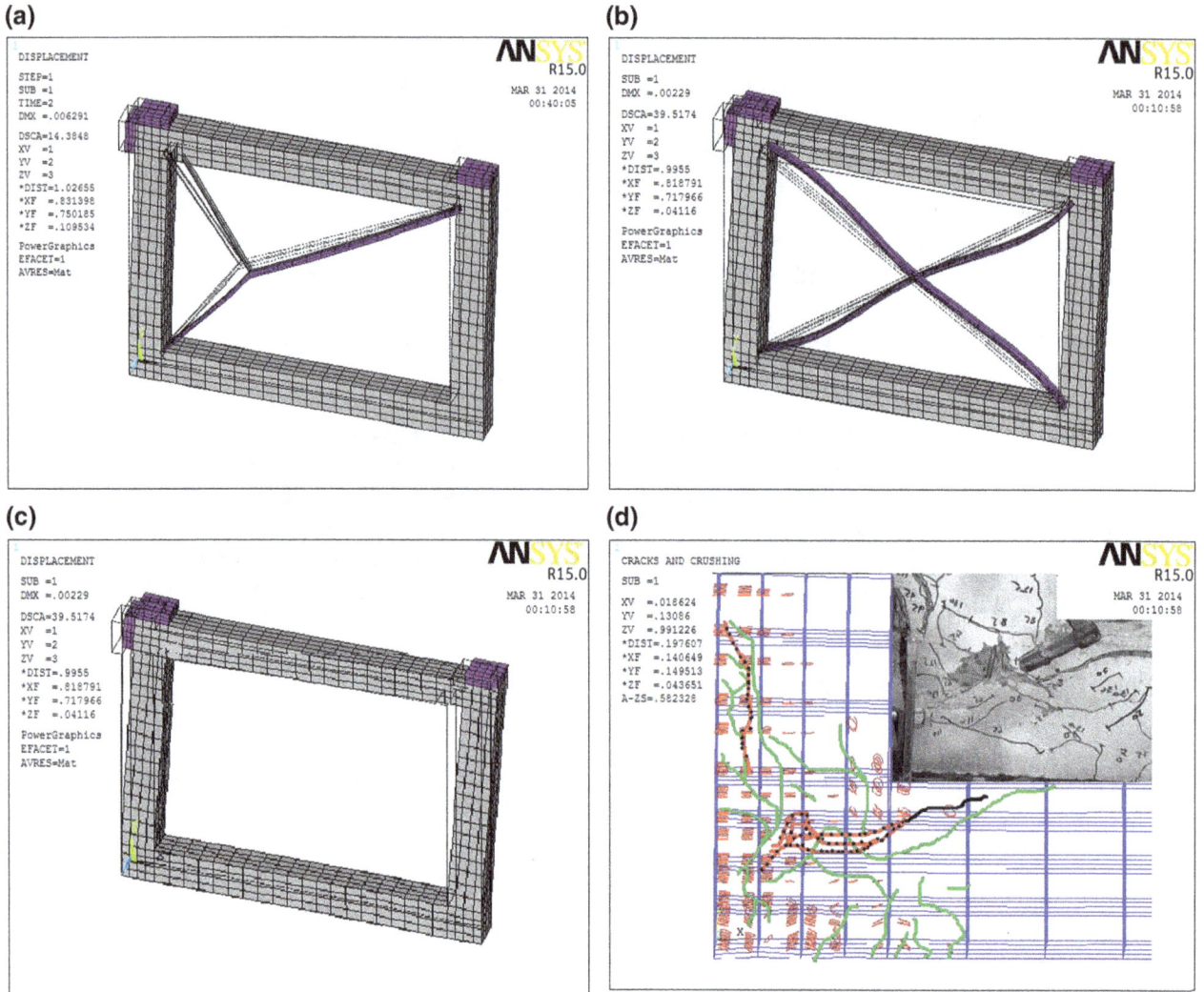

Fig. 4 Numerical model properties, **a** ODBS deformed shape; **b** X-Braced deformed shape; **c** flexural frame deformed shape and **d** flexural frame crushing comparison.

Braced frame that constructed in laboratory (Maheri and Hadjipour 2003).

5.2 Comparison Between FE and Experimental Results

As explained before, after performing the finite element 3-D modeling, for the next phase, the models were being analyzed and some results about them are obtained. The important status is concerned about results calibration for both of comparative models, flexural and X-braced frames where their results are indicated in Fig. 5, in the form of pushover diagrams calculated by nonlinear static analyses on reinforced concrete frame. The results are converged within the acceptance criteria for experimental and numerical investigations. As shown in Fig. 5, the flexural frame convergence is more than the x-braced frame, because in the existence of approximately modeling proceed in various elements such as gusset plates, expected length of reinforcement, steel brace section areas and also the other approximation are related to defining and modeling contact element. According the nonlinear static analysis, compared results of experimental and numerical models, have shown by Force–Displacement diagram and related control. Numerical

modeling is base on exact modeling and Real parameters. This diagram shows that errors in modeling were minimized and convergence condition was satisfied.

As consequences from the geometric properties of ODBS, a diagonal eccentricity is the one important parameters affected on this system's behavior. So the different kinds of models are simulated and tested by various eccentricities. Optimum eccentricity from the viewpoint of geometric characteristics of ODBS is about 0.3–0.4 for e_1 and 0.5 for e_2. The e_2 eccentricity is concerned about geometric properties, position of members, angle of braces, height and span dimensions and etc. Increasing ductility is directly in relation with increasing eccentricity but some ingredients such as allowable drift, crack width and deflection are from limitations applying eccentricity e_1. However if the structural purpose be proportional to large amount of stiffness, the off-diagonality should be decreased, but for the purpose of high ductility it was inverse. Some comparative results of possible eccentricities are indicated in Fig. 6.

Figure 7 illustrates a typical mode of failure of the tested reinforced concrete frame with and without bracing systems

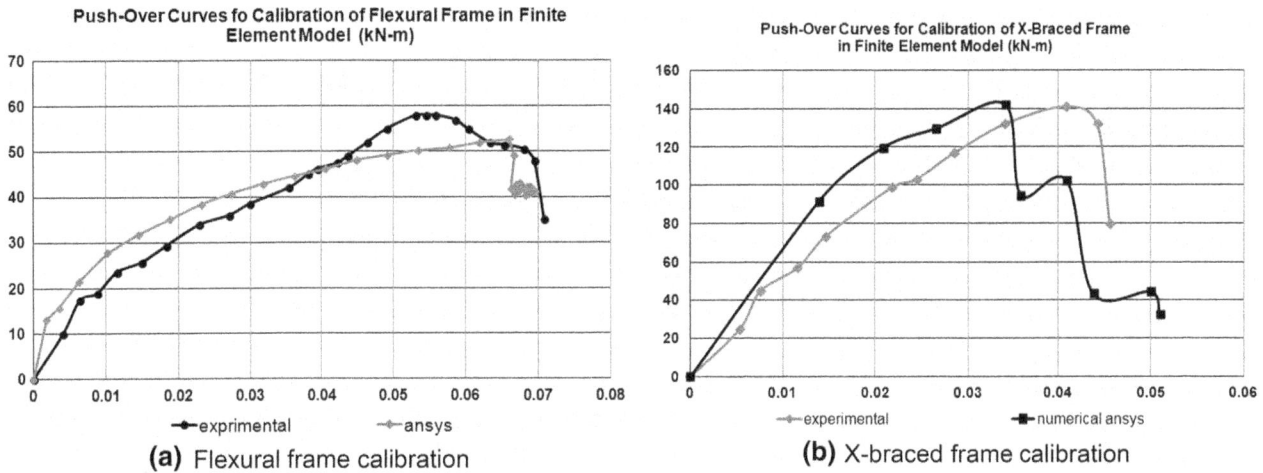

(a) Flexural frame calibration

(b) X-braced frame calibration

Fig. 5 Comparison Flexural frame calibration (**a**) and X-braced frame calibration (**b**) deformed Frames Pushover curves for performing calibration process between ANSYS model's obtained results and experimental results.

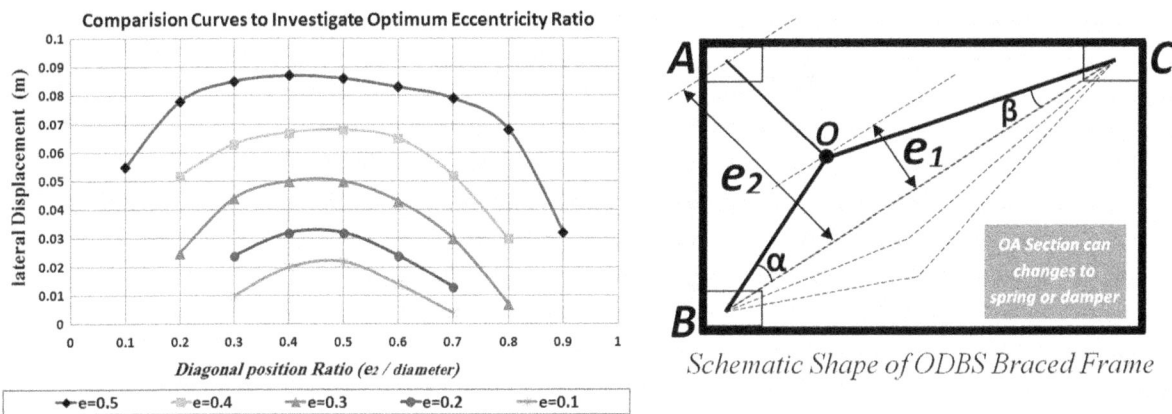

Fig. 6 Obtained lateral displacement in term of different eccentricities for Off-Diagonal bracing system (*left*) and parametric shape of ODBS system (*right*).

under static push-over loading. A typical deformed shape from the FE model is also provided in right side of figures, which perfectly matches the actual mode of failure.

Figure 7 shows a front view of this modeled RC frame. In Fig. 7a and b, the experimental and numerical force–displacement curves are compared for flexural frame and x-braced frame, respectively. Lateral displacements are measured at the base points at the top height of the frames. A comparison between these two figures shows the amount of calibration level in compare with the experimental modeling frames. The specific purpose of 3D finite element modeling is based upon the complex behavior of reinforced concrete nonlinearity and the geometric nonlinearity that applied for this paper.

The response modification factor (*R* factor) is the one of important factors for introducing the structural behavior specially to define ductility and stiffness of structures. To doing this aim, the separated calculations are performed to obtaining the factor of behaviour. Related force–displacement diagrams in form of bilinear pushover curves and the samples of finite element modeling are shown also in left side of Fig. 7.

After generating the pushover curves, the response modification factors are obtained for each model. The highest ductility and the large amount of energy dissipation are from ODBS system results.

5.3 The Response Modification Factor for ODBS Braced RC Frame

According to Fig. 8; Table 1 and by considering Pushover curves, to calculate the response modification factor, some basic formula is needed. As shown in Fig. 8, usually real nonlinear behaviour is idealized by a bilinear elastic perfectly plastic relationship. The yield base shear coefficient of structure is shown by V_y and the yield displacement is Δ_y. In this figure, V_e corresponds to the elastic response strength of the structure. The maximum base shear ratio in an elastic perfectly behaviour is Δ_y (Uang 1991). The ratio of maximum base shear coefficient considering elastic behaviour V_e to maximum base shear coefficient in elastic perfectly behaviour V_y is called force reduction factor.

$$R = R_\mu \cdot R_s \cdot Y, \quad R_\mu = V_e/V_y \tag{1}$$

The overstrength factor is defined as the ratio of maximum base shear coefficient in actual behaviour V_y to first significant yield strength in structure V_s.

$$R_s = V_y/V_s = \Omega \tag{2}$$

The concept of overstrength, redundancy and ductility, which are used to scale down the earthquake forces need to

(a) Force-Displacement curve of Flexural frame and deformed shape and sum of elastic & plastic strain contour.

(b) Force-Displacement curve of X-Braced frame and deformed shape and sum of elastic & plastic strain contour.

(c) Force-Displacement curve of steel ODBS braced frame and deformed shape and sum of elastic & plastic strain contour respectively.

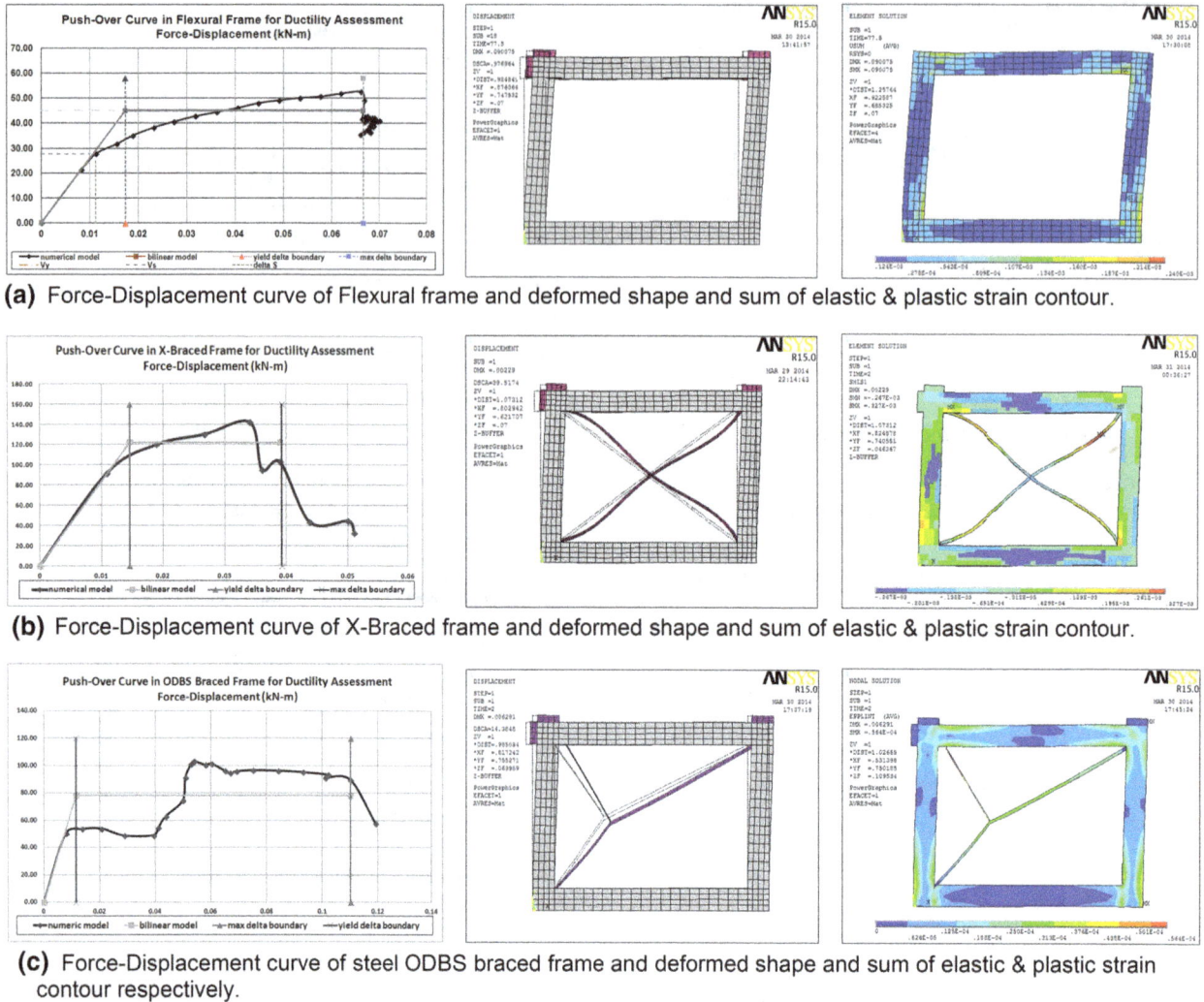

Fig. 7 Pushover Diagrams by specified bilinear equivalent curves in various Types of **a** RC Flexural Frame and **b** X- Steel Braced RC Flexural Frame and **c** ODBS Steel Braced RC flexural Frame (Plus results of simulated models in ANSYS).

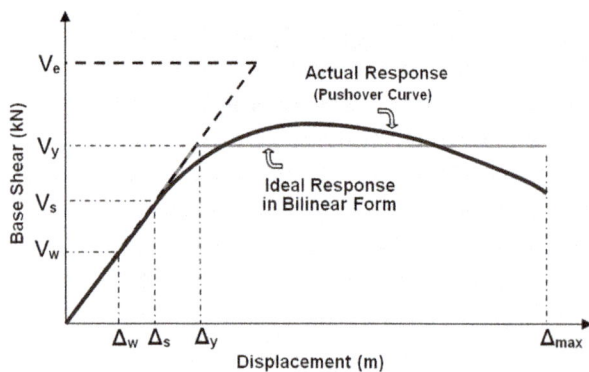

Fig. 8 Response modification factor evaluation along Pushover curve and its equivalent bilinear EPP curve.

be clearly defined and expressed in quantifiable terms. To design for allowable stress method, the design codes decrease design loads from V_s to V_w. This decrease is done by allowable stress factor Y (Uang 1991).

$$Y = V_s/V_w \qquad (3)$$

The range of this factor is about 1.4–1.5. In this paper allowable stress factor Y was considered as 1.4 (ATC 1995a).

$$R(R_w) = (V_e/V_y) \cdot (V_y/V_s) \cdot (V_s/V_w) = (V_e/V_w) \qquad (4)$$

$$R = R_\mu \cdot R_s = (V_e/V_s) \qquad (5)$$

Equation (4), shows the seismic response modification factor (R_w) in ultimate strength design method. Also, Eq. (5), indicates seismic response modification factor in allowable stress design method. Structural ductility, μ, is defined in terms of maximum structural drift (Δ_{max}) and the displacement corresponding to the idealized yield strength (Δ_y), as given in Eq. (6).

$$\mu = \Delta_{max}/\Delta_y \qquad (6)$$

The response modification factor (R) is included of the inherent ductility and ductility and overstrength effects of a structure and the difference in the design methods and limitations about related manual. Also Ductility reduction factor $R\mu$ is a function of both of the characteristics of the structure including ductility, damping and fundamental

Table 1 The maximum drift values (regardless of drift limitations) and ductility and R factor for floor.

Model	Δ_y (m)	Δ_{max} (m)	Δ_s (m)	Δ_w (m)	V_e (kg)	V_y (kg)	V_s (kg)	V_w (kg)	μ	R
(flexural RC frame) F_1	0.0173	0.0665	0.0112	0.067	17,272	4,500	2,875	1,740	3.84	9.92
(X-Braced frame) F_X	0.0103	0.039	0.0058	0.0547	33,034	12,200	9,147	5,480	2.69	6.04
(ODBS-Braced frame) F_{ODBS}	0.0125	0.1101	0.0082	0.0781	26,177	7,800	5,032	3,012	9.58	23.2

period of vibration (T), and the characteristics of earthquake ground motion. Figure 8 explains the schematic behaviour and its corresponding parameters to calculating response modification factor.

Where R_s is the overstrength factor and Y is termed the allowable stress factor (Maheri and Akbari 2003).

The results indicate the highest amount of response modification factor (R) is obtained for ODBS proportional to the other systems. By considering the presented formulation, the obtained results are gathered in Table 1. The R factor for ODBS is same as the system of displacement and vibration control, particularly same as the systems isolator and friction dampers. The author has other research paper about the ODBS innovation considering a new friction dampers substituted by third member. The system of composed damper-spring can be a suitable subject for further researches.

5.4 Crack Evaluation Through Finite Element Models

The next step of ODBS evaluation is concerned about control and compare of the existing and progressive cracks at the different time steps by the various FE models. Before any concluding for ODBS model, at the first, the flexural and X-braced models are compared along their cracking pattern with experimental models. To verifying the numerical model's cracks for this meaning, all happened cracks at the last phase of loading are represented and then they are compared with cracks on experimental model. Verifying assessments are performed on cracks place, directions, cracks width, time of occurrence and their geometry, that, they are matched by experimental results and fortunately they were similar up to 85 %. Figure 9 indicates comparative cracks in flexural and X-bracing frames.

After confirming the compared cracks, for the next stage, the crack analysis is performed on ODBS braced RC frame for further assessments. The ODBS braced frame is analyzed through 94 time steps by displacement control method. The cracking analysis is checked for three times in steps of 28, 65 and 92 from imposed displacement of ODBS model. The reasons for selecting the mentioned time steps along main cracking investigation are at first, effective parameters about crack development and performing plastic hinges along different rotations and at second existed deformation in steel

bracing members concerned about important stage of behavior such as yielding and plastic behavior for steel material.

As observed in Fig. 10 in 28th time step, the lateral displacement level is notated about 23 mm that it was equivalent to imposing 51 kN lateral loads on the ODBS braced frame. Until this step only the flexural cracks were observable and have not shown the major weakness along the structural concrete. Almost the third member is yielded and its behavior is investigated simultaneously with plastic behavior, but entirely does not exhibit any plastic behavior till this stage.

The time step 65, where flexural cracks developed densely and the shear cracks are started in some regions. Along 65th step, the lateral displacement is reached to 37 mm in upper level of the ODBS braced frame that this stage's displacement was equivalent to 101.5 kN in term of lateral force. Most of the longitudinal reinforcements were yielded and the relevant RC section is experienced the nonlinear strain perpendicular to section's surface. Members near the diagonal axis are yielded and also prepared to endure the large strain. The third member of steel bracing system achieved the maximum deformation and its plastic behavior was clearly obvious. And also some points of RC frame are experienced the plastic deformation. Crack pattern of step 65 is indicated in Fig. 11.

According to last considered time step, 92th step, adjacent to the collapse limit, the upper level's lateral displacement of the ODBS braced frame is reached about 100 mm that it was equivalent to 96.43 kN in term of lateral force. The second stage of yielding phenomenon and the second stage of plastic behaviour are created within the steps of 65–92 and maximized the displacement according to the last pages schematic shape. In this stage, sustainable lateral force is lower than the last stage because of plastic behavior happened in both of RC frame and steel ODBS members, conjugately. In this last stage many plastic zones are performed on the RC frame and the plastic strain is observed along three steel members. The third member of ODBS is given necking and lost its strength entirely. Crack pattern of step 92 is indicated in Fig. 12.

Through the stages of consideration, the ODBS braced RC frame is experienced various behaviours as elastic, elastic–plastic, secondary elastic, secondary elastic–plastic and plastic, then it was in threshold of collapse and also bracing

(a)

(b)

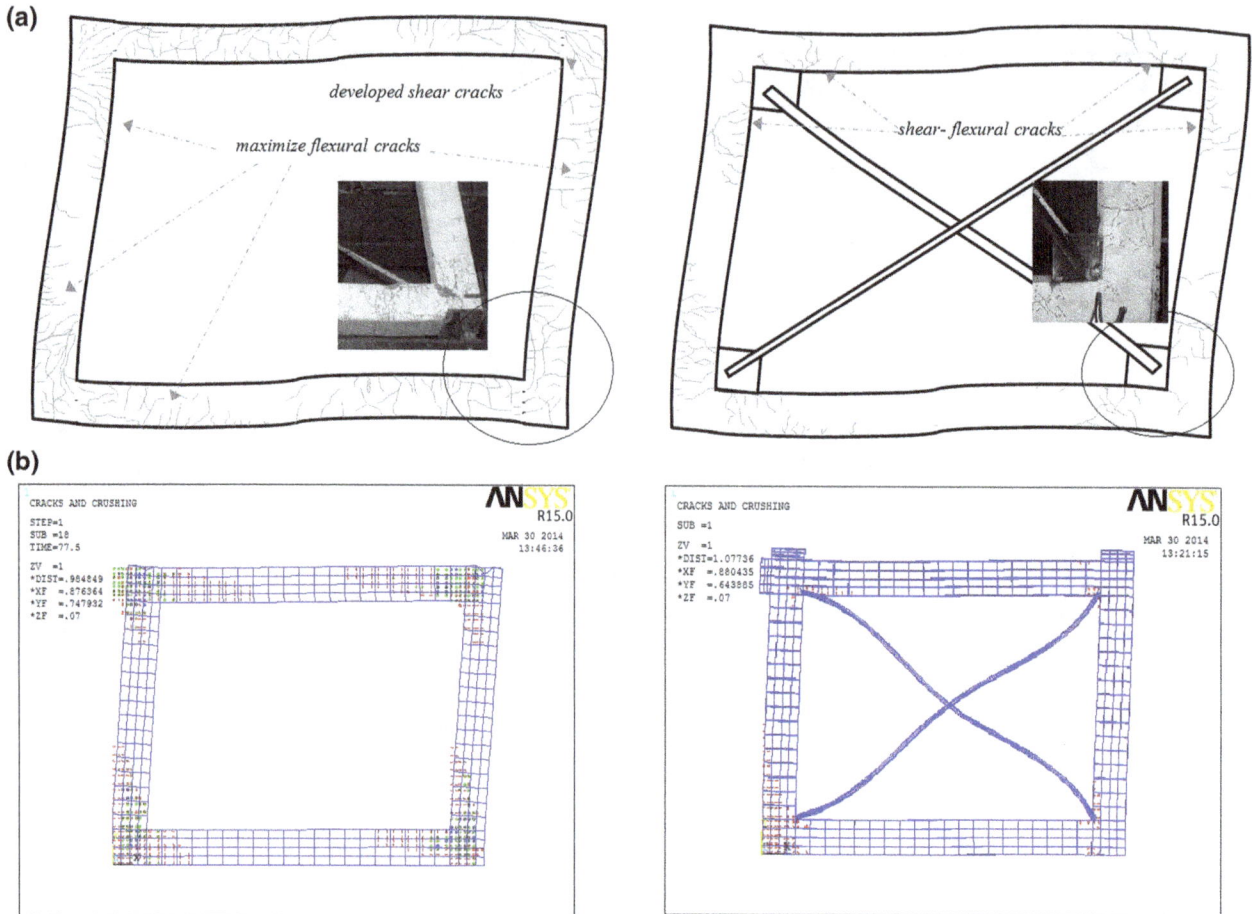

Fig. 9 Comparison between experimental and FE crack patterns for flexural frame (*left*) and x-braced frame; **a** happened cracks in experimental models (Maheri et al. 2006; Youssefa et al. 2007) and **b** happened cracks in numerical models.

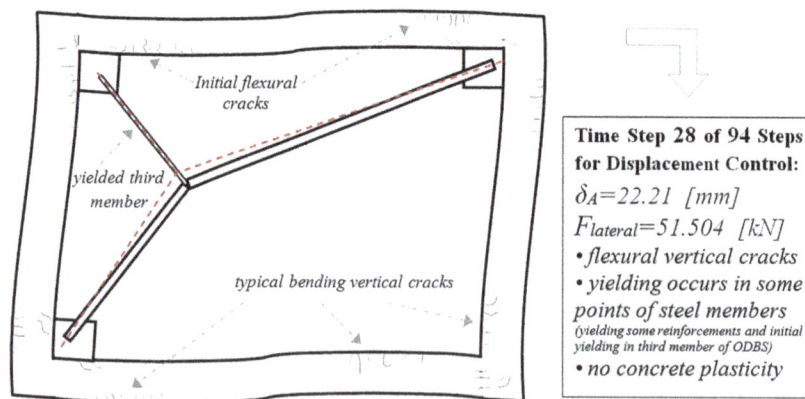

Fig. 10 Flexural and shear cracks development at the front side of the ODBS braced frame, deformed shape and cracks pattern observed at step 28.

members are formed parallel to each other and they were on diagonal axis. In this position, if the diagonal members lost their strength, then the whole frame was being collapsed.

Comparison between ODBS and other frames is not only about high ductility results of ODBS proportional to flexural and specially x-bracing system, but also the ODBS braced frame entirely used from flexural capacity of RC frame.

Whereas the x-braced RC frame did not use more than 20 percent from its flexural capacity. So the energy dissipation in the system of flexural and x-braced frame is lower than the ODBS braced frame. Figure 13 indicates the behaviour of ODBS for time step 28 in ANSYS finite element software. In another figure the plastic behaviour is concerned about concrete strain energy as shown.

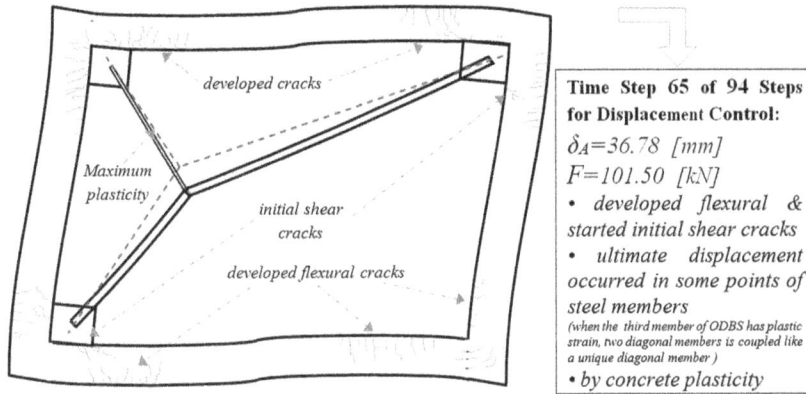

Fig. 11 Flexural and shear cracks development at the front side of the ODBS braced frame, deformed shape and cracks pattern observed at step 65.

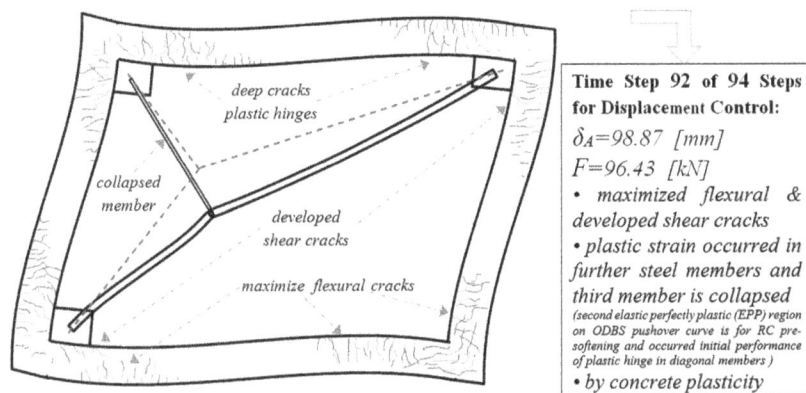

Fig. 12 Flexural and shear cracks development at the front side of the ODBS braced frame, deformed shape and cracks pattern observed at step 92 just before the collapse.

6. Numerical Models

In the next phase, various systems of bracing frame by different stories is studied for further investigation of ODBS compare with the other types of bracing. Figure 14 shows the different models under nonlinear dynamic analysis. Therefore the present model had the greater dimensions and also more stories, then using ANSYS program for these properties simulation was impossible, so for further modeling by raising height, spans and number of stories the SAP2000 (2010) software is used to continuing this research paper for optimizing the time consuming. This software is adapted to advanced analysis in the type of static and dynamic.

Almost, this software is calibrated and scaled by experimental results according to last scaling for ANSYS software. Finally, the results are verified for flexural and x-braced frame as recent models. The results of calibration along the numerical and experimental models are indicated in Fig. 15. By confirming the modeling proceeds in recent software, the pushover curves are drawn in term of force–displacement as indicated. These comparative results explained the convergence accuracy higher than 92 % between recorded numerical results and recorded experimental results.

7. Time History Analysis Methodology

7.1 Time History Records

Nonlinear dynamic analysis for this research has been imposed of three types of frame that mentioned before. Tabas, Naghan and Elcentro, the three scaled records of ground motion are considered for the dynamic time history analysis. Simplified form of these records is evaluated for response spectrum, so performed response spectrum analysis on ODBS to study about modal characteristics. Earthquake characteristics are indicated in Table 2. The load cases are defined at several conditions. Zero condition is concerned about applying the dead load and live load. Secondary condition is concerned about static and dynamic nonlinear analysis.

Macro modeling method was used to analyze the nonlinear behaviour of reinforced concrete frame strengthened by steel bracing system (macro element by lowest accuracy related to micro modeling elements). The model was calibrated using existing laboratory work results and then, larger number of floors and openings were analyzed. The SAP2000 (2010) software was employed to modeling the reinforced concrete frame braced with ODBS system. Dynamic time history analysis is done for modeling high rise concrete frames by

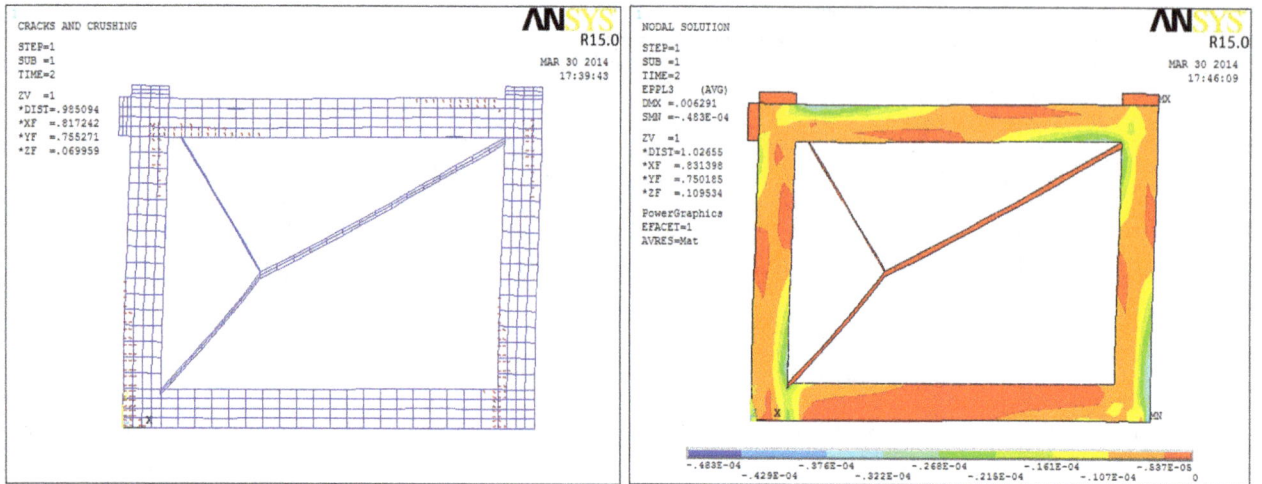

Fig. 13 FE crack pattern of ODBS braced frame obtained in step 28 of imposing displacement (*left*) and plastic strain energy contour along the RC frame (*right*).

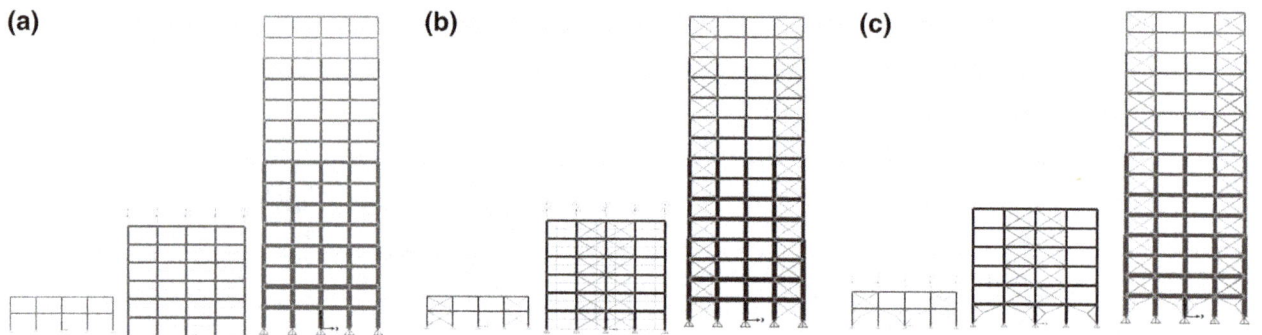

Fig. 14 Numerical models by various load bearing systems used for dynamic analysis by types of two-story, six-story and fifteen-story models respectively from *left to right* for flexural frame (**a**), x-braced frame (**b**) and ODBS braced frame (**c**).

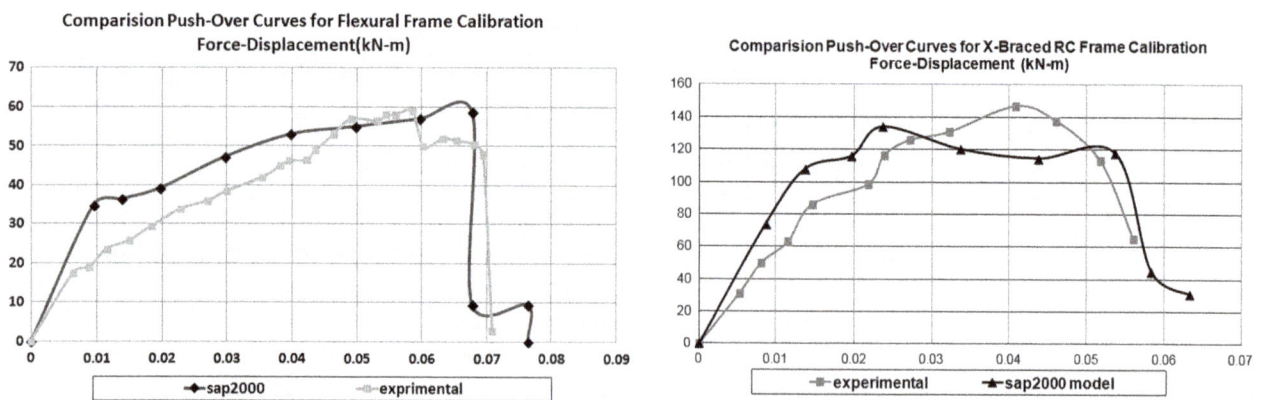

Fig. 15 Convergence between SAP2000 (2010) numerical results and experimental results for verifying the calibration process in term of their pushover curves.

increased height and width. Time history analysis using earthquake accelerograms is one of the suggested methods by most regulations to investigate the seismic behaviour of structures. In this study is used the three accelerograms of Naghan, Tabas and Elcentro. Their general characteristics are listed in Table 2.

The various maximum ground acceleration after scaling is set to 0.3 g. Three groups of records are selected based on two parameters; the closest distance to a fault rupture surface

[greater than 50 km (far field), nearer than 10 km (near fault)] and the moment magnitude in every scales (Berberian 1977 and Jamison et al. 2000). The other characteristics of the real accelerograms such as directivity, fault mechanism, and etc. are the same. The peak ground acceleration of all accelerograms is greater than 0.1 g. These accelerograms are selected from strong ground motion records. The specifications and classification of each group before doing matching procedure are tabulated in Table 2.

Table 2 Characteristics of scaled accelerograms used for time history analysis.

Records	Duration	PGA	Time	Country	Date of	Station	Position	Components	
	(s)	m/sec²	Step(s)	of event	event			Latitude	Longitude
Tabas	50	3.42	0.01	Iran	1978/09/16	Deyhook	33.3'N, 57.52'E	3.27	4.1
Naghan	5	7.09	0.001	Iran	1977/04/06	Central	31.98'N, 50.68'E	7.61	7.61
Elcentro	53.7	3.49	0.01	USA	1940/05/18	E06 array	32.44'N, 115.3'W	3.35	4.03

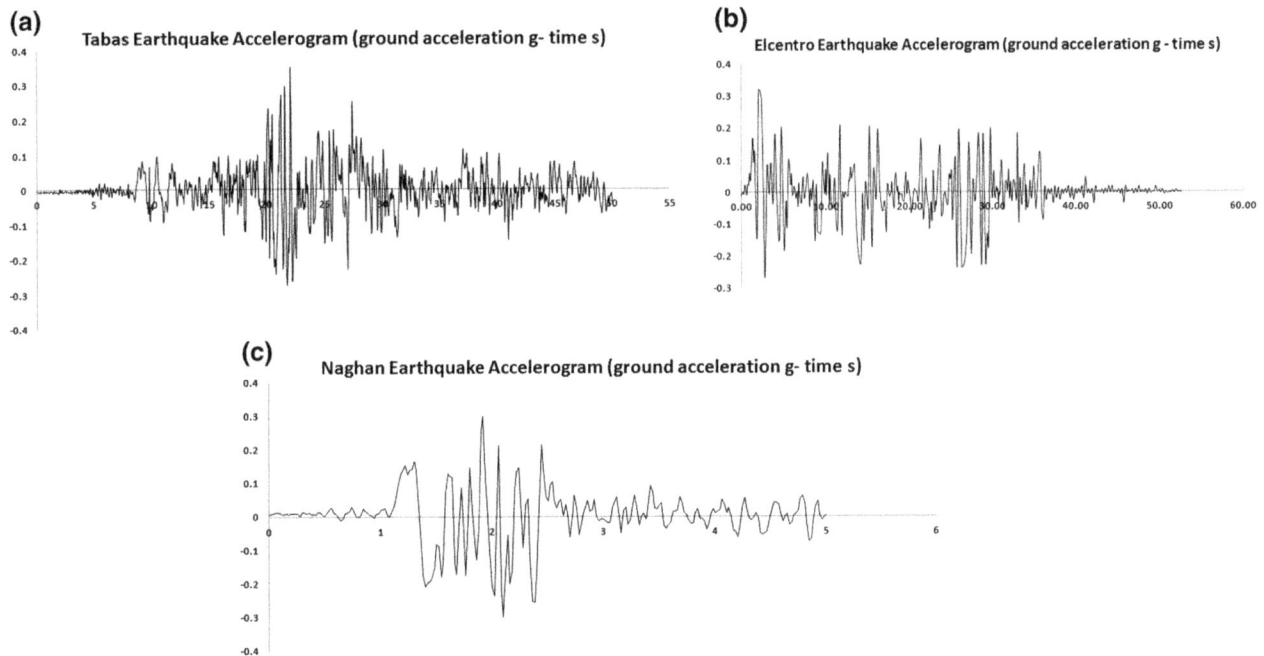

Fig. 16 Scaled ground acceleration used for time history analysis; **a** Tabas record accelerogram in scaled form, **b** Elcentro record accelerogram in scaled form and **c** Naghan record accelerogram in scaled form.

Accelerograms of various time histories, the earthquake excitations, in the scaled form are indicated in Fig. 16 as below.

7.2 Structural Modeling

Since the characteristics of these earthquakes are different from other places, they have to be scaled to a scale before using them for non-linear dynamic analyses of the studied models. To scale accelerograms using UBC-97 method, the values of natural oscillation period were initially calculated for three models. These three models included the models with three spans and 2, 6, and 15 floors. Models were divided to three groups of short, medium, and tall buildings and their natural period were considered to be in three categories of short, medium, and long periods (Razavian Amrei et al. 2011). Models dimensions have considered 3 m for height of stories and 4 m for uniform spans length. Suitable structural periods were selected to be smaller than 1.5T.[1] In the next stage, the acceleration spectrum were determined for all three accelerograms and knowing the spectrum suggested by regulations, the ratio of regulation spectrum to

acceleration spectrum of recent accelerograms were found to vary from 0.2 to 1.5 T. The arithmetic averages of the ratios were calculated for this range. In addition, scale factors were also investigated for taller buildings. Longitudinal and transverse reinforcements, characteristics and dimensions of three models flexural, x-braced and ODBS frame are assigned according to Tables 3, 4 and 5. The assigned plastic hinges at capable plastic points and also the acceptance criteria for rotational and translational displacements are defined to models as Tables 3, 4 and 5. The design criteria to selecting beam and column dimensions are corresponded to linear equivalent static analysis. Next analyses are considered for models and sections verification to doing their works properly.

According to the last table, the design sections of reinforced concrete frame and steel brace members are shown as below. The reinforced concrete sections are selected from 5 types. Each stories column is considered corresponding to Fig. 17 (left). Designed sections according to ACI manual are from maximum stress ratio or in the other word according to optimum state of strength and ductility. The ODBS sections are indicated in Fig. 17 (right). The schematic model of ODBS braced RC frame is presented in term

---

[1] The major vibration's period of models.

Table 3 Design sections characteristics and component properties for two-story frame's members.

Components	Floor	Type	Dimensions ($b \times h$)	Bars and Stirrups	Hinges	Acceptance criteria/ type
Beams	First floor		Rectangular 25 × 25 (cm)	3 Ø 18 Ø8 @10,20 cm (top &bottom)	Flexural (M3)	0.01 rad plastic rotation
	Second floor		Rectangular 25 × 20 (cm)	3 Ø 16 Ø8 @10,20 cm (top &bottom)	Flexural (M3)	0.01 rad plastic rotation
Columns	First floor		Rectangular 30 × 30 (cm)	8 Ø 18 Ø8 @10,20 cm	Flexural +Axial (P-M3)	0.012 rad plastic rotation
	Second floor		Rectangular 25 × 25 (cm)	8 Ø 16 Ø8 @10,20 cm	Flexural +Axial (P-M3)	0.012 rad plastic rotation
Braces	First floor	X	Box 8 × 8 × 0.5 (cm)	–	Axial (P)	7 Δ_T Plastic deformation
	Second floor		Box 8 × 8 × 0.5 (cm)	–	Axial (P)	7 Δ_T Plastic deformation
	First floor	ODBS	8 × 8 × 0.5: 1,2 (Members) 2 × 2 × 0.3: 3 (Members)	–	Axial (P)	7 Δ_T 9 Δ_T
	Second floor		8 × 8 × 0.5: 1,2 (Members) 2 × 2 × 0.3: 3 (Members)	–	Axial (P)	7 Δ_T 9Δ_T

of using spring element instead of steel element for third member of bracing system. The composed form of ODBS by damper, spring and steel bracing member could be state of the art characteristics for lateral bearing system with passive energy dissipation.

One of the important analyses for investigating the seismic behaviour of a structure is the time history analysis. So for supplementary analysis, the nonlinear dynamic analysis (NDA) is performed for various time histories.

8. Analysis for Results

8.1 Inter-Story Drift Investigation

The obtained inter-story drift results are indicated the amount of dissipated displacement and energy dissipation is increased in a system by more ductility and deformability. The flexural frame is contributed more than x-braced frame in strain energy absorption, because of its ductile characteristics. Construction of ODBS system is divided by two patterns. First, adding off-diagonal steel brace to the first story RC frame, since the x-brace system is used for other stories (composed bracing system by ODBS at first story and x-bracing system for other stories). As second pattern, adding the off-diagonal bracing system to all stories of RC frame (without any composed bracing system). If the system of bracing selected from first pattern, the first story treats as a ductile system and other stories were being treat as a semi

rigid body.

The first pattern advantage was a performance same as base isolation system that it is absorbed the vibrations of ground motion and the other stories had a minimum proportional displacement (limits of inter-story drift is checked always). What is too important by this pattern is the limitation of first inter-story's drift in term of seismic manuals. This system is behaved exclusively, if the regarded drifts being in use. The decision making about second pattern is too important because of its complicated behaviour. When the ODBS is performed in all stories, the energy dissipation potential is created in each story of the structure. By considering the basic lateral resisting system along flexural frame for ODBS system, by adding steel braces, the lateral displacement decreased strongly at each story. In the other word each story equipped by a type of damper and vibrations are controlled and dissipated specially for stories up to fifth or sixth.

One another investigation is about recorded displacements in special base points. The base points are usually selected along the point nearest to center of mass and center of lateral stiffness. Some single base points are chosen at the last floor's upper level, such as for pushover analysis, on each type of models, these points were monitored along the incremental static loads. Also in dynamic time history analysis, the central base points in each story are defined and monitored to control proportional displacement and/or in the other word, inter-story drift controlling.

Table 4 Design sections characteristics and component properties for six-story frame's members.

Components	Floor	Type	Dimensions ($b \times h$)	Bars and Stirrups	Hinges	Acceptance Criteria/Type
Beams	First and second floors		Rectangular 35 × 35 (cm)	4 Ø 18 Ø10@10,20 cm (top & bottom)	Flexural (M_3)	0.01 rad Plastic rotation
	Third and fourth floors		Rectangular 35 × 30 (cm)	4 Ø 16 Ø8 @10, 20 cm (top & bottom)	Flexural (M_3)	0.01 rad Plastic rotation
	Fifth and sixth floors		Rectangular 35 × 25 (cm)	3 Ø 16 Ø8 @10, 20 cm (top & bottom)	Flexural (M_3)	0.01 rad Plastic rotation
Columns	First and second Floors		Rectangular 40 × 40 (cm)	12 Ø 20 Ø8 @ 8,16 cm	Flexural +Axial (P-M_3)	0.012 rad Plastic rotation
	Third and fourth floors		Rectangular 35 × 35 (cm)	8 Ø 20 Ø8 @ 8,16 cm	Flexural +Axial (P-M_3)	0.012 rad Plastic rotation
	Fifth and sixth floors		Rectangular 30 × 30 (cm)	8 Ø 18 Ø8 @ 10,20 cm	Flexural +Axial (P-M_3)	0.012 rad Plastic rotation
Braces	Second and third floors	X	Box 10 × 10 × 0.7 (cm)	–	Axial (P)	7 Δ_T Plastic deformation be
	Fourth, fifth, & sixth floors		Box 8 × 8 × 0.5 (cm)	–	Axial (P)	7 Δ_T Plastic deformation
	First floor	ODBS	Components 1 and 2: Box 10 × 10 × 0.7 (cm)	–	Axial (P)	7 Δ_T Plastic deformation
			Component 3 Box 3 × 3 × 0.3 (cm)			9 Δ_T Plastic deformation

Table 6 gives the obtained values of proportional deformation (drift) of two adjacent floors. The obtained drift ratio, defines the amount of displacement in each floor. The Table 6 illustrates the amount of inter-story drift for each model of second, sixth, and fifteenth stories under the defined earthquakes for types of flexural, x-braced and off-diagonal bracing systems.

Corresponding inter-story values are indicated in Fig. 18. The inter-story drifts shown in three types of flexural, x-brace and off-diagonal braced frame. Differences between various simulated models are compared for their behaviours under three records Naghan, Tabas and Elcentro that they are scaled previously. Figures show the ODBS braced frame has maximum inter-story drift between the base and the first floor but, in upper floors this drift is minimized higher than the other systems. According to Fig. 18 the maximum displacement of ODBS structures in the first story, demonstrate large amount of energy absorption in this system.

If the results of various types of bracing system have been considered, the maximum inter-story drift is happened in the medium height of flexural frame and x-bracing frame wherever their fractural mode was happened there, at the same levels. Whereas in the ODBS braced frame, the maximum inter-story drift was happened in the first story and the other inter-story drifts were occurred proportional to the first story and almost decreased. This is the optimal and the ideal behaviour in a structure but, in the other systems the pattern of stiffness sorting that, it is from high to low stiffness, have not been respected and this is the reason of increasing the cost of construction, especially for structural skeleton. ODBS system has not only decrease of drift in upper story levels but also decrease of stiffness. In an approximately estimation of cost, the ODBS system has lowest cost for construction compare with the other RC bracing systems.

8.2 Plastic Hinges in terms of Levels of Performance

Subjoining the ODBS to concrete flexural frame is not only the cause of more plastic hinges formation in components of beam and column, but also the system ductility increases especially for three to ten story structures, as a result of producing axial plastic hinges of ODBS components. Considering the design of frame sections based on linear static analysis, a limited number of plastic hinges are formed in flexural frame and all of the members will not be

Table 5 Design sections characteristics and component properties for fifteen-story frame's members.

Components	Floor	Type	Dimensions ($b \times h$)	Bars and Stirrups	Hinges	Acceptance criteria/type
Beams	First, second, and third floors		Rectangular 70 × 55 (cm)	7 Ø 20 Ø16@12,25 cm (top & bottom)	Flexural (M_3)	0.01 rad Plastic rotation
	Fourth, fifth, sixth, and seventh floors		Rectangular 65 × 45 (cm)	8 Ø 22 Ø12@10,20 cm (top & bottom)	Flexural (M_3)	0.01 rad Plastic rotation
	Eighth, ninth, tenth, and eleventh floors		Rectangular 55 × 40 (cm)	7 Ø 18 Ø12@10,20 cm (top & bottom)	Flexural (M_3)	0.01 rad Plastic rotation
	Floors from twelve to fifteen		Rectangular 40 × 30 (cm)	6 Ø 18 Ø10@10,20 cm (top & bottom)	Flexural (M_3)	0.01 rad Plastic rotation
Columns	First floor and Second floor		Rectangular 70 × 70 (cm)	44 Ø 25 Ø16@ 10,20 cm	Flexural +Axial (P-M_3)	0.012 rad Plastic rotation
	Floors from three to six		Rectangular 65 × 65 (cm)	36 Ø 25 Ø14@ 10,20 cm	Flexural +Axial (P-M_3)	0.012 rad Plastic rotation
	Floors from seven to eleven		Rectangular 55×55 (cm)	28 Ø 22 Ø12@ 10,20 cm	Flexural +Axial (P-M_3)	0.012 rad Plastic rotation
	Floors from twelve to fifteen		Rectangular 40 × 40 (cm)	20 Ø 22 Ø10@ 10,20 cm	Flexural +Axial (P-M_3)	0.012 rad Plastic rotation
Braces	Floors from One to four	X	Box 20 × 20 × 1.0 (cm)	–	Axial (P)	7 Δ_T Plastic deformation
	Floors from five to nine		Box 15 × 15 × 0.9 (cm)	–	Axial (P)	7 Δ_T Plastic deformation
	Floors from ten to fifteen		Box 12 × 12 × 0.75 (cm)	–	Axial (P)	7 Δ_T Plastic deformation
	First floor	ODBS	Components 1&2: Box 20 × 20 × 1.0	–	Axial (P)	9 Δ_T Plastic deformation

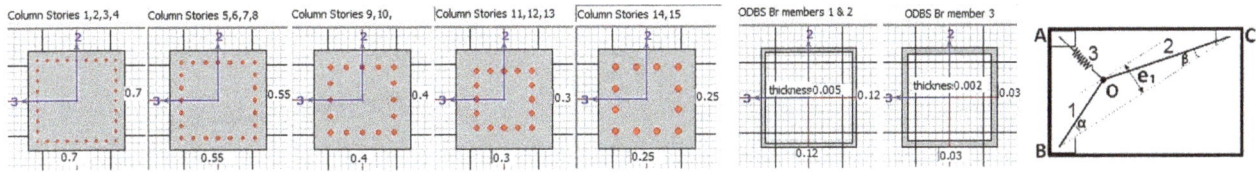

Fig. 17 Properties for applied sections used for this research models.

Table 6 Inter-story drift values for each model by second, sixth, and fifteenth stories under the earthquakes Naghan, Tabas, and Elcentro respectively (from *left to right*) and also in types of flexural, x-braced and off-diagonal bracing systems.

Floor	Inter story drift for Naghan, Tabas & Elcentro respectively (cm)											
	2- Stories Model											
	Flexural frame			x-braced frame			ODBS braced frame (all stories)			ODBS braced frame (first story)		
1	0.35	0.32	0.38	0.17	0.16	0.21	0.23	0.26	0.28	0.37	0.29	0.56
2	0.28	0.21	0.24	0.11	0.09	0.12	0.21	0.18	0.25	0.17	0.12	0.21
Floor	6- Stories Model											
1	2.95	1.68	2.93	0.62	0.6	1.4	1.08	1.67	2.63	3.28	2.96	3.95
2	1.05	1.98	2.85	0.7	0.83	1.71	2.76	1.24	3.12	0.94	0.91	1.08
3	2.48	1.85	2.33	0.8	1.15	1.93	2.05	2.93	3.34	0.85	0.63	0.79
4	1.49	1.08	1.45	1.45	2.41	1.68	1.46	1.25	1.68	0.72	0.85	1.13
5	1.2	0.9	1.21	0.72	1.52	0.79	0.77	0.72	0.9	0.43	0.52	0.54
6	0.8	0.58	0.86	0.51	0.89	0.46	0.68	0.55	0.88	0.08	0.29	0.42
Floor	15- Stories Model											
1	1.49	0.47	0.98	0.95	0.6	0. 8	3.11	2.12	3.43	3.43	4.05	5.93
2	3.13	0.51	1.83	0.9	0.45	0.4	3.66	3.03	4.21	4.21	1.03	1.21
3	2.45	1.37	2.93	0.75	0.39	1	3.71	3.47	3.63	3.63	0.96	0.99
4	2.17	1.21	3.91	1.47	0.51	1.1	3.43	3.42	3.36	3.36	0.89	0.93
5	2.96	1.53	4.85	1.68	1.41	2.1	2.84	2.98	3.02	3.02	0.98	0.76
6	3.85	1.92	3.72	1.93	1.5	2,6	1.98	2.18	2.45	2.45	0.85	0.85
7	4.61	2.13	3.53	0.76	0.95	3.01	2.45	1.97	3.38	3.38	0.77	1.45
8	3.42	1.94	2.98	0.93	2.1	1.76	3.53	2.63	3.23	3.23	0.63	1.23
9	3.23	2.73	2.45	0.81	2.05	2.51	2.81	2.51	3.87	3.87	1.51	0.82
10	3.12	2.92	2.95	0.75	2.5	1.12	1.71	1.33	2.67	2.67	1.84	0.79
11	2.31	3.43	1.95	0.6	1.63	0.93	1.58	1.04	2.32	2.32	1.51	0.71
12	1.86	2.95	1.43	0.72	0.95	0.81	1.62	0.73	2.13	2.13	1.06	0.62
13	1.03	2.08	1.32	0.58	0.71	0.6	1.08	0.66	1.48	1.48	0.43	0.51
14	0.68	1.81	1.11	0.49	0.45	0.48	0.78	0.51	0.94	0.94	0.41	0.43
15	0.34	0.71	0.95	0.33	0.2	0.23	0.54	0.36	0.67	0.67	0.32	0.36

able to produce plastic hinges, but when the steel off-diagonal bracing system is added to flexural frame, increasing the rotational capacity of RC members and also increasing the number of composed plastic hinge are some indices of increased ductility of ODBS braced RC flexural frame. Many results are generated along this analysis in ODBS braced RC frame before any damages, the third member of this system has been rotated and deflected near the plastic limit. In this hand, the initial sever vibrations have been damped through the flexibility of ODBS system and also its members elongation and energy absorption. Figure 19 illustrates the formation of plastic hinges and their rotational capacity.

Generated plastic hinges is indicated in Fig. 19. As shown in this figure, the off-diagonal system has the highest level of deformation not only in RC frame members but also in steel

(a) 2st Flexural Frame

(b) 2st X-Braced Frame

(c) 2st ODBS braced frame (all stories)

(d) 2st ODBS braced frame (first story)

(a) 6st Flexural Frame

(b) 6st X-Braced Frame

(c) 6st ODBS braced frame (all stories)

(d) 6st ODBS braced frame (first story)

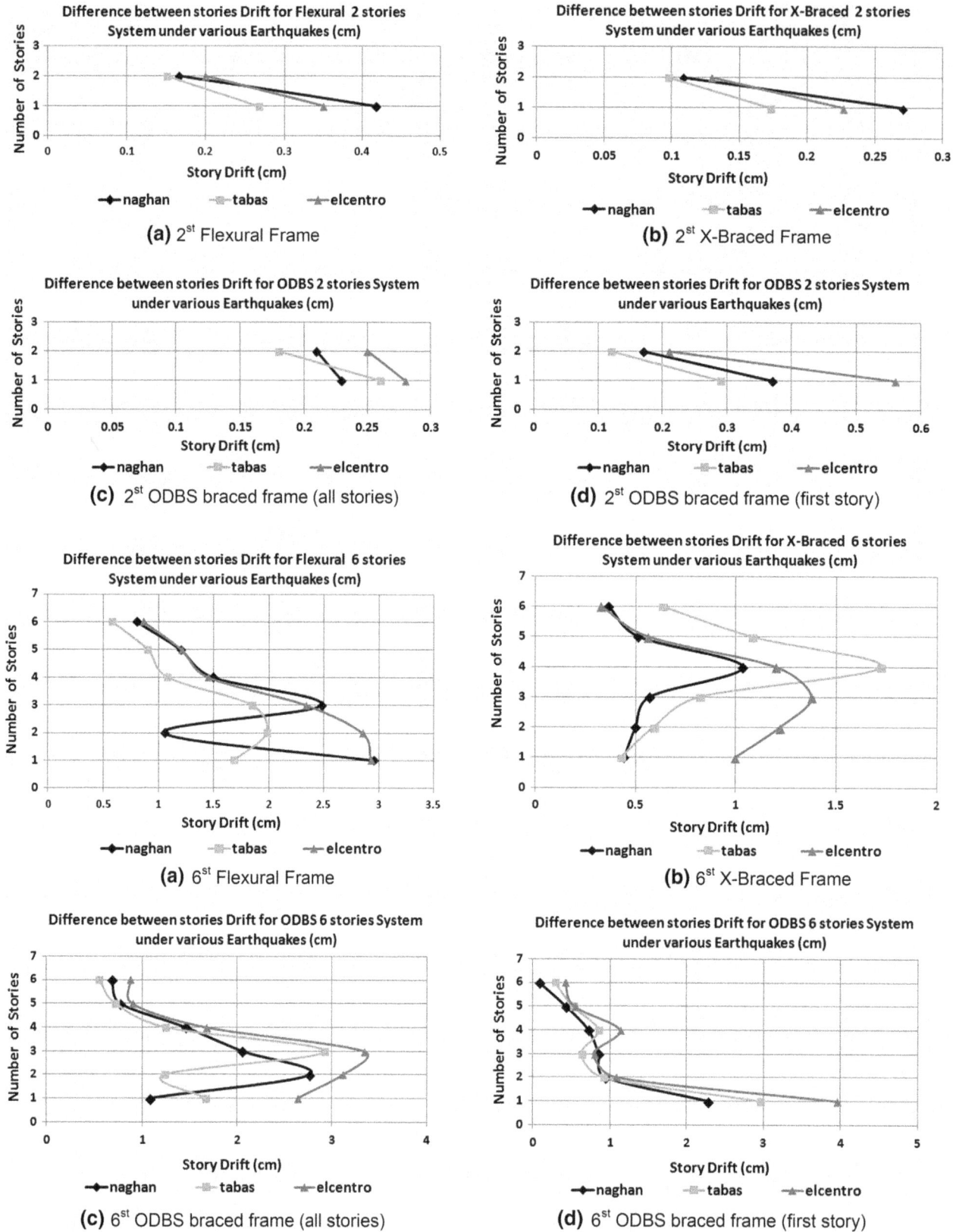

Fig. 18 Variation of inter-story drifts (cm) under various earthquakes by three types of 2, 6 and 15 stories frame respectively for **a** flexural frame, **b** x-braced, **c** ODBS system used in all stories and **d** ODBS system used only in the first story.

members, especially in third member of steel ODBS. The rotational capacity is increased in ODBS, the most ductile system. Performance levels of flexural and X-braced frames are limited to Life Safety (LS) level, but for ODBS, level of performance has been extended to higher ductility about

related design criteria. Six and fifteen stories flexural frames have plastic hinges more than X-braced frame. On the other hand in ODBS, more members are contributed in absorption of defined existing energy by nonlinear ductile behavior of plastic rotation and deformation proportional to flexural frame. Non-linear static analyses as well as dynamic step by step seismic analyses are performed and special purpose elements are employed for the needs of this study. Results showing the influence of the maximum rotation of the multi-storey frame members in terms of ductility requirements and rotational requirements of the frame members (Karayannis et al.).

Acceptance criteria for flexural frame of LS level is 0.02 for primary components and also the acceptance criteria for ODBS braced frame of CP level is 0.025 and 0.05 for primary and secondary components respectively. A more detailed scrutinizing of the results reveals that the hinges formed in ODBS system endure the maximum deformation and earn the structure a very high performance level along

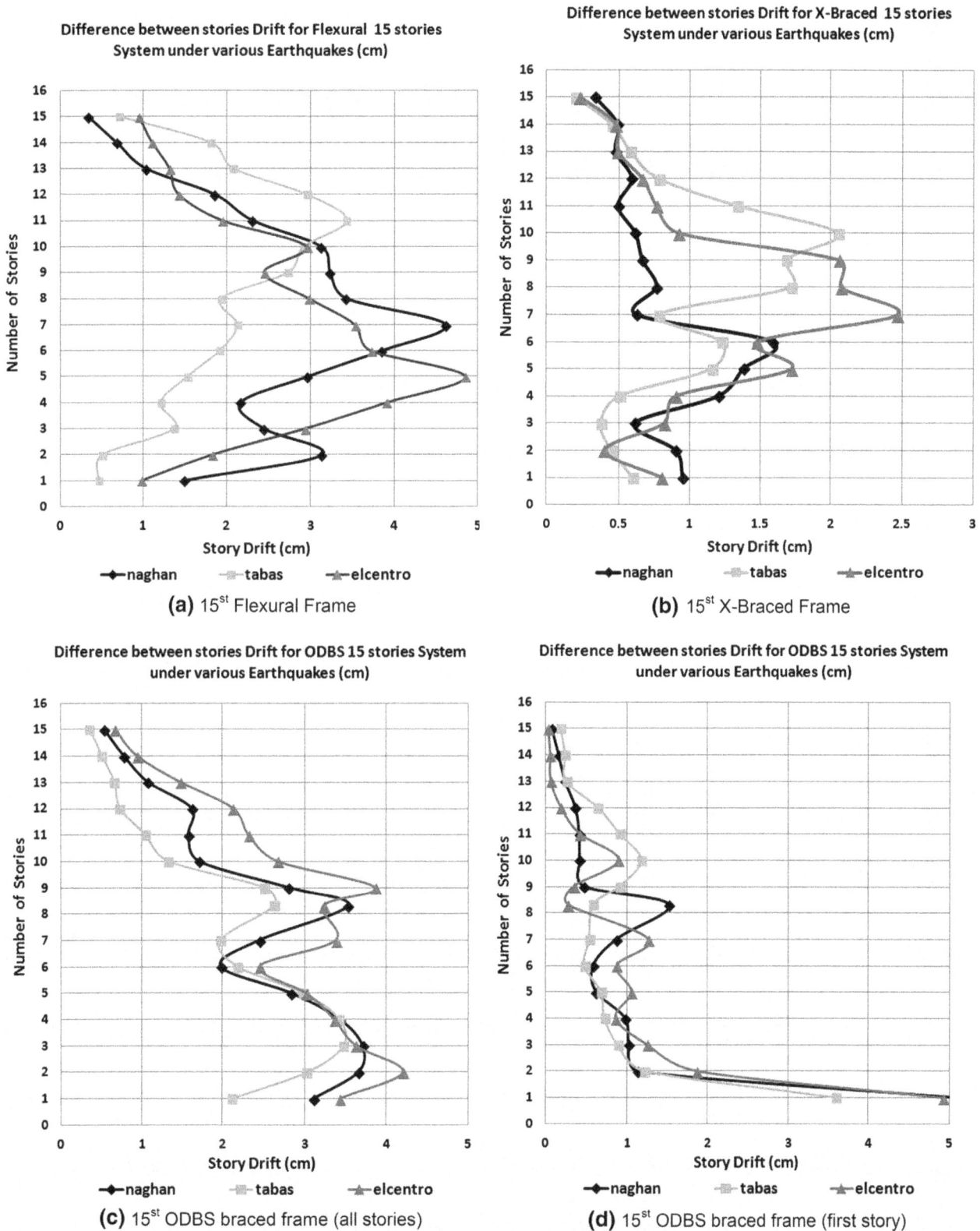

(a) 15st Flexural Frame

(b) 15st X-Braced Frame

(c) 15st ODBS braced frame (all stories)

(d) 15st ODBS braced frame (first story)

Fig. 18 continued

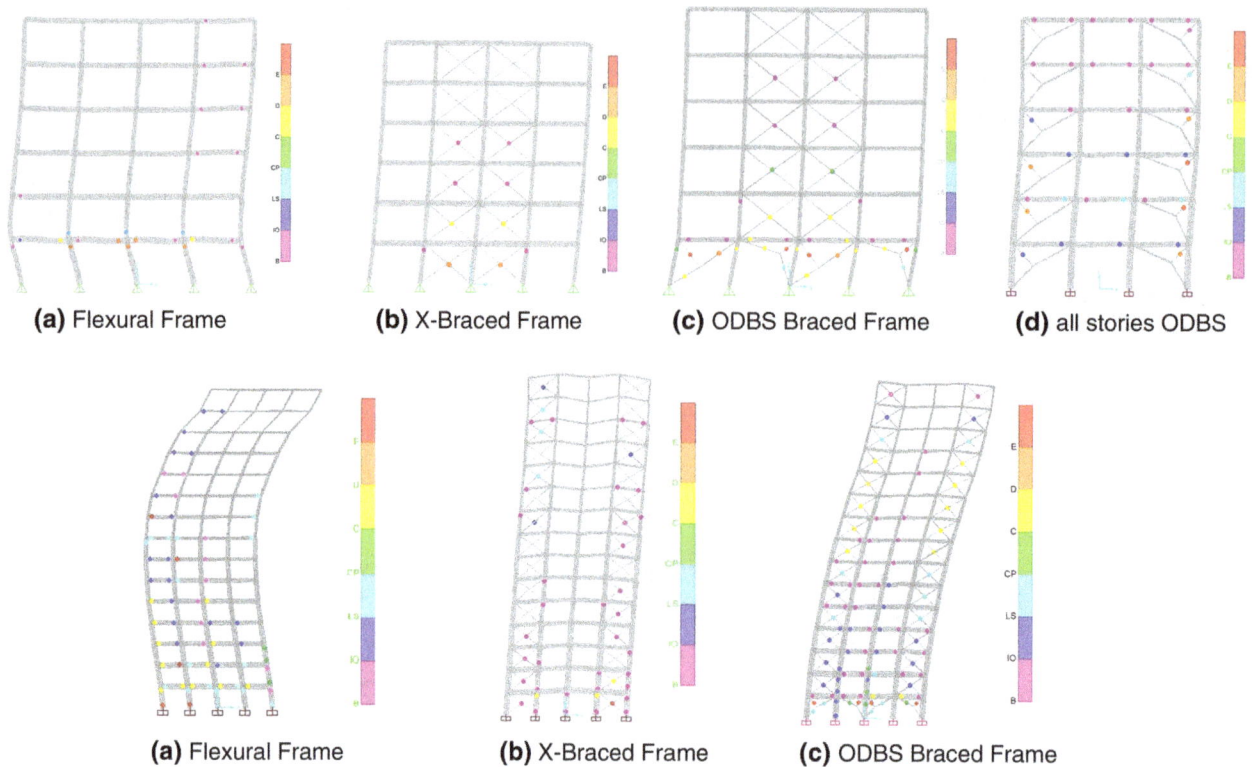

(a) Flexural Frame   (b) X-Braced Frame   (c) ODBS Braced Frame   (d) all stories ODBS

(a) Flexural Frame   (b) X-Braced Frame   (c) ODBS Braced Frame

Fig. 19 Sequence of performed plastic hinges under Tabas ground motion by considering geometric nonlinearity for six & fifteen stories RC frames, **a** flexural frames, **b** x-braced frames, **c** ODBS braced frames consisting and **d** all stories ODBS braced frame.

Table 7 The values of maximum response of ODBS braced frame under various eccentricities.

Models & index	Elcentro time-history acceleration response				General characteristics		
	Max. Ecc (e_1)	Max. Acc (g)	Max.Vel (cm/s)	Max. Displ (cm)	Initial stiffness (kN/cm)	Period (1st mode) (s)	Δ at first yield point (cm)
ODBS (p0)	0.00	0.342	46.08	3.93	491.78	0.47	2.83
ODBS (p1)	0.1	0.261	40.14	4.67	315.59	0.59	2.49
ODBS (p2)	0.2	0.175	37.72	5.10	158.51	0.83	2.31
ODBS (p3)	0.3	0.073	33.3	6.06	77.49	1.19	2.1
ODBS (p4)	0.4	0.034	34.71	6.81	37.77	1.71	1.7
ODBS (p5)	0.5	0.028	36.95	6.22	17.36	2.52	1.6
ODBS (p6)	0.6	0.023	38.71	6.16	6.50	4.12	1.3
ODBS (p7)	0.7	0.018	40.08	6.00	1.23	9.48	1.1
X-Bracing	–	0.581	78.35	2.04	894.39	0.39	2.96

ductile behavior. In addition, the more performance levels of plastic hinges are gathered in the structure, so by this level of ductility, the structure will be absorbed more quantity of energy. These results are deduced based on non-linear dynamic step by step analyses. The time steps for this analysis is considered less than $\Delta T = 0.02$ s. Future research will be dedicated to the full time history analysis and investigate the proportional hysteresis curves. Assessing the stiffness and/or the strength degrading is the most important to diagnosing the exact behavior of this system.

According to Fema-356 (FEMA-356 2000, Table 6–7 and Table 6–8), the plastic rotation of mentioned beams and columns of flexural frame are 0.025 and 0.02 rad respectively. These quantities are 0.05, 0.03 rad in the system like ODBS by high ductility and therefore, the structural damage can be prevented to a great extent. This is why the structure's ductility and its capacity of energy absorption decreases considerably when the structural performance is limited to the formation of first crack. As it is obvious, when the hinges occur in the beam and columns' concrete elements, the drift

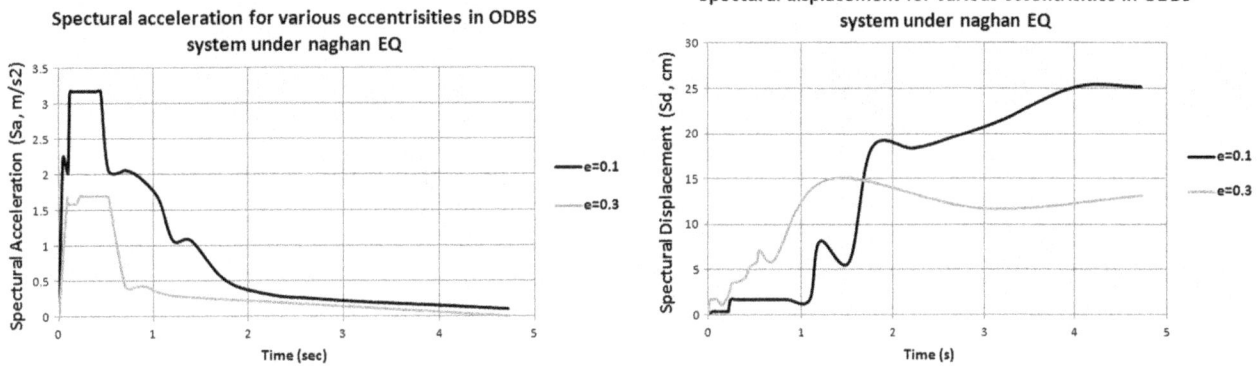

Fig. 20 Comparison response of ODBS system along Naghan earthquake for different eccentricities, spectral acceleration (*left*) and spectral displacement (*right*).

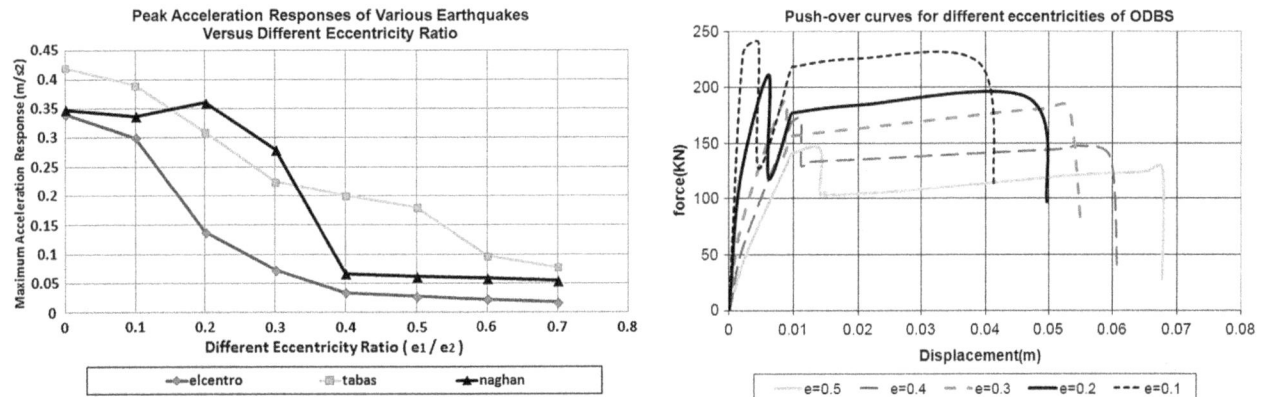

Fig. 21 Maximum Acceleration versus eccentricity ratio under Elcentro, Tabas and Naghan earthquakes (**a**), Ductility assessment of various eccentricities corresponding to pushover curves comparison (**b**).

and rotation of the frame attain to C and D points in the ODBS pushover curve, which will satisfy the safety requirements versus any deterioration related to its hazard and risk level. Thus, it is concluded that the application of steel ODBS in concrete flexural frame is quite suitable and earns the structural outstanding characteristics from the viewpoint of the economy and ductility of reinforced concrete frame.

8.3 Effects of Eccentricity on ODBS System

In other investigation, to recognize the effects of eccentricity on the seismic behaviour and the spectral response of ODBS, a six-story frame by previous properties is performed under spectral analysis. To this modeling, various eccentricities are considered from e = 0.0 to e = 0.7 and also the response of spectral acceleration, velocity and displacement is investigated for a base point on the upper level of sixth story floor. The ODBS system along primary spectral analysis is investigated under effect of Naghan earthquake and the related results are obtained as Fig. 20. Spectral acceleration quantities are registered for two different eccentricities e = 0.1 and e = 0.3. Figure 20 (left) indicates that the result of ODBS by more eccentricity (e = 0.3) was lower than the other one. As regards that the time domain of vibration for Naghan earthquake (measured period 5 s), by increasing the eccentricity, the response of ODBS is decreased in a short

time. Also the oscillation's intensity is decreased saliently. According to right figure, the amount of spectral displacements is indicated for ODBS by the same eccentricity. The spectral displacement for eccentricity about 0.3 was lower than the same by 0.1 for its eccentricity as indicated in right side of Fig. 20. The initial velocity for model by higher eccentricity is greater than the other model by lower eccentricity.

By investigating the recent results, it seems that the energy dissipation is increased by increasing eccentricity and indicates the optimum amount of eccentricity is about 0.1–0.4 for ODBS systems. The eccentricities out of the range of 0.1–0.4 are not suggested to use in structures. Table 7 includes the recorded spectral response as acceleration, velocity and spectral displacement for ODBS system by various eccentricities. By continuing the assessments, the spectral analysis is performed on x-braced frame too. The spectral response of ODBS in compare with x-braced frame indicated the advantage of ODBS system. The stiffness and displacement characteristics of models under Elcentro earthquake are illustrated in Table 7. The intense velocity of x-brace model proportional to ODBS is generated imposing loads by impulsive tendency within the time domain of acceleration. This phenomenon may be the cause of structural concrete deteriorations. The minimum amount of spectral velocity is concerned about ODBS braced frame by

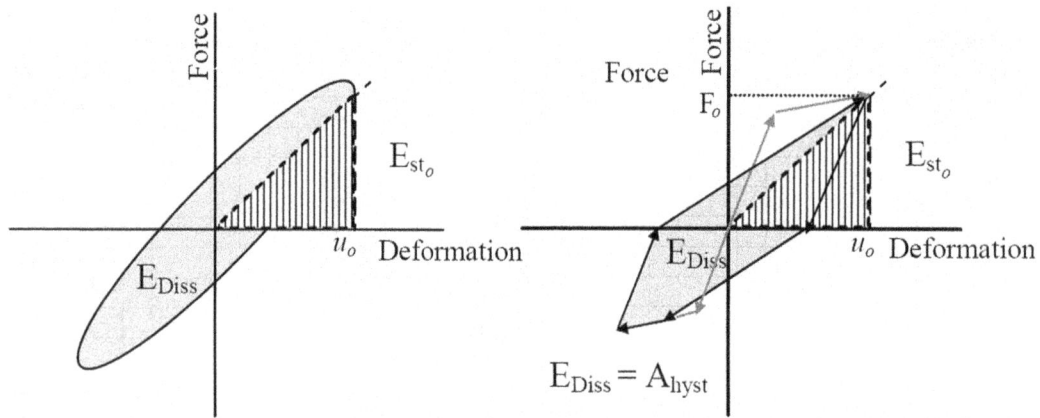

Fig. 22 Comparison between dissipated energy and stored energy for investigate viscous damping and hysteresis model.

eccentricity equal to 0.3. The lowest amount of spectral displacement in concerned about x-braced frame and ODBS braced frame, respectively. Finally the lowest response acceleration is recorded through 0.7 for eccentricity in ODBS braced frame.

Also the dynamic behaviour of the 6-story frames, retrofitted by Off-Diagonal bracing system, under three records of Tabas, Naghan and Elcentro, has been compared for optimizing normalized eccentricity ratio. Overall result related to several eccentricities is shown in Fig. 21.

The optimum levels for displacement and acceleration is affected from the structural stiffness and the mass inertia respectively. Manual limits should be considered for minimum displacement and allowable rotation. By knowing the optimum eccentricity about 0.2–0.5, the optimum behaviour of ODBS is generated. The relation between maximum acceleration response of ODBS and eccentricities variations under mentioned earthquake is specified as Fig. 21 (left). The results are converged for eccentricity about 0.1–0.3 under every excitation. Pushover curve due to different eccentricities is shown in Fig. 21 (right). Results are explained as how much the ductility levels are increased.

Approximation of the equivalent viscous damping ratio is effective parameters to verify the dynamic behaviour and is considered to predict the structural damping treatment and its energy dissipation capacity. The dynamic characteristic for comparison the various load bearing systems is proportional to equivalent damping and effective stiffness. Possibility to solving a simple linear braced frame instead of nonlinear ODBS braced frame is the reason why the equivalent viscous damping is estimated considering both elastic and inelastic energy dissipation.

The previous methods to approximate the equivalent viscous damping for present structure that this system is the composed structural material has indicated just for some hysteretic models, deformability or ductility range and frequency. Time history analysis carried out correspondent some results by amount of difference in present research from exact equivalent viscous damping. The exact and effective equivalent viscous damping ratio is related about two dependent factors, ductility and frequency (or period), because of this estimated values of equivalent viscous

damping ratio has been changed along any hysteresis analysis. Then the average equivalent viscous damping should be considered to investigate the present research subject.

The average value has minimum variation compare with exact value. This means that it is not an essential object to have the exact evaluation for damping ratio.

9. Equivalent Viscous Damping

Estimation of the equivalent viscous damping ratio is the one of effective parameters to verifying the dynamic behaviour and is considered to predict the structural damping treatment and its energy dissipation capacity. The dynamic characteristic for comparison the various load bearing systems is proportional to equivalent damping and effective stiffness. Possibility to calculating a simple linear braced frame instead of nonlinear ODBS braced frame is the reason why the equivalent viscous damping is estimated considering both elastic and inelastic energy dissipation.

By considering the unique ODBS braced RC flexural frame like a system of single degree of freedom (SDOF) and taking a part equivalent viscous damping ratio in elastic level of general damping and hysteresis damping in term of nonlinear behaviour along hysteresis analysis can be find:

$$\xi_{eq} = \xi_{lin} + \xi_{nonlin} = \xi_{elst} + \xi_{hyst} \tag{7}$$

That ξ_{elst} is linear or primary damping in elastic level and ξ_{hyst} is the nonlinear or secondary damping in inelastic level used for effect of energy dissipated loops. To calculate these quantities, the model should be subjected under harmonic cyclic loading by constant period in each cycle. Elastic damping ratio is assumed 5 % in this study. Because of this part of Eq. (7) is outside of this research's aim and scope, this quantity is considered in constant form (assumption is based on laboratory conditions in recent researches).

Equivalent viscous damping is related to hysteretic response that they have been referred to dissipated energy E_{dissip} and stored energy E_{stord}. By parallel use of Fig. 22 and Eq. (8), the quantity of equivalent damping ratio achieved from the value of excitation frequencies in each type of natural and main harmonic excitation.

(a)

(b)

(c)

(d)

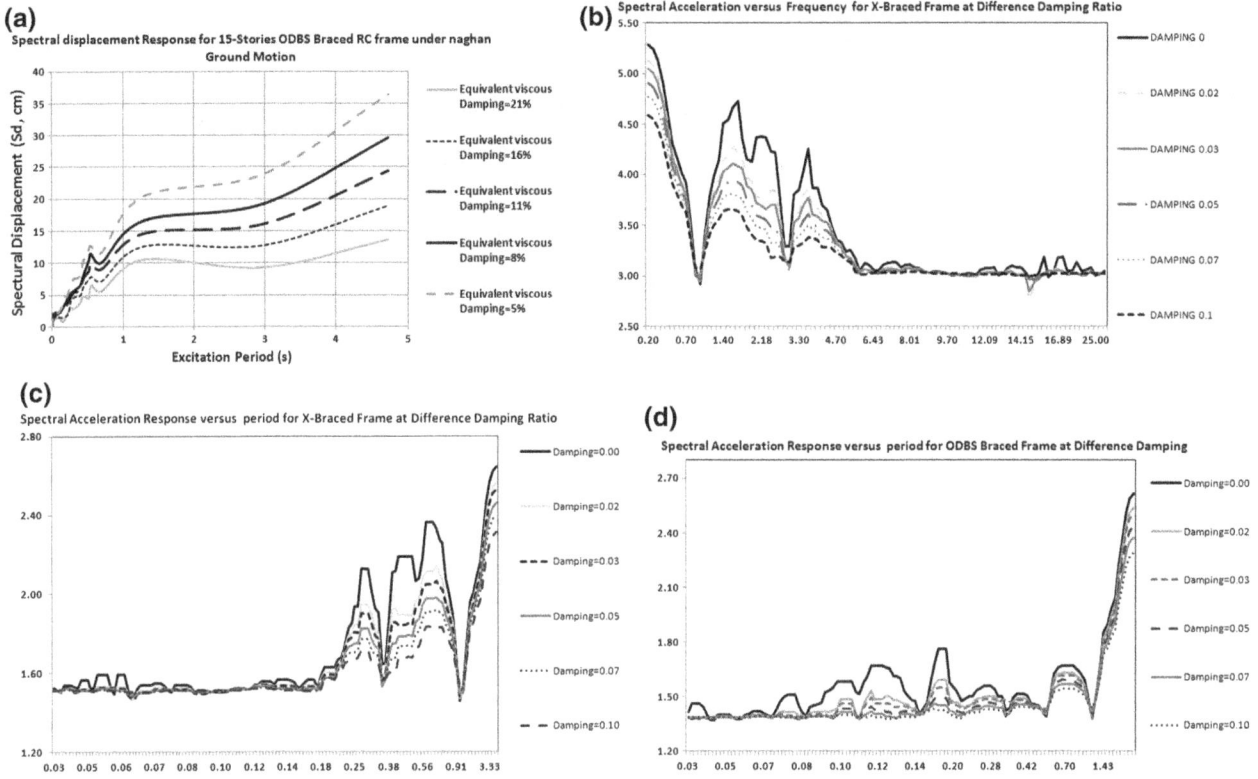

Fig. 23 Comparison spectral displacement response of ODBS frame model under Naghan excitation for different equivalent viscous damping ratios, **a** ODBS spectral displacement versus period, **b** flexural frame spectral acceleration versus frequency, **c** X-braced spectral acceleration versus period and **d** ODBS spectral acceleration versus period.

$$\xi_{hyst} = \left(\frac{1}{4\pi}\right)\left(\frac{\bar{\omega}_n}{\bar{\omega}}\right)\left(\frac{E_{dissip}}{E_{stord}}\right) \tag{8}$$

Also the equation of motion is assumed as below:

$$m\ddot{u} + c\dot{u} + ku = P_0 \sin \bar{\omega}t \tag{9}$$

The equation of motion in term of amplitude or maximum displacement by a phase angel along harmonic function is performed as:

$$u(t) = u_0 \sin(\bar{\omega}t - \varphi) \tag{10}$$

The phase angel in term of natural frequency and excitation frequency and almost the damping ratio is obtained from:

$$\varphi = \tan^{-1}\frac{2\xi(\bar{\omega}/\bar{\omega}_n)}{1 - (\bar{\omega}/\bar{\omega}_n)^2} \tag{11}$$

Since the energy stored calculated by:

$$E_{stord} = \frac{ku_0^2}{2} \tag{12}$$

The energy dissipation for the common motion in structure in considered as below. Substituting the functional parameter is related to each boundary condition and is performed distinguishably for any problems.

$$E_{dissip} = \int f_{damp}.du = \int_0^{\frac{2\pi}{\bar{\omega}}} (C\dot{u})\dot{u}.dt \tag{13}$$

This equation is defined by energy dissipated product by energy stored corresponding to characteristics of loading and time domain for imposed seismic or cyclic load. For example Eq. (8) is the one status of harmonic load substituted by u and $\dot{u}$ from Eq. (13).

Also the simple relation due to previous equations the equivalent period is substituted by equivalent damping ratio in form of Eq. (14). This equation explains the greater equivalent period for ductile systems and also the greater period is the result of the greater equivalent viscous damping ratio.

$$\frac{T_{eq}}{T} = 1 + 0.121(\mu - 1)^{0.939} \tag{14}$$

In general form, another equation is proposed by Priestley (2003) corresponding to equivalent viscous damping (ξ_{eq}), initial viscous damping (ξ_0) and the ductility level of the structures (μ).

$$\xi_{eq} = \xi_0 + \alpha\left(1 - \frac{1}{\mu^\beta}\right) \tag{15}$$

α and β constants are about the fatness and geometric property of hysteretic loops. These parameters assumption

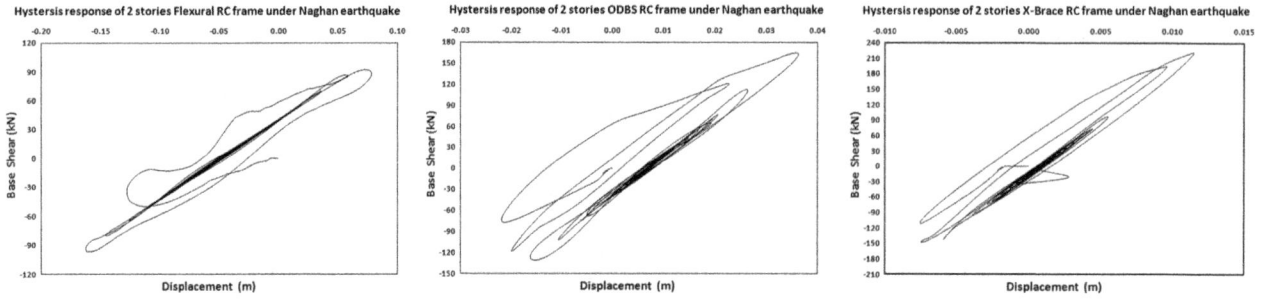

Fig. 24 Hysteretic response along Naghan seismic excitation for 2-stories Flexural frame (*left*), ODBS braced frame (*medium*) and X-braced frame (*right*).

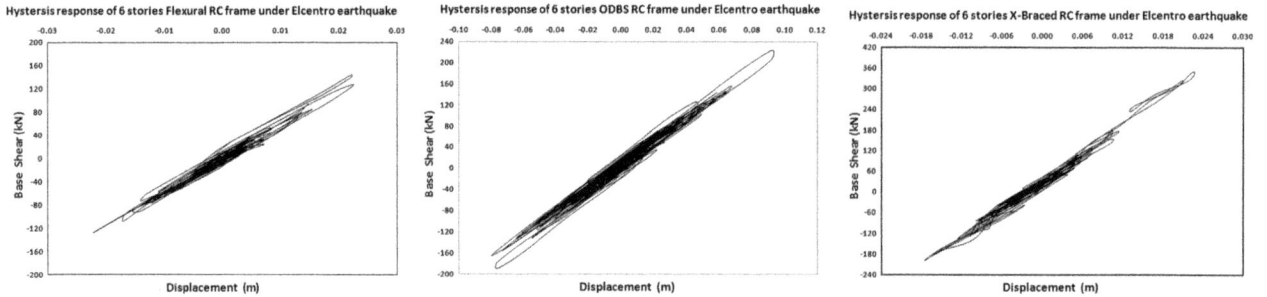

Fig. 25 Hysteretic response along Elcentro seismic excitation for 6-stories Flexural frame (*left*), ODBS braced frame (*medium*) and X-braced frame (*right*).

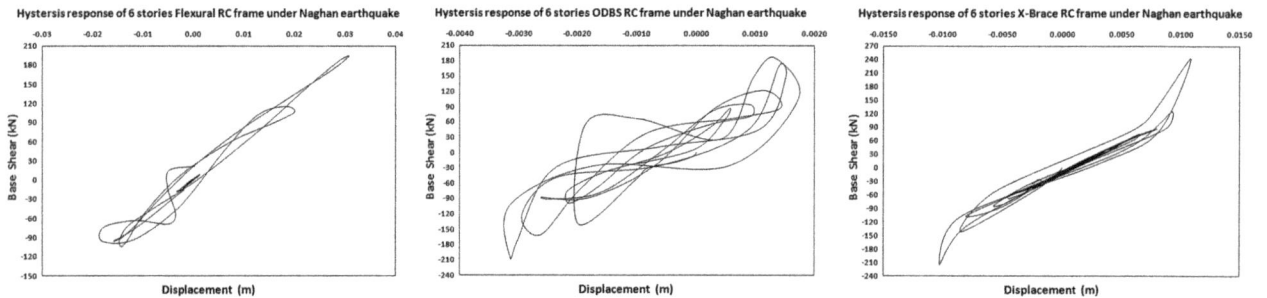

Fig. 26 Hysteretic response along Naghan seismic excitation for 6-stories Flexural frame (*left*), ODBS braced frame (*medium*) and X-braced frame (*right*).

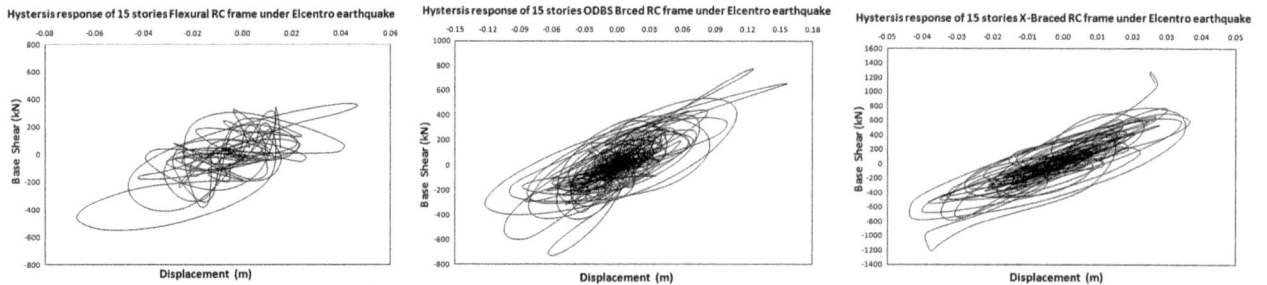

Fig. 27 Hysteretic response along Elcentro seismic excitation for 15-stories Flexural frame (*left*), ODBS braced frame (*medium*) and X-braced frame (*right*).

are considered by various types of selecting. Whatever seems important is increasing the equivalent viscous damping along increasing the ductility level. By consideration the α and β in the fix range, the ODBS system proportional to x-braced frame even to flexural frame, has more ductility and more equivalent viscous

damping ratio too. Although the equivalent damping ratio is considered to estimate the spectral displacement response for 15 stories ODBS braced frame as shown in Fig. 23. The under consideration model has 0.25 eccentricity that is the optimized quantity to ODBS investigation.

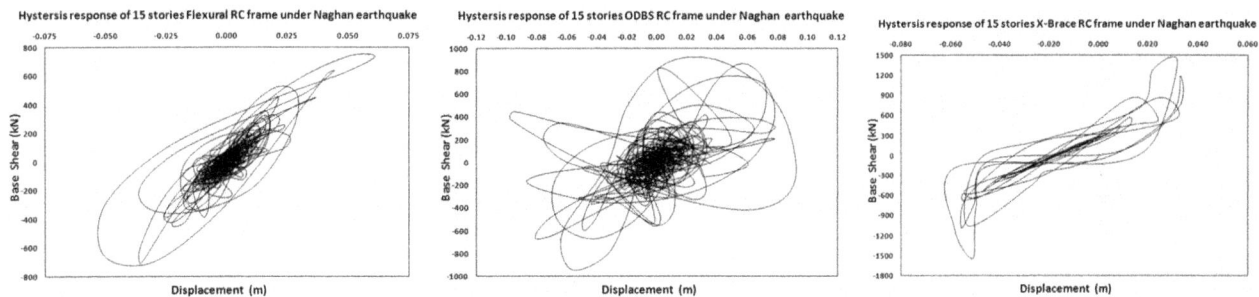

Fig. 28 Hysteretic response along Naghan seismic excitation for 15-stories Flexural frame (*left*), ODBS braced frame (*medium*) and X-braced frame (*right*).

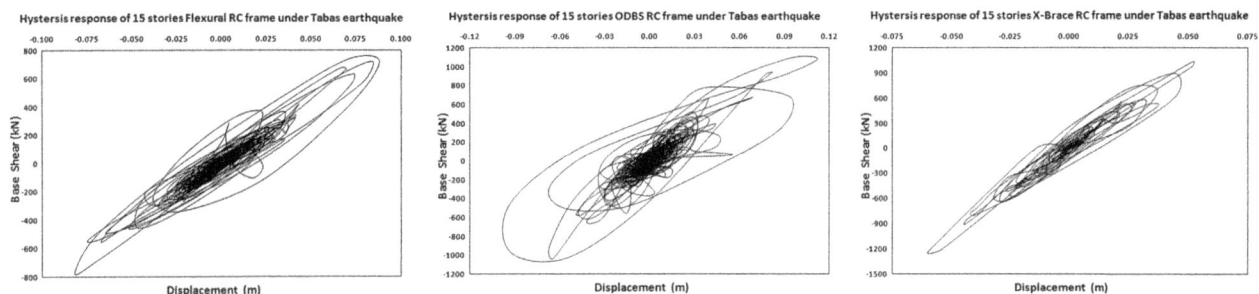

Fig. 29 Hysteretic response along Tabas seismic excitation for 15-stories Flexural frame (*left*), ODBS braced frame (*medium*) and X-braced frame (*right*).

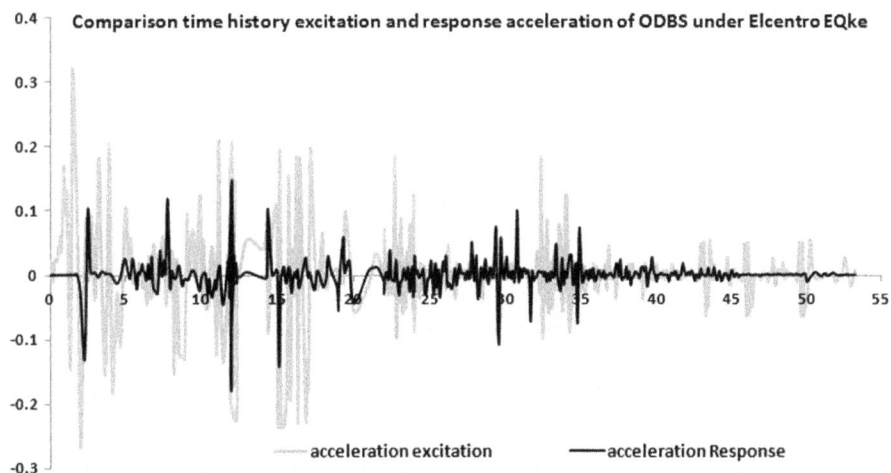

Fig. 30 Excitation of base level and ODBS response at first floor's level comparison for their accelerations under Elcentro time history.

As investigated for ODBS spectral response in last pages, the next assessment was about obtaining the equivalent viscous damping for ODBS braced frame. The obtained results according to Fig. 23 is illustrated the viscous damping characteristics for ODBS braced frame under Naghan earthquake. As indicated in recent figures and illustrative equations, the viscous damping ratio and the spectral displacements are related inversely. It means decreasing the spectral displacement is because of increasing the viscous damping for a system. In Fig. 23, the various amount of equivalent viscous damping is compared each model of ODBS and x-braced frame together. The effects of ODBS equivalent viscous damping ratios variations were more than other systems. In Fig.

23a, c and d, the response spectral displacements versus time periods are specified for various accelerograms assessment.

Figure 23 indicates the sequence of inelastic damping ratio for dissipating energy proportional to its exact treatment. This specific means ODBS system may be able to have 11–15 % of equivalent viscous damping ratio by comparing this status to real work. In minimum range of viscous damping, the ODBS gathers 6–10 % of damping ratio in term of hysteresis energy dissipation. On the other hand, according to extended analysis, these equivalent damping quantities are lower in the x-braced and flexural frame. In the other word the ductile capacity of ODBS to absorbing energy is higher than the other systems.

(a)

(b)

(c)

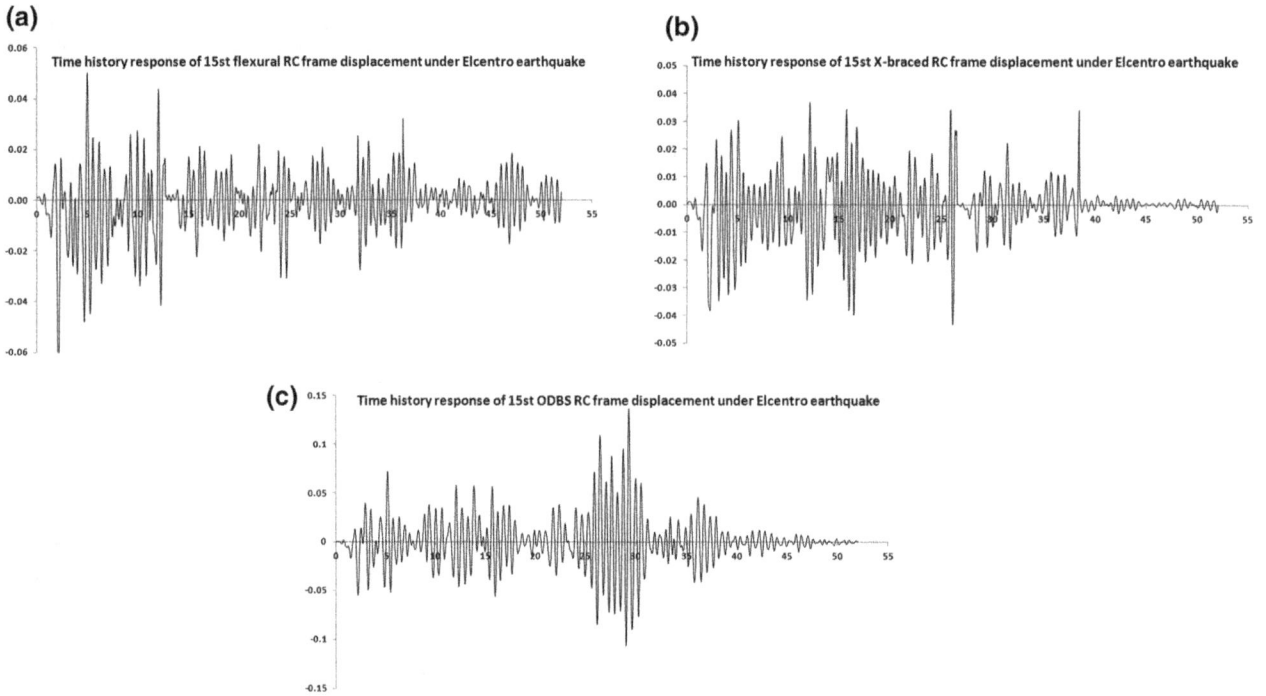

Fig. 31 Displacement response of a Flexural, b ODBS and c X-braced frame under Elcentro time history analysis.

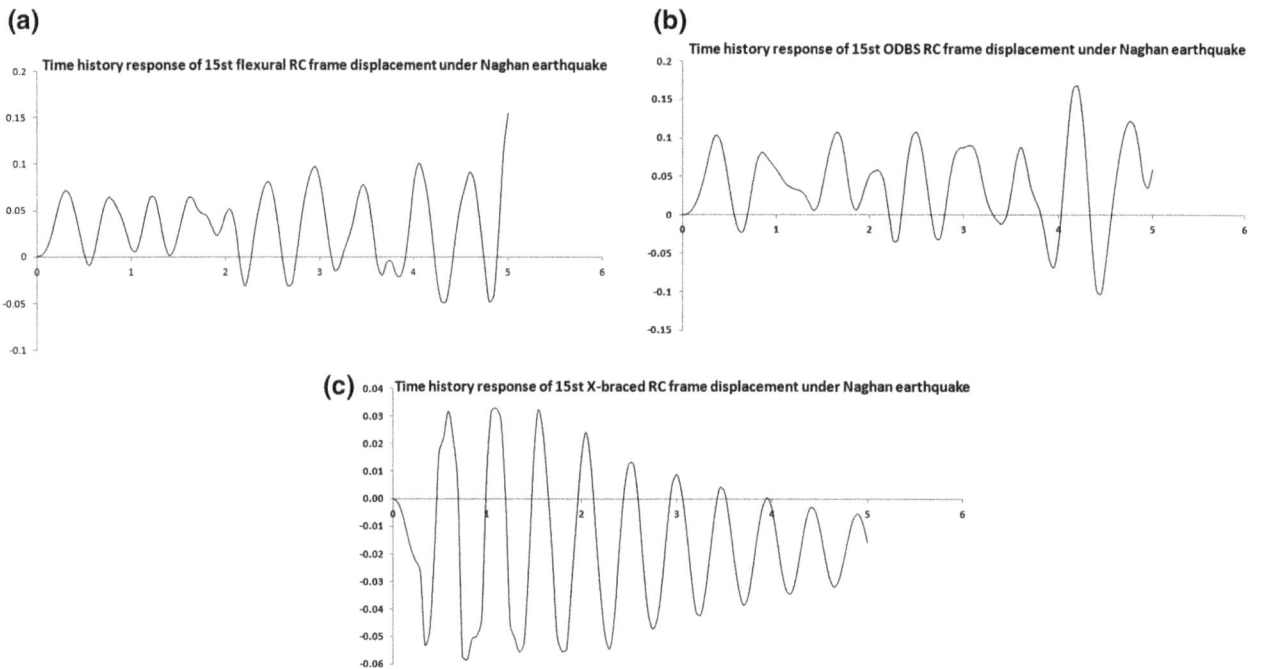

(a)

(b)

(c)

Fig. 32 Displacement response of a Flexural, b ODBS and c X-braced frame under Naghan time history analysis.

10. Energy Dissipation for Various Models

10.1 Hysteresis Response of Models Under Earthquake Excitations

Connecting the pick points of hysteresis loops obtained the spectral response that, it is the one of important parameters for control the structural response in the range of period domain. Figure 23 indicates the spectral response acceleration and spectral response displacement in some different case studies. The various models by 2, 6 and 15 stories for three types of flexural, x-brace and off-diagonal bracing

system are subjected to Elcentro, Naghan and Tabas accelerogram records.

It means that, after occurring large deformation (displacement or rotation), the flexural frame has been softened and the disability of load bearing in second stage, especially in term of strain hardening, may caused. The results indicate the ODBS system is the cause of adjustment in quantities of the energy stored and the hysteretic energy. This reason may be the cause of the upper equivalent viscous damping ratio for ODBS braced frame in compare with the others. Hysteresis analysis is performed for each system with their

(a)

Time history response of 15st flexural RC frame displacement under Tabas earthquake

(b)

Time history response of 15st X-braced RC frame displacement under Tabas earthquake

(c)

Time history response of 15st ODBS RC frame displacement under Tabas earthquake

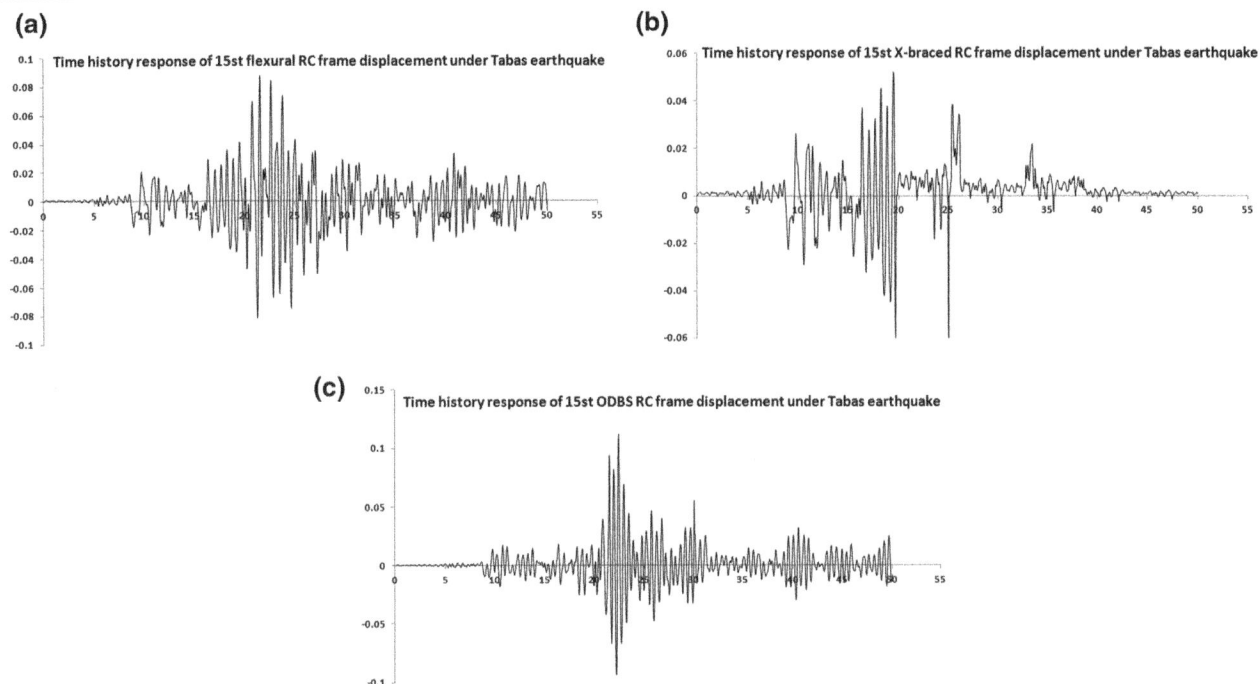

Fig. 33 Displacement response of **a** Flexural, **b** ODBS and **c** X-braced frame under Tabas time history analysis.

corresponding stories under earthquake records that are investigated one by one as shown in Figs. 24–29. Also as indicated in figures at center, another ODBS specification is about resisting along the earthquake loads up to last cycles of vibration or even to the moment of collapse, vice versa, sole use of flexural frame is not able to resist earthquake in the next cycles just similar to its initial behaviour.

10.2 Time History Analysis Response

Ultimately in last step, the diagrams of time history responses are investigated for different systems of load bearing and under the various seismic records. According to mentioned scaled records, the sightly accelerograms are defined for time history analysis within this research article. As indicated in comparative Fig. 30, the levels of vibration in ODBS system are compared for exciting acceleration and response acceleration.

The level of ODBS response acceleration (in each story by off-diagonal bracing system) is about 41 percent of exciting acceleration. Similar treatment of ODBS has been observed in friction damper device and/or in isolators. Figures 31, 32 and 33 indicate the variation of acceleration and displacement response for various systems and under different earthquakes. For the ODBS system, the number of mode and their vibration's amplitude could be decreased, that this behaviour is suitable for a structure.

For the future, the exact modeling and analysis of ODBS by experimental investigation could be performed and verified by highest level of accuracy. According to obtained results, the influence factor for variation control and energy dissipation of ODBS system is more than other lateral resisting systems, specially compare with flexural and x-braced frame. The ODBS system was very advantages to

controlling deteriorations due to loading by short time period just like an isolation damper or viscous damper and/or the composed system of them.

11. Conclusion

This research has several results about applying off-diagonal bracing system in reinforced concrete frame. The ODBS is considered in single story and multi stories frame to investigating its behaviour under nonlinear static and dynamic analysis. The results can be explained as below:

(1) FE models are treated properly as verified by experimental investigations. Nonlinear analysis of FE models is indicated as converged response along its ductility confirmations. The finite element analyses are performed by comparative models as Flexural, X-braced and ODBS braced frames where their results indicate the highest ductility for ODBS. Pushover diagrams are also performed for investigating the capacity of various mentioned frames.

(2) The acceptance criteria for rotations and displacements are compared with related quantities of the ATC manual. The ODBS models had the most rotation at nodal elements, before the collapse level. Each of occurred plastic hinges is deformed elastically and plastically two times and it may be the cause of more energy dissipation.

(3) Also the crack analysis is performed according to finite element models in a time domain consisted of several time steps. The time steps were the steps of imposing displacement. The ODBS system was fully cracked pattern in term of flexural and shear crack. The

differences between flexural and ODBS systems are indicated in capacity of bearing for both of deformations and loads.

(4) High-rise frames in 2, 6 and 15 floors are modeled numerically for considering ODBS effects on a simple flexural frame. The optimum number of stories to use ODBS system in first story is about 4–12 stories but if the ODBS system utilized for all stories, it has no limitation for number of stories. The only disadvantage of using ODBS in all stories is that it's uneconomically.

(5) Application of ODBS system is considered for different regions and their given results are concentrated along its better behaviour under strong ground motion, because the natural period of ODBS system is high and for preventing the occurrence of resonance phenomenon, it is better that, this system being under strong ground excitation by low excitation's period.

(6) The specific property of ODBS system is about its damping and its ability for dissipating received energy. Along analytical investigation of ODBS system, the equivalent viscous damping ratio is calculated about 11–15 % according to the results. The ODBS is absorbed 6–10 % of damping ratio in term of hysteresis energy dissipation. The ODBS equivalent damping is higher than related quantities for x-braced and flexural frame. The ODBS damping ratio is about 1.8–2.5 times of flexural frame damping.

(7) Mentioned numerical models are also considered to assessing the ODBS hysteretic behaviour. Many hysteresis diagrams are obtained to various models under several earthquake records. Hysteresis curves indicate the ODBS system had less pinching and the most strain energy. The obtained curves are geometrical comparison to calculating its energy dissipated quantities. The absorbed energy for ODBS model was several times of x-bracing system. The strength degradation for flexural frame is more than the other systems. The ODBS system is a sustainable system versus the large number of cycles.

Acknowledgments

I would like to express my deep gratitude to professors of Shiraz University for their life pattern specially thanks to Profs. H. Seyyedian; S. A. Anvar & A. R. Ranjbaran. Finally, I wish to specially thanks to Mehri Zoroufian for her support and encouragement throughout my study.

References

Abou Elfath, A., & Ghobarah, A. (2000). Behaviour of reinforced concrete frames rehabilitated with concentric steel bracing. *Canadian Journal of Civil Engineering, 27*, 433–444.

ACI Committee 318 (2002). Building code requirements for reinforced concrete (ACI 318-02). Detroit, MI: American Concrete Institute.

Altun, F., & Birdal, F. (2012). Analytical investigation of a three-dimensional FRP-retrofitted reinforced concrete structure's behaviour under earthquake load effect in ANSYS program. *Natural Hazards and Earth System Sciences., 12*, 3701–3707.

Amiri, G., Gh, R., et al. (2008). Evaluation of performance behaviour of reinforced concrete frame rehabilitated with coaxial steel braces. *The Journal of Structure and Steel, 4*(4), 17–25.

Amiri, A., & Tabatabaei, R. (2008). Earthquake risk management strategy plan using nonparametric estimation of hazard rate. *American Journal of Applied Sciences, 5*(5), 581–585.

Anam, I & Shoma, Z. N. (2011). Nonlinear properties of reinforced concrete structures, Department of Civil and Environmental Engineering, The University of Asia Pacific, tech_ bulletin n & journal.

ANSYS (2009&2015). ANSYS manual, ANSYS, INC, Canonsburg, PA 15317, USA.

ATC. (1995a). Structural response modification factors. ATC-19 Report. Applied Technology Council. Redwood City, CA, 1995.

ATC. (1995b). A critical review of current approaches to earthquake-resistant design. ATC-34 Report. Applied Technology Council. Redwood City, CA, 1995.

Badoux, M., & Jirsa, J. (1990). Steel Bracing of RC Frame for seismic retrofitting. *ASCE Journal of Structural Engineering, 116*, 55–74.

Bathe, K. J. (1996). *Finite element procedures.* Upper Saddle River, NJ: Prentice Hall.

Bazant, Z. P., & Becq-Giraudon, E. (2002). Statistical prediction of fracture parameters of concrete and implications for choice of testing standard. *Cement and Concrete Research, 32*, 529–556.

Berberian, M., & Navai I. (1977). Naghan (chahar mahal bakhtiari-high zagros, Iran) earthquake of 6 April 1977: A preliminary field report and a seismotectonic discussion. *Geological Survey Iran, 40*, pp. 51–77.

Bozorgnia, Y., & Bertero, V. (2004). *Earthquake engineering: From engineering seismology to performance-based engineering.* Boca Raton, FL: CRC Press.

Bush, T. D., Wyllie, L. A., & Jirsa, J. O. (1991). Observations on two seismic strengthening schemes for concrete frames. *Earthquake Spectra, 7*(4), 511–527.

Constantinou, M. C., Soong, T. T., & Dargush, G. F. (1998). Passive energy dissipation systems for structural design and retrofit. Monograph no. 1. Buffalo, NY: Multidisciplinary Center for Earthquake Engineering Research.

Cotsovos, D. M. (2013). Cracking of RC beam/column joints: Implications for the analysis of frame-type structures. *Engineering Structures, 52*, 131–139.

El-Metwally, S., & Chen, W. F. (1988). Moment-rotation modelling of reinforced concrete beam–column connections. *ACI Structural Journal, 85*(4), 384–394.

FEMA-356, & FEMA-270. (2000). *Pre-standard and commentary for seismic rehabilitation of buildings.* Washington, DC: Federal Emergency Federal Agency.

Ghaffarzadeh, H., & Maheri, M. R. (2006). Cyclic tests on the internally braced RC frames. *Journal of Seismology and Earthquake Engineering (JSEE), 8*(2), 177–186.

Ghobarah, A., & Abou Elfath, A. (2001). Rehabilitation of a reinforced concrete frame using eccentric steel bracing. *Engineering Structures, 23*, 745–755.

Gourabi, A., & Yamani, M. (2011). Active faulting and quaternary landforms deformation related to the nain fault. *American Journal of Environmental Sciences, 7*(5), 441–447.

Guan, Y., Cheng, X., & Zhang, Y. (2011). Study on the earthquake disaster reduction information management system and its application. *International Journal of Intelligent Systems and Applications, 1*, 51–57.

Kawamata, S., & Ohnuma, M. (1981). Strengthening effect of eccentric steel braces to existing reinforced concrete frames, In *7WCEE Conference, Proceedings of 2nd Seminar on Repair and Retrofit of Structures.* Ann Arbor, MI: NSF.

Kent, D. C., & Park, R. (1971). Flexural members with confined concrete. *Journal of the Structural Division ASCE, 97*, 1969–1990.

Maheri, M. R., & Akbari, R. (2003). Seismic Behaviour factor, R, for steel x-braced and knee-braced RC building. *Engineering Structures, 25*, 1505–1513.

Maheri, M. R., & Ghaffarzadeh, H. (2008). Connection overstrength in steel-braced RC frames. *Engineering Structures, 30*(7), 1938–1948.

Maheri, M. R., & Hadjipour, A. (2003). Experimental investigation and design of steel brace connection to RC frame. *Engineering Structures, 25*(13), 1707–1714.

Maheri, M. R., Kousari, R., & Razzazan, M. (2003). Pushover tests on steel X-braced and knee-braced RC frames. *Engineering Structures, 25*(13), 1697–1705.

Maheri, M. R., & Sahebi, A. (1995a). Experimental investigation of the use steel bracing in reinforced concrete frames. In *Proceedings of 2nd International Conf. on Seismology and Earthquake Eng.*, Tehran, Islamic Republic of Iran, Vol. 1, pp.755–784.

Maheri, M. R., & Sahebi, A. (1995b). Use of Steel bracing in reinforced concrete frames. *Engineering Structures, 9*, 545–552.

Masri, M., & Goel, S. C. (1996). Seismic design and testing of an RC slab-column frame strengthening by steel bracing. *Earthquake Spectra, 12*(4), 645–666.

Mastrandrea, L., & Piluso, V. (2009a). Plastic design of eccentrically braced frames, I: Moment–shear interaction. *Journal of Constructional Steel Research, 65*(5), 1007–1014.

Mastrandrea, L., & Piluso, V. (2009b). Plastic design of eccentrically braced frames, II: Failure mode control. *Journal of Constructional Steel Research, 65*(5), 1015–1028.

Moghaddam, H. A., & Estekanchi, H. (1994). On the characteristics of an off-diagonal bracing system. *Elsevier, Journal of Constructional Steel Research, 35*, 361–376.

Moghaddam, H. A., & Estekanchi, H. (1999). Seismic behaviour of off-diagonal bracing systems. *Elsevier, Journal of Constructional Steel Research, 51*, 177–196.

Mohyeddin, A., Helen, M., Goldsworthy, B., & Emad, F. Gad. (2013). FE modelling of RC frames with masonry infill panels under in-plane and out-of-plane loading. *Engineering Structures, 51*, 73–87.

PEER. (2005). Strong motion database http://peer.berkeley.edu Retrieved 4 Apr 2009.

Providakis, C. P. (2008). Pushover analysis of base-isolated steel–concrete composite structures under near-fault excitations. *Soil Dynamics and Earthquake Engineering, 28*(4), 293–304.

Ramin, K. (2009). Seismic investigation and numerical analysis of RC frame retrofitted by off diagonal steel bracing system (ODBS) Master science of Shiraz university (Thesis in Farsi).

Rastiveis, H., Samadzadegan, F., & Reinartz, P. (2013). A fuzzy decision making system for building damage map creation using high resolution satellite imagery. *Natural Hazards and Earth System Sciences, 13*, 455–472.

Ravi Kumar, G., Kalyanaraman, V., & Kumar, S. (2007). Behaviour of frames with non-buckling bracing under earthquakes. *Journal of Constructional Steel Research, Elsevier, 63*, 254–262.

Razavian Amrei, S. A., Ghodrati Amiri, G., & Rezaei, D. (2011). Evaluation of horizontal seismic hazard of Naghan, Iran World Academy of Science, Engineering and Technology, Vol. 5.

Roeder, C., Popov, E. (1977). Inelastic behaviour of eccentrically braced steel frames under cyclic loading, Univ. of Calif. Berkeley, CA.

SAP2000 (ver11.0). (2010). *Software and manual.* Berkeley, CA: CSI: Computers and Structures Inc.

Scott, B. D., Park, R., & Priestley, M. J. N. (1982). Stress–strain behaviour of concrete confined by overlapping hoops at low and high strain rates. *ACI Journal, 79*, 13–27.

Steidl, J. H., & Lee, Y. (2000). The SCEC phase III strong-motion database. *Bulletin of the Seismological Society of America, 90*, S113–S135.

Sugano, S., & Fujimura, M. (1980). Seismic strengthening of existing reinforced concrete building. In *Proceeding of the 7th World Conference on Earthquake Eng.*, Turkey, Part 1, Vol. 4, pp. 449–459.

Tasnimi, A., & Masoumi, A. (1999). Study of the behaviour of reinforced concrete frames strengthened with steel brace. In *Third international conference on seismology and seismic engineering.* Tehran, Iran.

Taucer, F., Spacone, E., & Filippou, F. C. (1991). *A fiber beam-column element for seismic response analysis of reinforced concrete structures.* Berkeley, CA: Earthquake Engineering Research Center, University of California at Berkeley.

Uang, C. M. (1991). Establishing R (or Rw) and Cd factors for building seismic provisions. *Journal of structural Engineering, ASCE, 117,* 19–28.

Vatani Oskouei, A., & Rafi'ee, M. H. (2009). Damage modeling in reinforced concrete bending frames originated by quake and its restoration using X-bracing system. *Esteghlal, 28*(1), 49–73.

Willam K.J., & Warnke E.P. (1974). Constitutive model for the triaxial behaviour of concrete. In *Proceedings of the international association for bridge and structural engineering,* Vol. 19, (pp. 1–30). Bergamo, Italy.

Yaghmaei Sabegh, S. (2011). Stochastic finite fault modeling for the 16 september 1978 Tabas, Iran, earthquake. *IJE Transaction A Basics, 24*(1), 15–24.

Youssefa, M. A., Ghaffarzadehb, H., & Nehdia, M. (2007). Seismic performance of RC frames with concentric internal steel bracing. *Engineering Structures, 29,* 1561–1568.

2

Compression Behavior of Form Block Walls Corresponding to the Strength of Block and Grout Concrete

S. Y. Seo[1],*, S. M. Jeon[1], K. T. Kim[2], M. Kuroki[3], and K. Kikuchi[3]

Abstract: This study aimed to present a reinforced concrete block system that reduces the flange thickness of the existing form block used in new buildings and optimizes the web form, and can thus capable of being used in the seismic retrofit of new and existing buildings. By conducting a compression test and finite element analysis based on the block and grouted concrete strength, it attempted to determine the compression capacity of the form block that can be used in new construction and seismic retrofit. As a result, the comparison of the strength equation from Architectural Institute of Japan to the prism compression test showed that the mortar coefficient of 0.55 was suitable instead of 0.75 recommended in the equation. The stress–strain relation of the block was proposed as a bi-linear model based on the compression test result of the single form block. Using the proposed model, finite element analysis was conducted on the prism specimens, and it was shown that the proposed model predicted the compression behavior of the form block appropriately.

Keywords: form block, prism test, finite element analysis, grout concrete, mortar coefficient, compression behavior.

1. Introduction

In the recent seismic retrofit of school buildings in South Korea, whose construction is susceptible to damage from earthquakes, the most widely used retrofit method for frame buildings is installing a damper in the openings such as windows in tandem with the expansion of infill walls. Infill walls are usually cast-in-place concrete, but another possible construction method involves the use of reinforced block walls.

Compared to cast-in-place concrete, reinforced block walls have a somewhat lower structural capacity but offer excellent constructability, and thus, if they satisfy the required capacity, they can be recommended for use in a seismic-strengthening method. Usually, blocks used as reinforced block walls make it difficult to fill the hollow block zones with a sufficient amount of grouted concrete. Furthermore, their hollow areas are relatively small, and as such, the amount of grouted concrete is also small. In addition, bar arranging in construction is difficult, and thus, they provide little structural integrity. If the amount of grouted concrete increases, the blocks become thinner, so that when casting,

their resistance to lateral pressure decreases, and they can be easily broken during delivery or construction. Therefore, the size of the hollow zones should be determined based on a comprehensive review of the constructability and strength of the blocks. The form block used for a new construction has wider hollow zones than the existing blocks. In other words, it increases the volumetric ratio of grouted concrete, which increases its strength and thus improves the structural capacity of the wall after the completion of the construction.

Accordingly, this study aimed to present a reinforced concrete block system that reduces the flange thickness of the existing form block used in new buildings and optimizes the web form, and can thus capable of being used in the seismic retrofit of new and existing buildings. By conducting a compression test and finite element analysis based on the block and grouted concrete strength, it attempted to determine the compression capacity of the form block that can be used in new construction and seismic retrofit.

2. Compression Behavior of the Reinforced Form Block Wall

2.1 Construction Process of the Reinforced Form Block Wall

The construction process of reinforced form-block-walls follows the Standard Specifications for architectural construction (2013), where the vertical mortar joint of the block wall is set as the continuous joint to reinforce the hollow zones with bar and grouted concrete. In the case of the infill wall, as a seismic retrofit method for a building, as shown in Fig. 1, the vertical bar is fixed onto the base plate and the block corners,

---

[1]Department of Architectural Engineering, Korea National University of Transportation, Chungju, Korea.
*Corresponding Author; E-mail: syseo@ut.ac.kr

[2]B&K Construction Technology, Chungju, Korea.
[3]Department of Architecture and Mechatronics, Oita University, Oita, Japan.

(a) Settlement of vertical bars

(b) Staking blocks

(c) Casting grout concrete

(d) Completion of wall construction

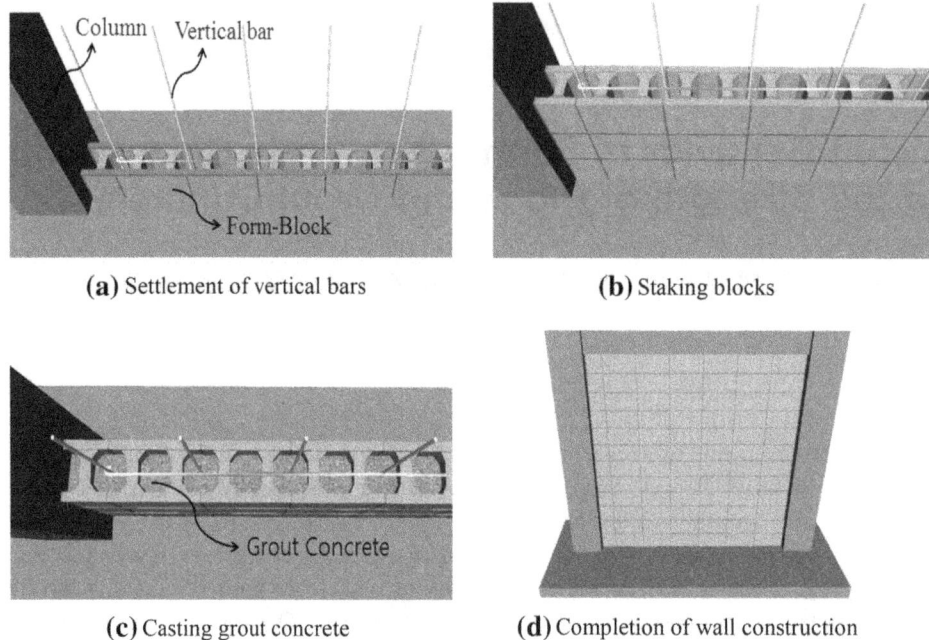

Fig. 1 Construction process of form-block-wall.

and the other standard parts are first constructed, and then the horizontal bar is arranged and blocks are stacked on top of the mortar. On top of the walls are gaps for grouted concrete, which are finished using non-shrinking mortar.

For the design of the reinforced form block walls, the structural capacity of the walls against compression and shear force should be determined. In particular, the compression capacity of the walls whose hollow block zones are filled with grouted concrete allows the walls to function as braces on frames with the compression strut under the horizontal load, and therefore becomes the most important structural capacity. Therefore, for the design of the new construction of the form block and for seismic retrofit, it is essential to predict the appropriate compression structure of block walls.

2.2 Previous Researches

Masonry walls and concrete blocks have long been used in construction, and much research has been conducted on them in other countries. Recently, Jonaitis and Zavalis (2013) conducted a test to determine the compression behavior of an empty concrete block and of another block filled with grouted concrete. A fracture was found on the mortar joint of the empty concrete block, and as such, an increase in compression stress caused lateral deformation. Furthermore, it was discovered that the concrete block filled with grouted concrete showed a compression behavior similar to that of cast-in-place concrete. Shing et al. (1989) evaluated the fracture mechanism, ductility, and energy dissipation capacity of shear block walls through the cyclic loading test to determine their nonlinear behavior. Based on the test results, he reported that through seismic retrofit by using shear block walls, a resistance capacity to shear and ductility could be acquired to some degree. Zhai and Stewart (2010) presented a new theoretical equation with the

material strength, types of live load, the ratio of dead load to live load, and the combination of the eccentricity and concentric load as parameters to establish the safety guideline for reinforced block walls in China, and verified the appropriateness of the equation through tests. In Japan, form blocks are widely used in new construction projects where only grouted concrete used to fill the hollow zones inside the blocks would be considered to provide sufficient strength, and blocks would be considered to function only as a form (Architecture Institute of Japan 2006).

As opposed to Japan, in South Korea, three-story or lower buildings are constructed with unreinforced block, and those higher than three-story buildings are usually frame building or wall-type apartment. Therefore, few new buildings are constructed with form block. Accordingly, little research has been conducted on the form block, and some studies focused on its use for the seismic retrofit of frame type building. Yun et al. (2005) conducted an experiment on the use of blocks manufactured with recycled aggregate as infill walls, which showed that if block walls made with recycled aggregate are used in the construction of infill walls, the initial stiffness and shear strength can be improved. Compared to the cast-in-place infill walls, however, the reinforcement with block walls made with recycled aggregate is less effective, and due to constructability issues, it was shown to be difficult to acquire a reliable reinforcement effect. Kim et al. (2004) established a concept of a reinforced masonry wall with stacked hollow form blocks with a bar arranged inside and filled with concrete, and conducted a test on the wall reinforced with form block walls in a concrete frame. The test results showed that such a concept could produce an excellent seismic retrofit effect, and when reviewed based on the equation for the shear strength of reinforced masonry walls in Japan Code, the shear strength of the wall was found to be relatively under-evaluated.

As for the design of masonry wall construction in the Korean Building Code (2009), the cement-to-sand ratio (1:2.0–3.0) for bearing mortar is codified instead of the design strength; it is known that the value is for developing the compression strength of 30–40 Mpa. In the case of the grout, as with the bearing mortar, cement-to-sand ratio is used, but the minimum strength is set to be 1.3 times more than the compression strength of the masonry unit. If it is satisfied, after 28-day curing, the compression strength of the masonry can be decided through the prism test or unit strength test of block.

In International Building Code (2012) of USA, also, the strength of bearing mortar is decided by the mix ratio of cement-to-sand. The grout design also is achieved by the cement-to-sand mix ratio and its minimum strength and slump flow are codified. The compression strength of the grout should be over the design compression strength of masonry after 28-day curing, between 13.79 and 34.47 MPa (in the case of the concrete block). The slump flow should be ranged from 610 to 762 mm. As with KBC, the compression strength of the masonry wall can be verified through either the prism test or the unit strength of masonry, and it should meet the scope of 10.34–27.58 MPa (in the case of the concrete block).

From the above, in the case of KBC and IBC, if the bearing mortar and grout are mixed based on the standard, and the compression strength is over the standard strength, either the result of the prism test or that of the masonry unit strength can be used so that in the end, the strength of the form bock wall is determined by the masonry unit. When the strength of the grout, however, is lower than the masonry unit strength, neither code does not clearly identify how the strength can be calculated.

On the other hand, Japan (JASS 7), which shows considerable use of the form block, which is often used in seismic retrofit, offers equations for the compression strength, elastic module, and shear module of block walls using the material test results of the bearing mortar, grout, and masonry unit.

Seismic retrofit using blocks needs sufficient explanation of the structural capacity of the concrete-filled block walls against compression, based on which the resistance capacity against lateral force should be determined. Accordingly, this study aimed to present a design method for form block walls with optimized web and flange dimensions by considering the constructability and structural capacity and by conducting a compression test and a prism compression test on a single block as part of the research on examining the efficiency of reinforcement against compression. Furthermore, it aimed to examine an analysis method with which to predict the compression behavior of form block walls using nonlinear finite element analysis.

3. Compression Test of the Single Form Block

3.1 Compression Test Layout

As shown in Fig. 2, the size of the form block that was used in this study was 390 × 200 × 185 mm. The shape and dimension of the block was decided from the consideration to improve the quality control during the production of the blocks as well as the fabrication of those at the site. Based on the concept, the center of webs was moved to litter bit upper from the horizontal center line and the end of web was enlarged to increase the rigidity of both web and flange especially at the bottom region. Finite element analyses were performed to find best dimension.

This block was developed to be used as an in-filled wall in seismic retrofit of frame structure. As a construction process shown in Fig. 1, at first, the bars for vertical continuity shall be vertically anchored in the beams and then the form blocks shall be laid on the beams. In order to have horizontal continuity, bars also shall be horizontally anchored to columns. Those bars shall be connected to other bars by lap splicing. All blocks shall be laid on bearing mortar. After laying three blocks vertically, grout concrete shall be poured into the void holding bars vertically.

The compression-test parameters of the single form block were four mixing design ratios, as shown in Table 1, and the load was applied on the specimen, as in Fig. 3, based on the compression test on the whole section of the hollow concrete block according to KS F 4002 (2011).

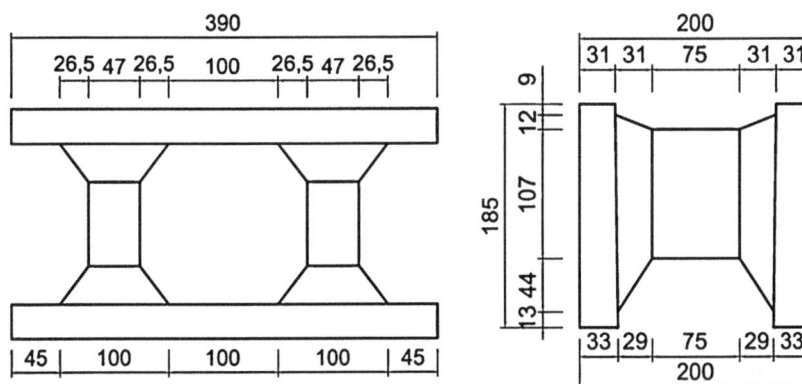

Fig. 2 Dimension of developed form-block (Unit: mm).

Table 1 Mixing design of form-block.

Type	Designed compressive strength (MPa)	Cement (%)	Sand (%)	Water (%)	Water/Cement (W/C:%)
B1	23.0	12.07	80.45	7.48	62.11
B2	25.0	13.70	78.31	7.99	58.48
B3	34.0	15.48	77.33	7.19	46.51
B4	38.0	19.35	72.53	8.12	42.02

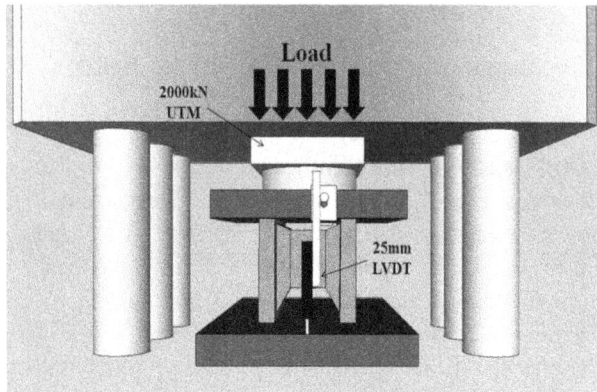

Fig. 3 Test setup for testing compressive strength of form-block.

3.2 Compression Test Result

Shown in Fig. 4 is the load–displacement curve measured as the result of the single form block test. It shows a linear curve initially, and the compression displacement rapidly increases at a certain point when several vertical cracks occur on flange from top to middle, leading to the crushing failure of upper part of both flanges without any failures in the webs. This pattern was observed in most blocks. Shown in Table 2 are the stress and strain at the maximum load and the strain value at the point when the strain suddenly increases after the elastic zone showing vertical cracks on flanges. This shows that as the amount of the cement increases, the compression strength of the block also increases, exceeding the predicted design strength based on each mixing ratio (23, 25, 34, and 38 MPa) shown in Table 1.

Figure 5 shows the strains at both yield and ultimate state corresponding to the compression strength. The strain at ultimate state decreases corresponding to the increase of compression strength while the yield strain is constant regardless the change of that. From Fig. 6 representing the ratio $\varepsilon_{by}/\varepsilon_{bu}$, it is found that the ratio is linearly dependent on compressive strength. The ratio can be expressed as a function of ultimate strength of block such as $0.02f_{bu}$ (dimensionless). This means the plastic deformation decreases gradually corresponding to the increase of ultimate strength of block due to brittle behavior as shown in Fig. 7. Also, the whole average value of elastic limit

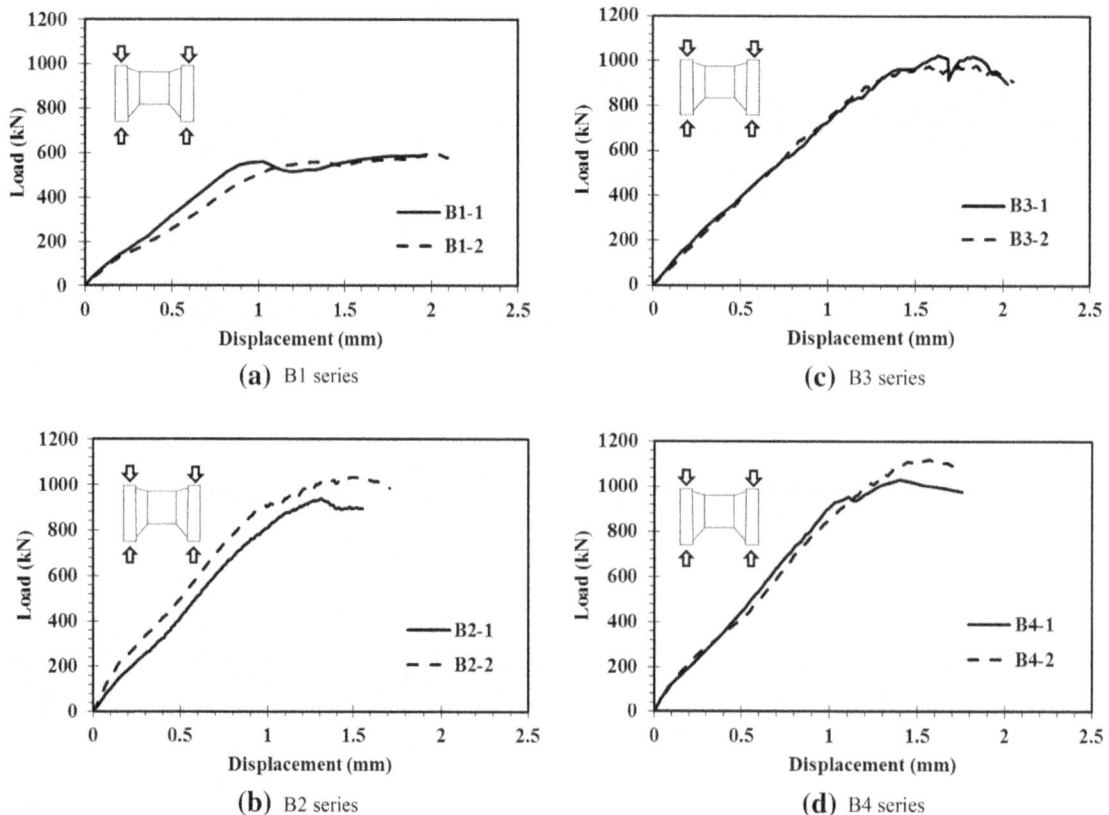

(a) B1 series

(c) B3 series

(b) B2 series

(d) B4 series

Fig. 4 Load-displacement curve of single form-block.

Table 2 Strain variation of single form-block.

Block type		B1 series			B2 series			B3 series			B3 series		
		B1-1	B1-2	Avg.	B2-1	B2-2	Avg.	B3-1	B3-2	Avg.	B4-1	B4-2	Avg.
At ultimate state	Load, P_{bu} (kN)	588.06	599.29	593.67	937.50	1031.60	984.55	1025.1	977.93	1,001.52	1030.10	1,119.21	1,074.66
	Stress*, f_{bu} (MPa)	23.56	24.01	23.79	37.56	41.33	39.45	41.07	39.18	40.13	41.27	44.84	43.06
	Displacement δ_{bu} (mm)	1.868	2.030	1.95	1.556	1.482	1.52	1.809	1.853	1.83	1.399	1.575	1.49
	Strain+, ε_{bu} (%)	1.01	1.10	1.05	0.84	0.80	0.82	0.98	1.00	0.99	0.76	0.85	0.80
At elastic end	Displacement δ_{by} (mm)	1.133	1.373	1.25	1.305	1.006	1.16	1.628	1.235	1.43	1.064	1.349	1.21
	Strain+, ε_{by} (%)	0.61	0.74	0.68	0.71	0.54	0.63	0.88	0.67	0.77	0.58	0.73	0.65
Strain ratio	$\varepsilon_{by}/\varepsilon_{bu}$	0.607	0.677	0.643	0.839	0.679	0.761	0.900	0.667	0.782	0.760	0.857	0.811

* Stress = Load/Flange area, + Strain = Displacement/height.

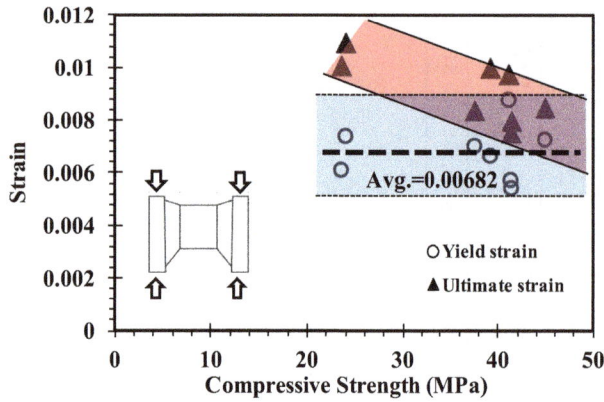

Fig. 5 Variation of strains corresponding to compressive strength of single form-blocks.

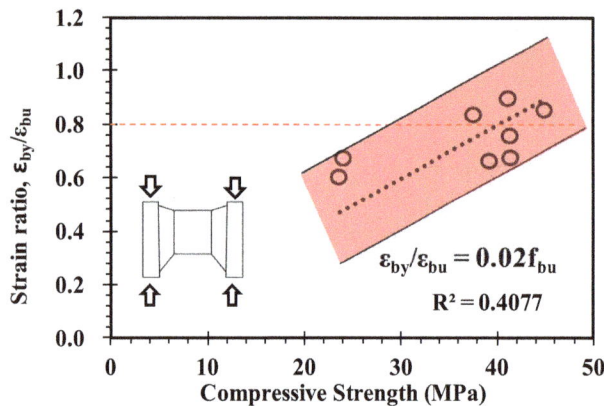

Fig. 6 Relationship between strain ratio and compressive strength in single form-block.

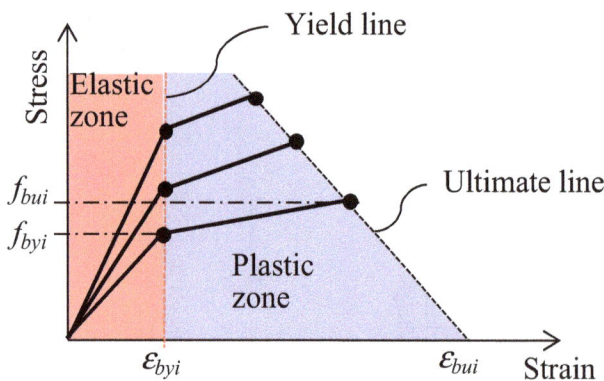

Fig. 7 Variations of yield and ultimate lines corresponding to the ultimate strength of block.

strain without considering the compression strength change of the block was about 80 % of the strain at ultimate strength. Considering such a result, the stress–strain relation of the form block can be expressed in a bi-linear model, as shown in Fig. 8.

4. Prism Test

4.1 Overview and Method

To determine the compression strength of the block filled with grouted concrete, the prism test was conducted, where

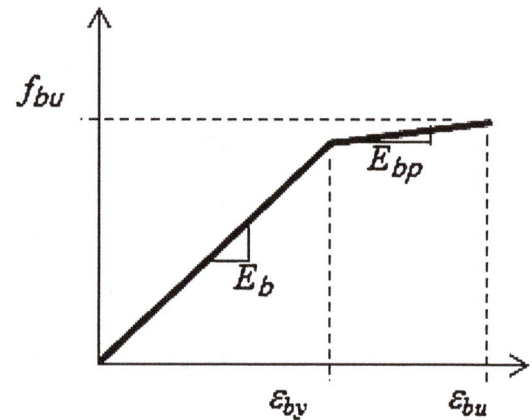

Fig. 8 Idealized bi-linear stress–strain model of form-block.

there were every three prism specimens for a total of eight parameters (two grouted concrete parameters and four single block parameters; total of 24 specimens). Shown in Table 3 are the dimension of the specimens, the compression strength of the single form block, and the material test strength of the grouted concrete and bearing mortar. Referring to KBC 2009 Prism Construction and Test (2009), the single block was cut in halves, as shown in Photo 1, and the blocks were stacked by three with the layering mortar, based on a 10 mm continuous joint thickness. Then concrete was poured inside the hollow zone attached with the form on the side to create prism specimens. After 20-day curing, compression force was applied on the specimens, using the Universal Testing Machine. To prevent specimen eccentricity, the compression force was applied to the specimens after attaching steel plates horizontally on the top and bottom of the specimens, using gypsum.

As shown in Fig. 9, the displacement was measured using the LVDT (linear variable differential transformer) installed in four corners, and a 60 mm concrete strain gauge was attached perpendicularly to the four central parts of the specimen to measure the strain on the initial elastic zone. The displacement gauge was used to measure the overall compression strain based on the increase in compression force, and the concrete strain gauge was used to observe the stress from the amount of compression strain, which was generated at the block flange and grouted concrete.

4.2 Prism Test Result
4.2.1 Load–Displacement Curve

Shown in Fig. 10 is the load–displacement curve that illustrates the effect of the block strength by the compression strength of grout concrete. If the strength of grout concrete is identical and the block strength differs, the larger the block strength is, the larger the prism strength becomes. This difference was shown to be larger with low grout concrete strength. In particular, in the case of the initial stiffness on the load–displacement curve, the stiffness change based on the block strength was found to be large for grout concrete with low strength whereas it was even regardless of the block strength if the grout concrete had high strength. This shows that if the strength of grout concrete is similar to or

Table 3 Dimension and material strengths of prism test specimens.

Specimen name	Width × length × height (mm × mm × mm)	f_{bu} (Mpa)	f_g (Mpa)
B1-C1	200 × 208 × 580	23.27	19.7
B1-C2	200 × 208 × 582	23.27	36.3
B2-C1	200 × 207 × 584	39.45	19.7
B2-C2	200 × 208 × 584	39.45	36.3
B3-C1	200 × 209 × 581	40.13	19.7
B3-C2	200 × 209 × 578	40.13	36.3
B4-C1	200 × 208 × 580	43.06	19.7
B4-C2	200 × 209 × 582	43.06	36.3

f_{bu} and f_g are compressive strength of block and grout concrete, respectively.
Bearing strength of mortar is 18.6 Mpa.
B Mixed property, C Compressive strength of grout concrete.

(a) Before grouting　　　**(b)** After grouting

Photo 1 Manufacturing specimens for prism test.

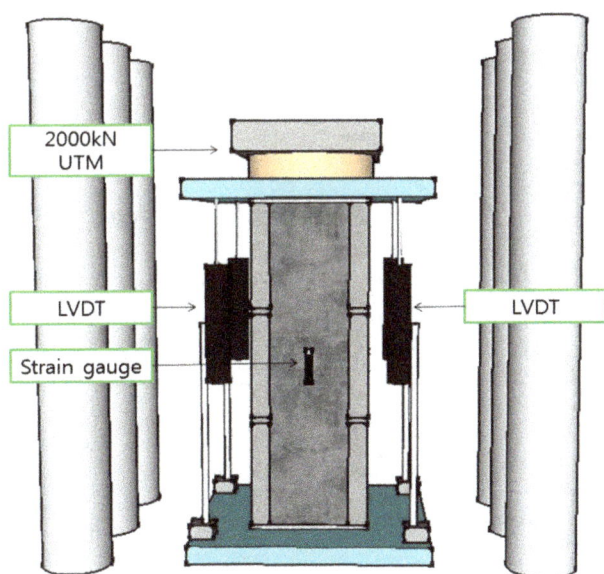

Fig. 9 Setup of prism test.

lower than the strength of the block, the block strength determines the prism strength, but if it is considerably higher than the block strength, the strength of grout concrete determines the prism strength. Figures 11 and 12 show the load variation corresponding to the strength of block and grout concrete, respectively. As mentioned above, the strength variation due to the change of block strength is large in the case of high strength of grout.

4.2.2 Strain and Failure Shape

The strain gauge, which was installed to determine the stress flow between the block and grout concrete against the gradually increasing compression stress, showed that stress was relatively evenly delivered to the block and grout concrete. Figure 13 shows graphs of representative specimens, which show the stresses resulting from the strain gauges, which were attached perpendicularly onto the center surface of the block and grout concrete, and the applied force divided by the total area. The strain pattern based on the stress of the strain gauge attached onto the surface of the block flange and grout concrete showed that the strain was low on the block while the stress was low, but the amounts of strain were similar to each other in the end. In a construction project where the form block is used, the surface of grout concrete is not exposed outside and is constrained so that there would be a constraint effect until the failure of the block flange wall.

At the end of the process, the tensile force on the block web due to the horizontal expansion of the grout concrete poured into the hollow zone of the block as well as the

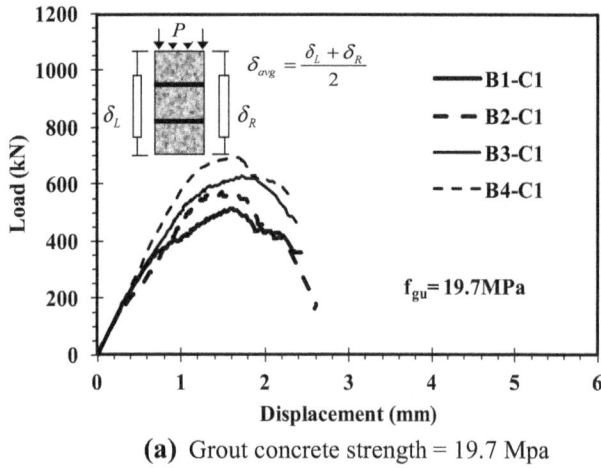

(a) Grout concrete strength = 19.7 Mpa

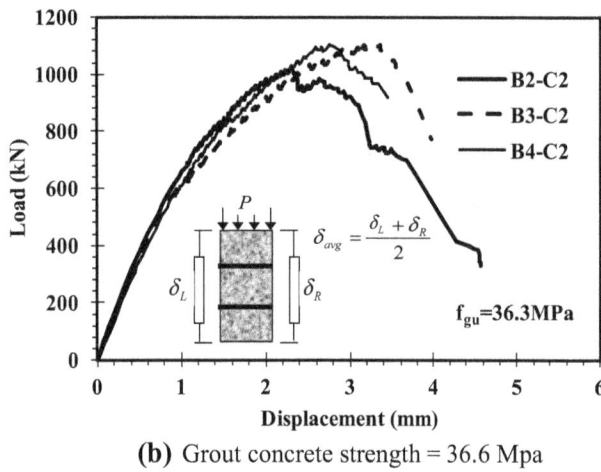

(b) Grout concrete strength = 36.6 Mpa

Fig. 10 Load-displacement curve corresponding to compressive strength of grout concrete.

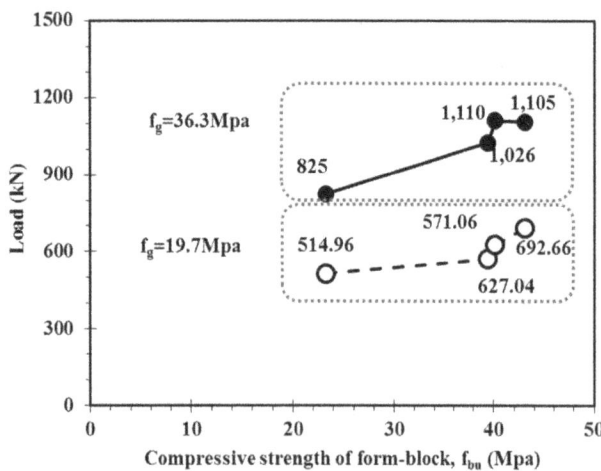

Fig. 11 Load variation due to compression strength of form-block.

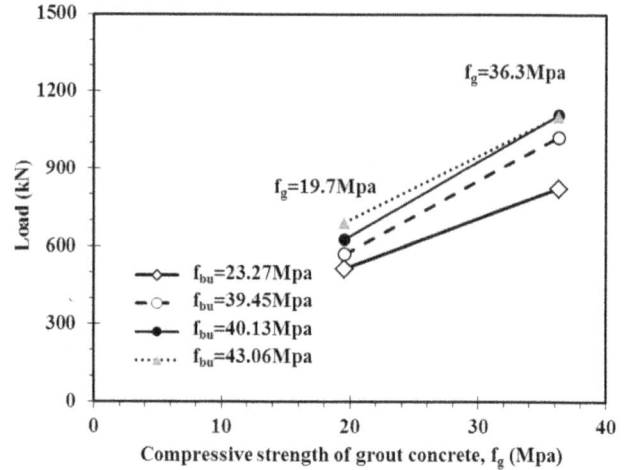

Fig. 12 Load variation due to compressive strength of grout concrete.

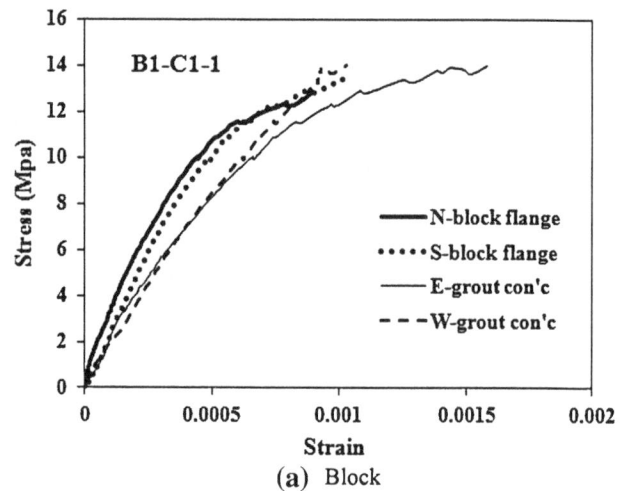

(a) Block

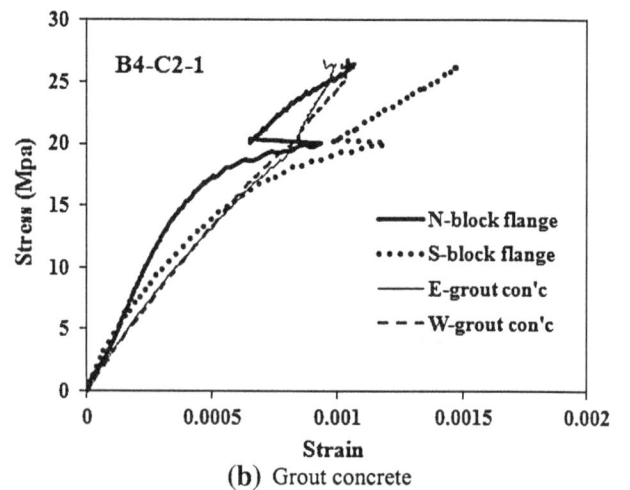

(b) Grout concrete

Fig. 13 Stress–strain curve of block and grout concrete.

compression displacement caused by the increased force resulted in the failure of the side flange of the form block. Shown in Photo 2 and Fig. 14 are the typical final failure and the failure behavior of the prism compression-strength specimens. There were few failures on the block flange whereas horizontal fractures were shown on the side of the grouted concrete. Also, as has been mentioned, the final

failure occurred as the flange was detached due to the failure of the block web shear.

4.2.3 Compression Strength Evaluation of the Form Block

In relation to calculating the compression strength of the form block, Architectural Institute of Japan Code (2006)

Photo 2 Typical failure shape of block wall after prism test (B3-C1-1)

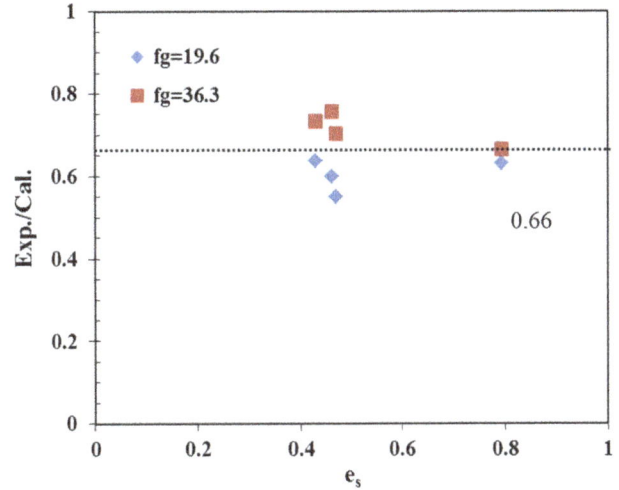

Fig. 15 Variation of ratios between test and calculation results corresponding to mortar factor.

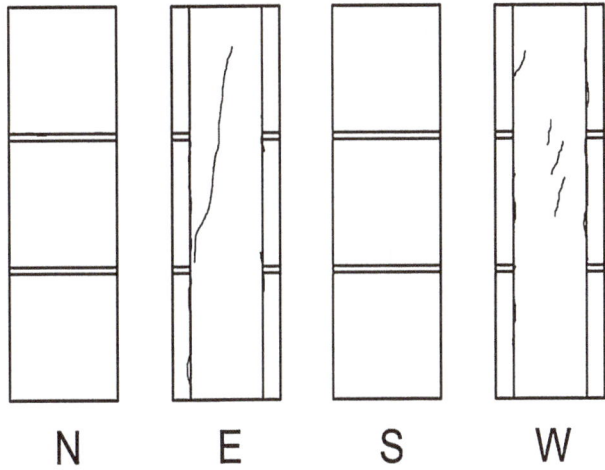

Fig. 14 Typical final crack shape of block wall after prism test (B3-C1-1).

defines, in their wall-structure design standard, the four-week compression strength through the test on the walls as the design strength, which can be calculated using Eq. (1) if there is sufficient information on the materials constituting the masonry walls. This equation is used to calculate the compression strength of the new form block, where the strength of the grout concrete and that of the block parts are simply added up to produce the prism strength. Also, by implementing the mortar coefficient, the strength is reduced based on the ratio of the bearing mortar strength to the block strength if the former is lower than the latter. The suggested usual value, however, is 0.75.

$$F_m = e_s \left[(1 - \beta') F_u + \beta' F_g \right] \tag{1}$$

where e_s is mortar factor, ration between strength of bearing mortar and block strength, F_u: strength of form-block (Mpa), F_g: grout concrete strength (Mpa), β': volumetric hollow ratio, and F_m: prism strength (Mpa).

Table 4 and Fig. 15 show the results of the calculation of the prism compression strength using Eq. (1), and of comparing the calculation result to the test result. The volumetric porosity of the form block was 65 %, and the strength of the form block and grout concrete was calculated based on the

Table 4 Comparison of prism test and calculation result.

Specimen name	Test result		Mortar factor e_s	$F_{m,c1}$ (MPa)	$F_{m,c2}$ (MPa)	$\dfrac{F_{m,\exp}}{F_{m,c1}}$	$\dfrac{F_{m,\exp}}{F_{m,c2}}$
	Ultimate load (kN)	Ultimate strength $F_{m,\exp}$ (MPa)					
B1-C1	514.96	13.20	0.79	20.88	11.49	0.63	1.15
B2-C1	571.06	14.64	0.47	26.55	14.60	0.55	1.00
B3-C1	627.04	16.08	0.46	26.79	14.73	0.60	1.09
B4-C1	692.66	17.76	0.43	27.81	15.30	0.64	1.16
B1-C2	824.58	21.14-	0.79	31.74	17.46	0.67	1.21
B2-C2	1025.5	26.29	0.47	37.40	20.57	0.70	1.28
B3-C2	1110.32	28.47	0.46	37.64	20.70	0.76	1.38
B4-C3	1105.1	28.34	0.43	38.67	21.27	0.73	1.33
Avg.						0.66	1.20

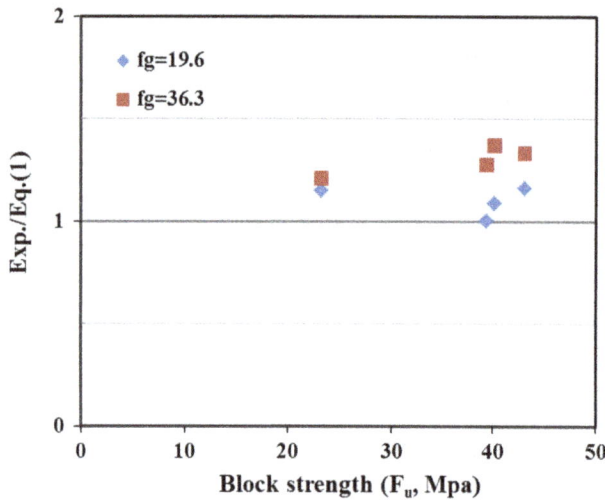

Fig. 16 Comparison of test and calculation results after applying new average mortar factor.

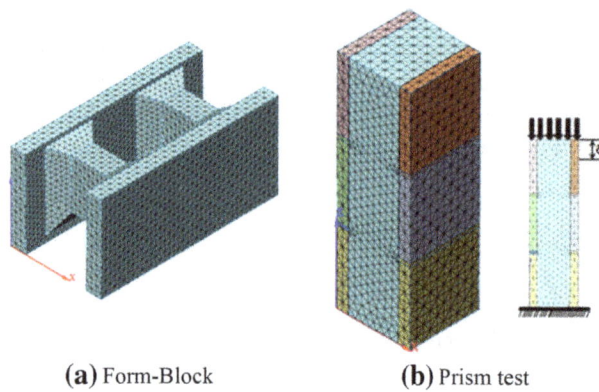

(a) Form-Block   **(b)** Prism test

Fig. 17 Modeling for finite element analysis.

material test results. In this study, the strength of the bearing mortar was constant while the block strength changed so that the aforementioned mortar factor changed. As the mortar coefficient which is horizontal axis changed from 0.46 to 0.79, there was no difference in the strength ratio. This means that the mortar coefficient does not change based on

the strength change of the bearing mortar or block, and is determined by the lower strength of the two. If the mortar coefficient is not considered, the ratio of the test result to the calculation was 0.66 on average, but the average strength of the B2-C1 specimen group was extremely low. As such, 0.66 was discarded, and 0.55 was used instead as the mortar coefficient to calculate the strength. The result of the calculation is shown in Fig. 16. Overall, the figure shows that the calculation result predicted the test result appropriately and conservatively.

5. Analysis of Compression Behavior Using Finite Element Analysis

5.1 Overview and Modeling

In this study, to analyze the compression-failure behavior of the single form block and prism specimens, nonlinear finite element analyses of the single form block in Fig. 17a and of the prism specimens in Fig. 17b were conducted by using Midas FEA program (2013). The compression material block model that was used in the analyses was a 3D solid element. The material model presented was used as the result of the compression test of the single form block, and its properties were based on the material test results. Shown in Fig. 18 are the tensile and compression material models of grout concrete and mortar, where the tensile strength was set to 1/10 of the compression strength.

The crack model of the prism specimens was the total strain crack model, which was used with the consideration of concrete cracks. In this analysis, the total strain crack model of the discrete crack model was used. And fixed crack model was applied.

To compare the parameters with the test results, the failure behavior was analyzed while considering the interfacial condition between the materials and the same strength of the form block and grout concrete as the test parameters. The interfacial condition model used two methods: the weld contact and the general contact. The weld contact method is used when the main and subsequent contact areas are attached to one another from the initial stage; it does not

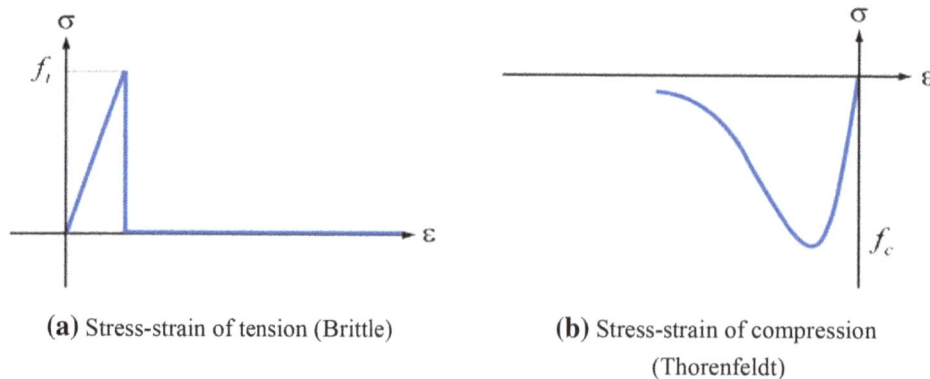

(a) Stress-strain of tension (Brittle)   **(b)** Stress-strain of compression (Thorenfeldt)

Fig. 18 Material model of grout concrete and mortar.

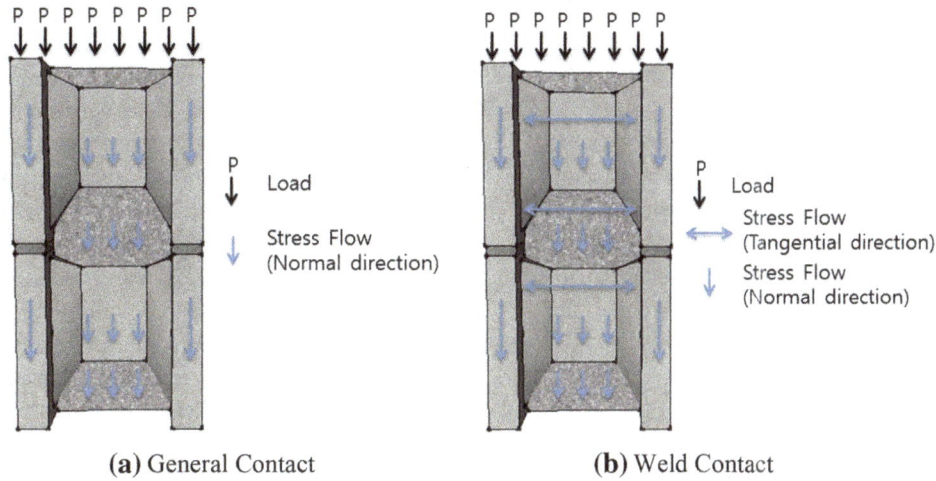

Fig. 19 Conceptual diagram of stress flow.

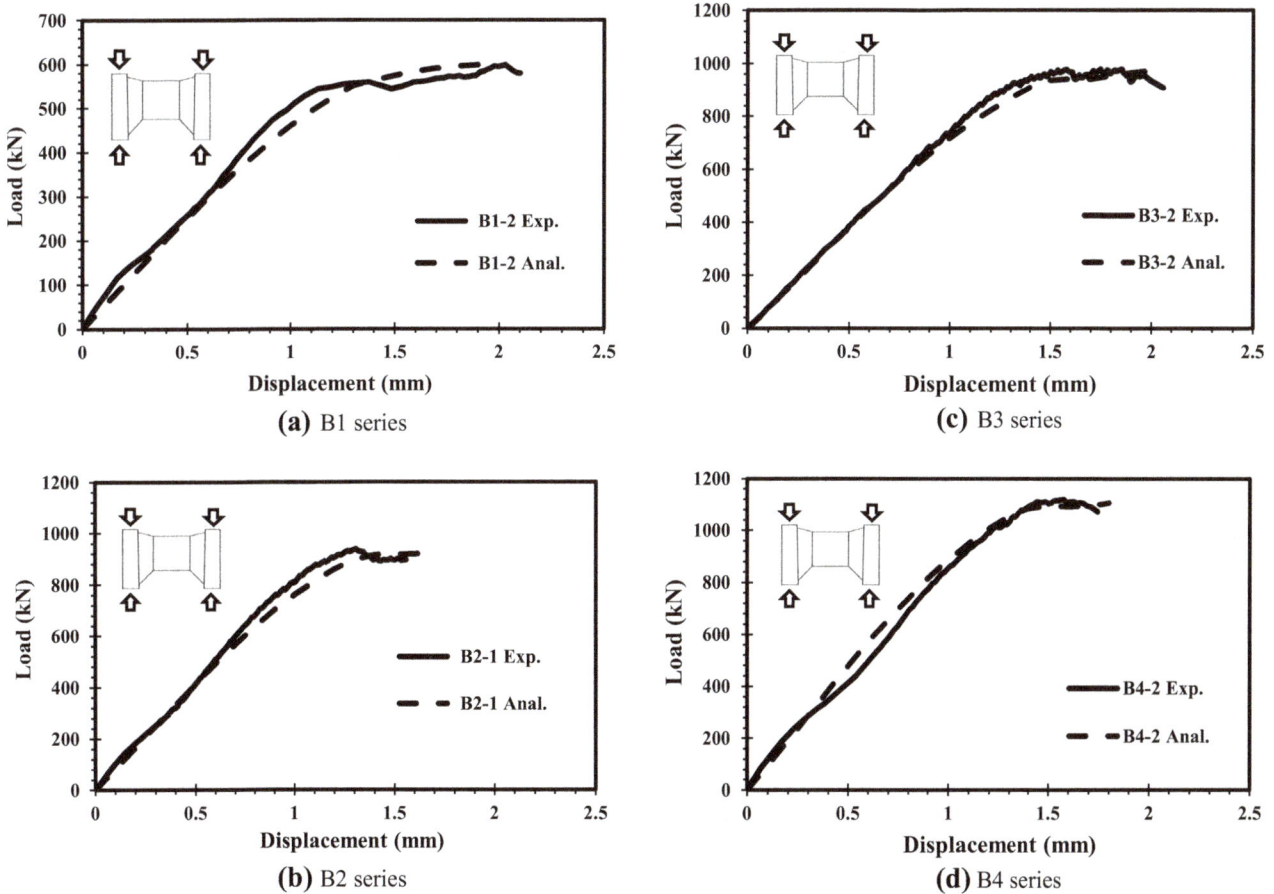

Fig. 20 Comparison of FE analysis and test results for single block walls.

allow the two surfaces to be detached during the analysis so that the load is delivered in the direction of the normal and contact directions of the model. The general contact method is used when the two surfaces were attached before the analysis and detached during the analysis as well as when the two surfaces continue to be attached and detached. The load is delivered only to the normal direction as shown in Fig. 19.

5.2 Analysis Result of the Single Form Blocks

Shown in Fig. 20 is the single form block analysis result, which was shown in the load–displacement curve by comparing it to the test result. The block compression-material model used the bi-linear model in Fig. 8, which was based on the single form block test result. The analysis result showed that the load–displacement curve pattern, the maximum load, the initial stiffness, etc. of the test results were appropriately demonstrated so that it was determined that the block behavior could be explained by using the bi-linear model in Fig. 8.

5.3 Analysis Result of the Prism Specimens

Figure 21 shows the flow of the force at the maximum load, which demonstrates that together with the compression displacement caused by the increase in the working load, the tensile force on the block web due to the horizontal expansion of the grout concrete poured into the hollow zone of the

block resulted in the failure of the specimens where the lateral flanges of the form block failed. The general and subsequent contacts, used as parameters, showed the same failure behavior so that the stress delivered by the contact direction had little effect on the specimens, and the stress on the normal direction determined the compression behavior.

Shown in Fig. 22a and b are the load–displacement curves of the prism specimens (B3 series) compared with the test results, and shown in Fig. 22c and d are those of the prism specimens (B4 series) compared with the test results. While the analysis result of the strength of prism specimens was similar to the test result, even if the strain at the maximum strength was little bit smaller than the test result.

6. Conclusions

(1) From the single form block, all the specimens showed bi-linear load–displacement relation. The strain ratio of the elastic limit to the strain at maximum strength increased according to the maximum compression strength of material. It reaches 2 % of the maximum compression strength of the block, and the total average value without considering the compression strength change of the block was about 80 % of the strain when the strain at the elastic limit was at the maximum strength.

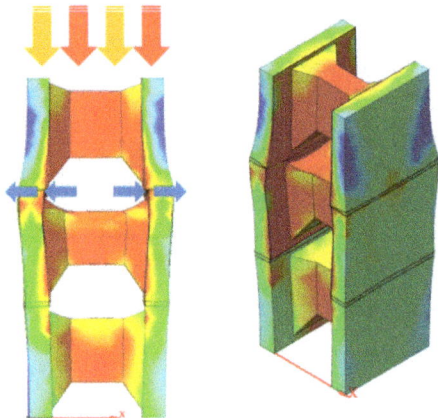

Fig. 21 Stress flow and deformation shape of prism test specimen at ultimate strength.

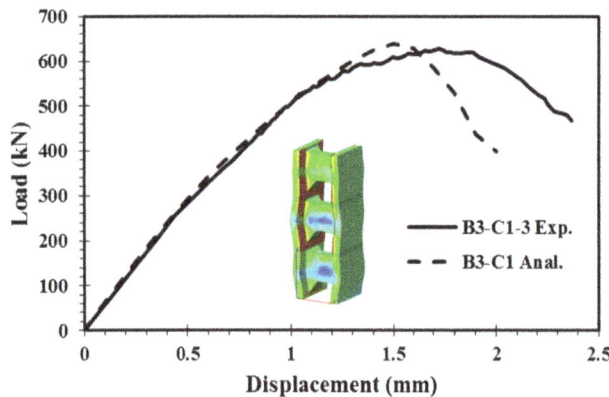

(a) B3-C1 series

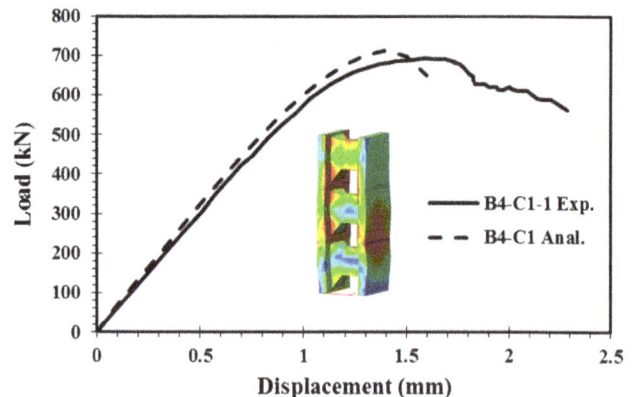

(c) B4-C1 series

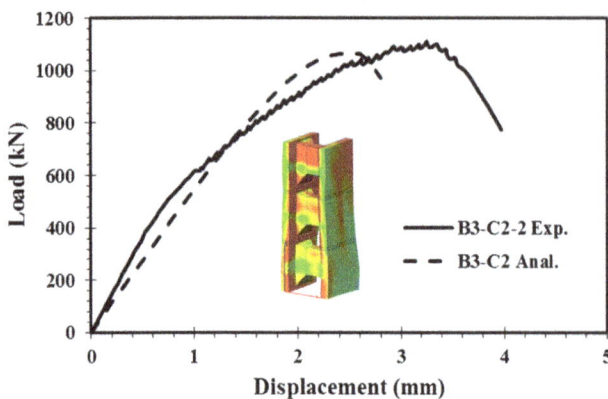

(b) B3-C2 series

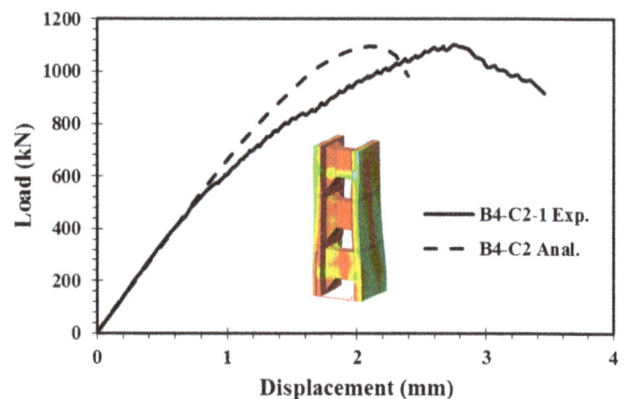

(d) B4-C2 series

Fig. 22 Comparison of FE analysis and test result of prism specimens.

(2) The prism test result showed that when the grout concrete strength was lower than or similar to the strength of the block, the block strength governed the prism strength. When the strength of the grout concrete was considerably higher than that of the block, however, the grout concrete strength governed the prism strength. In the case of the failure shape, due to the compression deformation generated by the load increment as well as the tensile force on the block web due to the horizontal expansion of the grouted concrete poured into the hollow zone of the block, the lateral flanges of the form block failed.

(3) The comparison of the strength equation from Architectural Institute of Japan to the prism compression test shows that the mortar coefficient is not variable value. If 0.75, which is usually implemented in the Japanese wall structure design standard, is used, the strength of the specimens is over evaluated by about 12 %. In this study, based on the test result, 0.55 was used as the mortar coefficient for safe design purpose. The calculation result based on this mortar coefficient was compatible with the test result.

(4) Based on the compression test result of the single form block, the stress–strain relation of the block is proposed as a bi-linear model. The finite element analysis result using the proposed model showed that the stress flow and the load–displacement curve were very similar to those of the test result. Also, using the proposed model, finite element analysis was conducted on the prism specimens, and it was shown that the proposed model predicted the compression behavior of the form block appropriately.

Acknowledgments

The authors acknowledge the support provided by Korea Association of Industry, Academy, and Research Institute (KAIARI) as one of the 2012 international business cooperation and technology development projects.

References

Architecture Institute of Japan (2006). *Standards for structural design of masonry structures.*

Jonaitis, B., & Zavalis, R. (2013). Experimental research of hollow concrete block masonry stress deformations. *Procedia Engineering, 57,* 473–478.

KS F 4002 (2011). *Hollow concrete block.*

Kim, K.-T., Seo, S.-Y., Yoon, S.-J., Yoshimura, K., & Sung, K.-T. (2004). Experimental study for higher seismic performance of confined masonry wall system, *Korea Concrete Institute,* (Vol.*12*(1), pp. 3–8) Proceeding of Fall Conference, November.

Ministry of Construction and Transportation. (2013). *Korean architectural standard specification,* pp. 7–53-7–55.

Ministry of Land, Transport and Maritime Affairs (2009). *KBC 2009, Production and Experiment of Prism,* 2009-1245, 0603.4.6.

Shing, P. B., Noland, J. L., Klamerus, E., & Spaeh, H. (1989). Inelastic behavior of concrete masonry shear wall. *Journal of Structural Engineering ASCE, 115*(9), 2204–2225.

Yun, H.-D., Han, M.-K., Kim, S.-W., Lee, G.-W., Park, W.-S., & Choi, C.-S. (2005). Structural performance of lightly reinforced concrete frame strengthened with infilled walls by concrete blocks made in recycled sands. *Journal of the Architectural Institute of Korea, 21*(4), 83–90.

Zhai, X., & Stewart, M. G. (2010). Structural reliability analysis of reinforced grouted concrete block masonry walls in compression. *Engineering Structures, 32*(1), 106–114.

3

Correlation Between Bulk and Surface Resistivity of Concrete

Pratanu Ghosh*, and Quang Tran

Abstract: Electrical resistivity is an important physical property of portland cement concrete which is directly related to chloride induced corrosion process. This study examined the electrical surface resistivity (SR) and bulk electrical resistivity (BR) of concrete cylinders for various binary and ternary based high-performance concrete (HPC) mixtures from 7 to 161 days. Two different types of instruments were utilized for this investigation and they were 4 point Wenner probe meter for SR and Merlin conductivity tester for bulk resistivity measurements. Chronological development of electrical resistivity as well as correlation between two types of resistivity on several days was established for all concrete mixtures. The ratio of experimental surface resistance to bulk resistance and corresponding resistivity was computed and compared with theoretical values. Results depicted that bulk and SR are well correlated for different groups of HPC mixtures and these mixtures have attained higher range of electrical resistivity for both types of measurements. In addition, this study presents distribution of surface and bulk resistivity in different permeability classes as proposed by Florida Department of Transportation (FDOT) specification from 7 to 161 days. Furthermore, electrical resistivity data for several HPC mixtures and testing procedure provide multiple promising options for long lasting bridge decks against chloride induced corrosion due to its ease of implementation, repeatability, non-destructive nature, and low cost.

Keywords: surface resistivity, bulk resistivity, HPC, corrosion, conductivity.

1. Introduction

Chloride induced corrosion is one of the major problems in concrete bridge decks, pavements, and other marine structures. For this reason, electrical resistance against corrosion protection needs to be evaluated for repair and retrofit of concrete structures. As a result, since the last decade, there is growing demand to develop and implement electrical response techniques as a health monitoring tool for concrete structures.

Electrical resistivity is noninvasive and non-destructive and can evaluate microstructure of concrete, it can be related to the volume fraction of the pores, conductivity of the pore solution and can be utilized to predict the diffusion coefficients of chloride ions and water permeability (Christensen et al. 1994). Typically chloride resistance of concrete by electrical response is determined by the rapid chloride ion penetration (RCPT) according to ASTM C1202-05 standard. The total charge passed in coulomb in 6 h of RCPT is considered a relative measure and indication of the concrete's resistance to chloride ingress. One major limitation of the RCPT is the high current flow through permeable concrete mixtures results in a "joule effect". The increase in

temperature effectively decreases the electrical resistance and encourages the electrical current to flow rapidly and produce more heat which further accelerates the current flow (Julio-Betancourt and Hooton 2004). However, the monitoring of the non-destructive methodology such as change of the electrical resistivity provides much more data about concrete properties in different conditions as a quality control tool in a simple way. One of the existing non-destructive methods of determination of concrete resistance to the chloride ion penetration is the surface electrical resistivity by Wenner 4-probe device. The FDOT has developed a method to standardize procedures for collection of resistivity readings (FDOT Standard 2004). Experimentation using the Wenner device on 529 samples was conducted by Kessler et al. (2005) at the FDOT to investigate whether resistivity can be used as a quality control measure in place of the RCPT. Tikalsky et al. (2011) completed a recent study on different binary and ternary based HPC mixtures electrical resistivity testing a 91 days and found that resistivity data is well correlated with RCPT data for different binary and ternary based HPC mixtures.

Previous research suggested that supplementary cementitious materials (SCMs) can be used as a partial cement replacement to increase the electrical resistivity of mortar and concrete (Katherine et al. 2010). Marriaga et al. (2010) studied the reliability of the RCPT and resistivity test on the basis of chloride resistance of ground granulated blast furnace slag (GGBFS) mixtures with different levels of cement replacements. They established that the electrical resistivity and the

---

Department of Civil and Environmental Engineering, California State University, Fullerton, Fullerton, USA.
*Corresponding Author; E-mail: pghosh@fullerton.edu

total charge passed is an indirect measure of the chloride penetration suitable for both ordinary portland cement (OPC) and GGBS mixtures. Icenogle et al. (2012) recently showed that the better precision of Wenner Probe resistivity meter from their experimental investigation of single laboratory and multi laboratory measurements and surface resistivity test shows lower variability than rapid chloride permeability test with different HPC mixtures. Paredes et al. (2012) conducted rigorous round robin program to document the repeatability and reproducibility of surface measurements data on 12 different PCC mixtures in several laboratories. Darren et al. (2011) established effectiveness of electrical resistivity technique for HPC to obtain a relationship with chloride diffusivity in order to evaluate the quality of the concrete. Their findings showed a high correlation coefficient in the range between 0.94–0.99, representing the suitability of using electrical resistivity technique to evaluate the quality control of high performance concrete and prediction of corrosion rate. Another possible method is to measure electrical resistance of concrete cylinder by using plate electrodes on the end of the sample (Polder et al. 2004; Newlands et al. 2008). This test can be performed by utilizing conductive medium and needs to be remembered that surface finish needs to be flat as much as possible for proper contact pressure and sponges were used between sample and plates to obtain better contact. Recently, Spragg et al. (2012) analyzed variability studies on 12 different cementitious mixtures for BR and SR and correlation was established at testing ages of 28, 56 and 91 days. Additionally, the effect of electrode resistance was discussed. It was noticed that the effect is not significant on high resistivity concrete.

Most of the previous study on electrical resistivity focused on measurement of a limited number of concrete mixtures over a shorter period of time. In this study, comprehensive investigation of bulk and surface electrical resistivity of thirty three different binary and ternary based HPC mixtures containing large numbers of supplementary materials have been conducted from 7 to 161 days and correlation was established between two different types of resistivity in different groups. Ratio of surface to bulk resistivity and corresponding resistance were calculated and compared with theoretical values at all ages starting from 7 to 161 days. It also highlights pattern of chronological development of surface and bulk electrical resistivity over a longer period of time and correlation of surface resistivity between 161 days with other time periods.

2. Experimental Investigation

2.1 Materials and Mixture Proportions

Thirty three different types of ternary and binary cementitious mixtures including the control mixture of 100 % ordinary portland cement with a water/cementitious materials ratio of 0.44 were designed to give a wide range of values for this experimental program. This water/cementitious materials ratio is typical of exposed bridge deck and substructure concrete. All mixtures contained 335 kg/m³

(564 lbs/yd³) of cementitious material with a coarse aggregate factor (CAF) of 0.67. Limestone coarse aggregate of size 19 mm (3/4 inch) meeting ASTM C33 No. 67 gradation and ASTM C33 silica sand were used. All the SCMs were replaced by mass. Tests were performed on mixtures using:

- Type II-V cement (TII-V)
- Ground granulated blast furnace slag of grade 120 (G120S)
- Ground granulated blast furnace slag of grade 100 (G100S)
- Class C fly Ash (C)
- Class F fly Ash (F)
- Silica fume (SF)
- Metakaolin (M)

Due to sulfate attack problems in California, it is mandatory to use Type II-V cement instead of Type I cement. The selection of mixture design was based on concrete mixtures meeting basic technical properties and also representing a diverse range of solutions to long term durability. The basic mixture parameters were coded into the names of the mixtures with percentage of each cementitious material, e.g. 75TII-V/20F/5SF means 75 % Type II-V Cement, 20 % Class F fly ash and 5 % Silica Fume. A high-range water reducing admixture (Glenium 7500) and an air entraining agent (MBVR AE90) were used to meet better workability and other durability performance specifications. All the mixtures were cast according to ASTM C192 practice and four cylinders 100 × 200 mm (4 × 8 in.) were prepared for both bulk and surface electrical resistivity testing at all ages. The cylinders were demolded after 24 ± 2 h and they were continuously cured in lime water tank. Electrical resistivity was measured on 7, 14, 28, 56, 91 and 161 days.

2.2 Measurement of Surface Electrical Resistivity

Surface resistivity measurement was performed by commercially available 4 point Wenner probe surface resistivity (SR) meter, manufactured by Proceq. In this study, Florida testing method was used for electrical resistivity measurement on 7, 14, 28, 56, 91 and 161 days for 100 × 200 mm (4 × 8 in.) cylinders except the curing condition and the probe spacing. All the cylinders were cast and then demolded within 24 ± 2 h. After demolding, the cylinders were placed in lime water tank. A multiplier of 1.1 is used for electrical resistivity data as suggested by AASHTO TP-95 specification for lime water curing condition. All the cylinders were removed from lime water tank on the specified testing days and tested in saturated surface dry (SSD) condition at 23 ± 2 °C by Resipod Wenner Probe meter. Readings were taken two times with 0, 90, 180 and 270 degree angles of circular face of each concrete cylinder. The data in this research were collected using a probe spacing of 50 mm (2 inches), instead of 38 mm (1.5 inches) as recommended by FDOT. The probe spacing could not be changed as it came from the manufacturer with 50 mm (2 inches) spacing. The whole experimental process took less than half hour to complete. Four cylinders were tested for each concrete mixture and altogether 32 data points (4 × 8 = 32 points) were collected for each mixture for surface electrical resistivity. The equipment measures the current flowing between the outer electrodes and the potential difference between the two inner electrodes. Assuming that the concrete cylinder has homogeneous semi-infinite geometry (the dimensions of the element are large in comparison of the probe spacing), and the probe depth is far less than the probe spacing, the concrete cylinder resistivity ρ can be computed as:

$$\rho_1 = 2\pi a \frac{V}{I} = 2\pi a R_1 \qquad (1)$$

where a is the probe spacing in mm; V is the applied voltage in volt; I is the current in ampere; and R_1 is the surface resistance in KOhm.

2.3 Measurement of Bulk Electrical Resistivity

The Merlin conductivity tester was used to measure the bulk electrical conductivity, or its inverse, the bulk electrical resistivity, of water saturated concrete cylinders of 100×200 mm (4×8 in) in lime water tank. Bulk resistivity measurement was performed on the same cylinder sample as of surface electrical resistivity measurement. This test is also non-destructive and simple to perform. A test result was obtained within 2 s, and sample preparation and testing altogether takes less than 30 min. The conductivity of a saturated concrete specimen provides information on the resistance of the concrete to penetration of ionic species by the diffusion mechanism. The curing criteria and number of specimens were same as of surface electrical resistivity. Additionally, before testing, the tester verified with Merlin verification cylinder for rapid calibration purpose. A cylinder was placed on the support and two ends were wet with spraying bottle. This test method consists of applying a potential difference to the cylindrical specimen, thereby producing a current flow through the cylinder. The potential difference and resulting current can be utilized to obtain the electrical resistance. Two readings were obtained from data logger for each cylinder specimen by swapping the two end faces of a cylinder. From the measured current I and voltage V, the bulk resistivity was calculated as follows:

$$\rho_2 = \frac{V}{I}\left(\frac{A}{L}\right) = R_2\left(\frac{A}{L}\right) \qquad (2)$$

where, A is the surface area, L is the length of the specimen and R_2 is the bulk resistance. Figure 1a and b shows experimental set up for Wenner 4-probe meter and Merlin conductivity tester.

For 100×200 mm (4×8 inch) cylinder, surface area $A = \pi\frac{d^2}{4}$, $d = 100$ mm (4 inches), $L = 200$ mm (8 inches), and probe spacing of $a = 50$ mm (2 inches). Finally, the ratio of theoretical surface and bulk resistance can be computed in Eq. (3).

$$\frac{R_1}{R_2} = \left(\frac{\rho_1}{\rho_2}\right)(1/8) \qquad (3)$$

Morris et al. (1996) developed the geometry correction factor for specific cylinder sizes and the ratio of two different

types of resistivity is computed as 2.63. As a result, the ratio of theoretical surface and bulk resistance can be calculated as:

$$\frac{R_1}{R_2} = 2.63/8 = 0.33 \qquad (4)$$

3. Results and Discussions

3.1 Bulk and Surface Resistivity Data Analysis and Correlation

Table 1 depicts classification of chloride ion permeability criteria of different concrete mixtures on the basis of surface electrical resistivity as suggested by FDOT specification. The correlation between bulk and SR on different days in different groups of mixtures is established and is shown in Figs. 2, 3, 4 and 5. Average values of surface and bulk resistivity data from four cylinders for all mixtures (moderate to low permeability class) at all ages are provided in Tables 2 and 3. They show that both surface and bulk electrical resistivity increases over time and increment is more prominent at later ages in case of ternary based HPC mixtures compared to binary and ordinary portland cement mixtures. It is also evident that high percentage replacement of Class C fly ash did not perform well in binary or ternary blends. The possible reason is due to its incompatibility with Class F fly ash or other SCMs. Similar problem of Class C fly ash was studied by Rupnow et al. (2007). Average standard deviation of measurement of SR and bulk resistivity ranges between 0.4 to 6.6 and 0 to 3 from 7 to 161 days from all cementitious mixtures. It should be noted that mostly the standard deviation increases over time. It is believed that this may be due to slight variations in temperature or configuration of the sample affected by manual labor which may have occurred at the laboratory which could change differences overtime.

Figures 2, 3, 4 and 5 show the correlation between bulk and SR for different group of mixtures. It is observed that for most group of mixtures, the coefficient of determination values for linear trend line are higher than 0.8 and sometimes close to 1 except some silica fume mixtures at early ages where the coefficient becomes less than 0.80. This proves that all binary and ternary mixtures are linearly well correlated in two different types of resistivity measurements at different test ages. In summary, it can be observed that the relationship between the bulk and SR of concrete is observed as linear and it follows same trend over time.

As mentioned above, the ratio of SR over bulk resistivity is computed as 2.63. In another word, the inverse ratio of bulk to SR is $1/2.63 = 0.38$. In Figs. 2, 3, 4 and 5, it can be observed that the slope of the linear trend line varies from 0.29 to 0.47. It is to be also noted that the variation of slope values of all group mixtures decreases over time as it varies from 0.29 to 0.45 at 7 days and 0.32 to 0.44 at 161 days. It can be concluded that the surface and the bulk resistivity has strong linear correlation and the trend line slope found in all group mixtures again justifies the ratio between surface and bulk resistivity.

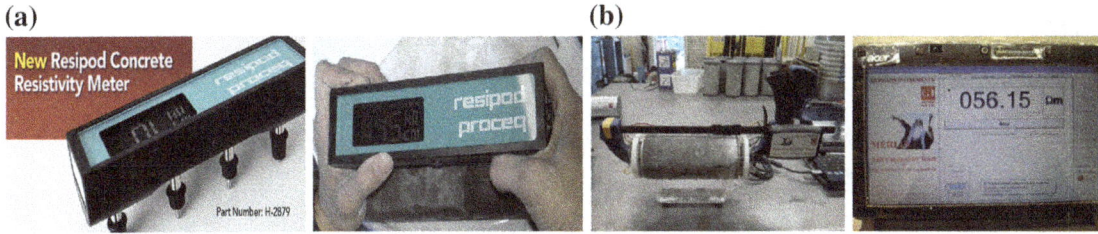

Fig. 1 Wenner 4-probe meter and Merlin conductivity tester. **a** Wenner 4-probe meter. **b** Merlin tester.

Table 1 Relationship between permeability class and surface resistivity (FDOT Standard 2004).

Chloride ion permeability classification	Surface resistivity test 28 day test Kohm-cm
High	<12
Moderate	12–21
Low	21–37
Very low	37–254
Negligible	>254

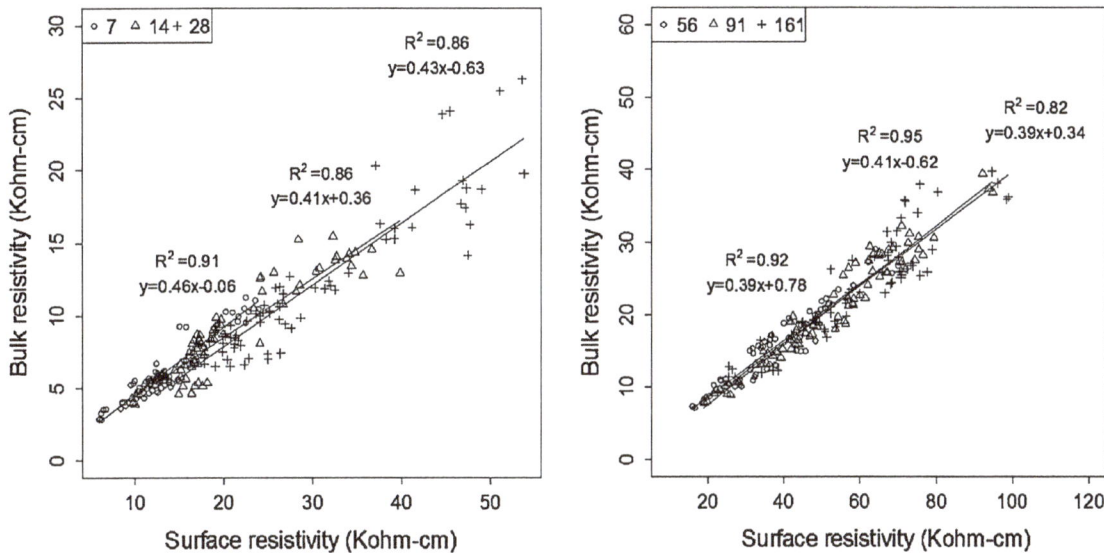

Fig. 2 Relationship between surface and bulk resistivity for fly ash mixtures.

3.2 Comparison of Experimental Resistance and Resistivity Values

Figure 6 shows distribution of ratio of experimental surface resistance to bulk resistance on different time periods. The box plot represents the range of average ratios between theoretical and experimental resistance those lay between the first quartile (Q_1) and the third quartile numbers (Q_3) for all cementitious mixtures. The horizontal line inside the box represents the median value of ratio. The vertical line below and above the box represent the remaining values of the ratios excluding the outlier. Outlier is shown as a symbol of (*) in their respective group obtained directly from statistical analysis software. An outlier is any data point that is more than 1.5 times the Inter Quartile Range (IQR = $Q_3 - Q_1$) from either end of the box. It is evident that for all mixtures the ratio falls in between 0.29 to 0.36 (in between first and third quartile numbers) from 28 to 161 days. Some lower

ratio in the range of 0.22–0.28 is observed at early ages especially on 7 and 14 days due to inadequate development of SR compared to bulk resistivity. For this reason, at 14 days an outlier is observed for minimum values on the basis of statistical analysis. At 14 days, the minimum and maximum outliers are 0.21 and 0.41 respectively. Similarly, for some mixtures development of SR is significantly higher compared to bulk resistivity at 28, 56 and 91 days and this trend causes some outliers for maximum values of experimental ratio at those specific days. The maximum outliers at 28, 56 and 91 days are 0.51, 0.41 and 0.45, respectively. However, the outliers did not exist consistently for a specific mixture except the mixture 60TII-V/35G120S/5SF as the ratio remains high from 28 to 161 days. The possible reason for maximum outliers is the testing condition. In order to measure the bulk resistivity accurately, several requirements must be satisfied during testing. In the beginning of testing,

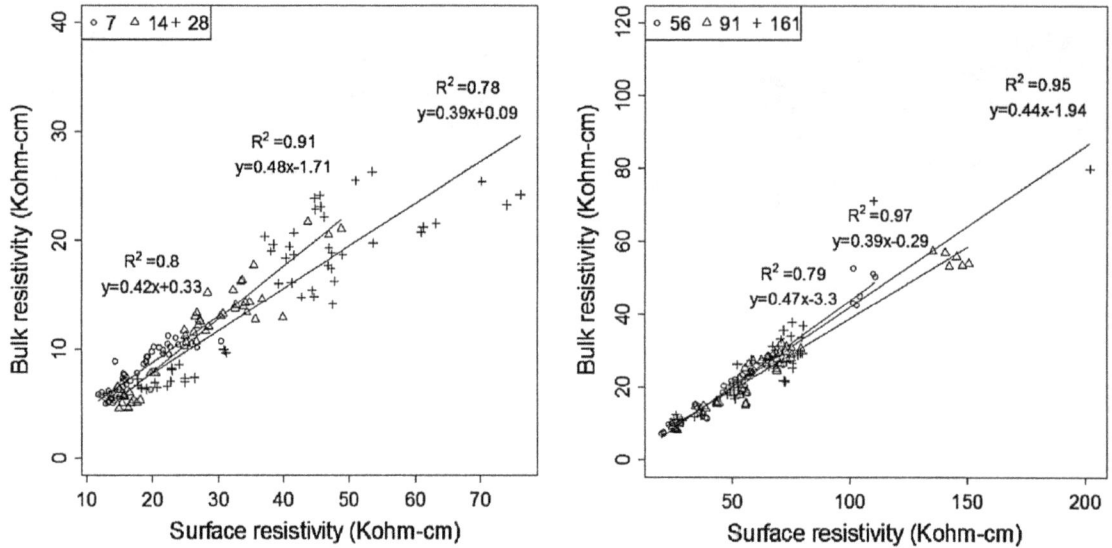

Fig. 3 Relationship between surface and bulk resistivity for slag mixtures.

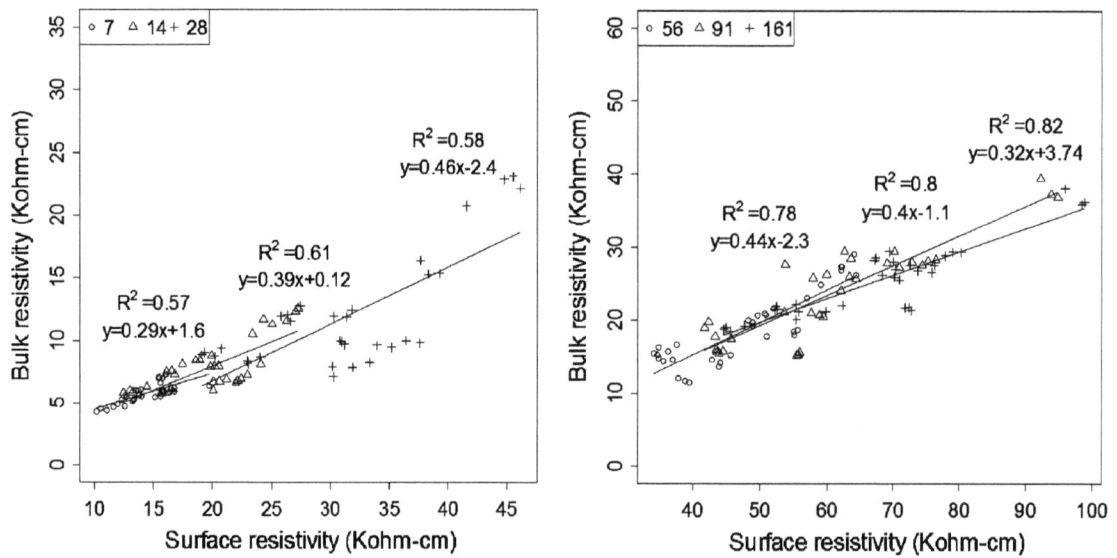

Fig. 4 Relationship between surface and bulk resistivity for silica fume mixtures.

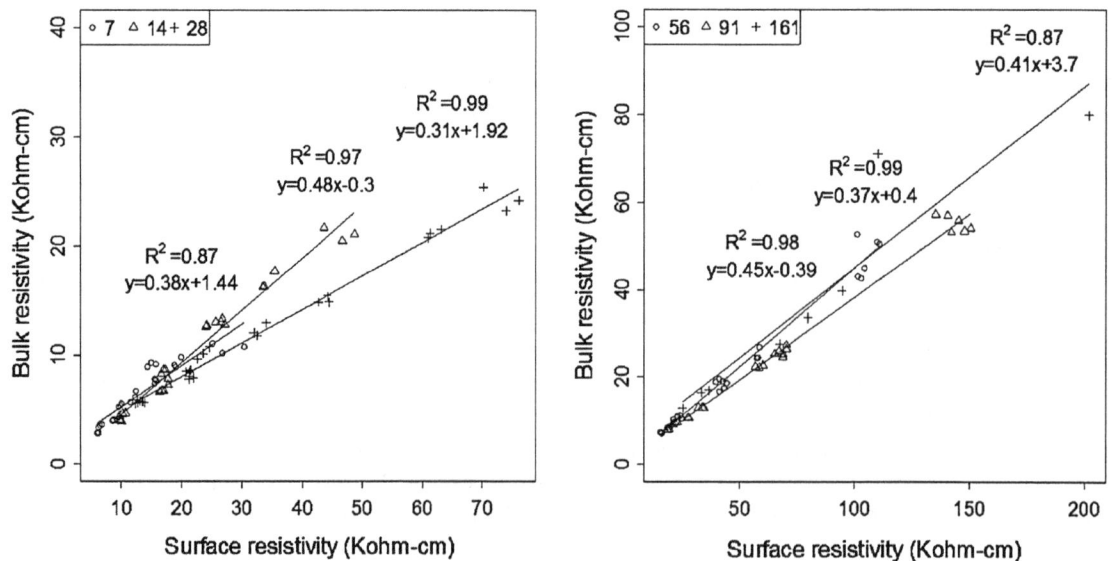

Fig. 5 Relationship between surface and bulk resistivity for metakaolin mixtures.

Table 2 Surface resistivity of different mixtures at all ages.

No.	Mixture ID	Classification	Surface resistivity (Kohm-cm)											
			7 days		14 days		28 days		56 days		91 days		161 days	
			Mean	Standard deviation	Mean	Standard deviation	Mean	Standard deviation	Mean	Standard deviation	Mean	Standard deviation	Mean	Standard deviation
1	100TII-V	Moderate	14.5	0.7	15.4	1.2	21.0	1.1	28.3	1.5	36.5	2.7	42.2	2.0
2	60TII-V/25C/15M	Moderate	6.5	0.3	10.2	0.4	12.8	0.4	19.8	0.7	22.4	1.0	33.2	1.3
3	60TII-V/30C/10M	Moderate	6.2	0.3	9.9	0.5	13.6	0.8	16.4	0.9	19.3	1.0	27.9	1.9
4	80TII-V/20C	Moderate	11.2	0.8	12.5	1.1	18.2	1.2	27.3	1.6	37.5	2.3	45.4	2.3
5	45TII-V/40G120S/15C	Moderate	13.4	0.6	15.3	0.7	18.3	1.1	24.5	1.3	25.3	1.2	26.5	1.1
6	65TII-V/35G120S	Moderate	12.1	0.6	14.8	0.6	18.6	1.1	20.9	1.1	26.6	1.4	28.6	1.9
7	65TII-V/5SF/30C	Moderate	10.8	0.7	13.3	0.9	19.8	0.9	35.8	1.9	43.3	1.8	45.7	2.1
8	60TII-V/20C/20F	Moderate	8.7	0.5	12.8	0.5	20.4	0.8	32.1	1.4	42.3	1.9	54.1	2.5
9	60TII-V/30F/10M	Low	8.8	0.4	16.8	0.5	21.2	0.7	24.1	4.2	33.8	1.2	66.2	3.4
10	65TII-V/28C/7M	Low	9.8	0.4	17.0	0.6	21.5	0.6	23.9	0.9	28.2	0.9	37.9	1.4
11	50TII-V/35G120S/15C	Low	13.5	0.9	15.7	1.0	21.6	1.5	27.3	2.1	36.6	2.3	37.0	2.1
12	55TII-V/5SF/40G120S	Low	13.5	0.4	15.9	0.5	23.3	0.7	35.6	1.4	43.9	1.2	53.2	1.5
13	60TII-V/25F/15M	Low	12.1	0.4	17.3	0.7	23.6	0.9	43.0	1.5	67.5	2.9	172.2	6.6
14	50TII-V/35G120S/15F	Low	14.3	0.6	17.4	0.7	25.7	1.0	37.0	1.4	55.0	1.8	51.6	2.1
15	60TII-V/30C/10F	Low	11.8	0.4	17.0	0.7	24.4	1.0	32.1	1.4	44.0	2.8	57.6	2.9
16	60TII-V/30F/10C	Low	12.3	0.6	19.3	0.8	26.1	1.1	36.1	1.5	48.2	1.9	68.4	4.1
17	65TII-V/5SF/30F	Low	12.4	0.6	16.7	0.7	26.6	1.2	48.9	1.8	63.0	0.7	68.6	2.9
18	80TII-V/20F	Low	10.8	0.5	15.9	0.8	27.3	1.2	45.5	2.3	58.8	2.7	73.8	4.2
19	60TII-V/35G120S/5SF	Low	13.7	0.5	20.1	0.8	31.0	1.2	38.7	1.7	55.8	2.0	72.4	1.8
20	75TII-V/20C/5SF	Low	13.3	0.6	19.1	0.9	31.2	1.2	51.2	1.7	73.9	2.7	71.6	2.9

Table 2 continued

No.	Mixture ID	Classification	Surface resistivity (Kohm-cm)											
			7 days		14 days		28 days		56 days		91 days		161 days	
			Mean	Standard deviation	Mean	Standard deviation	Mean	Standard deviation	Mean	Standard deviation	Mean	Standard deviation	Mean	Standard deviation
21	95TII-V/5SF	Low	16.1	0.7	20.8	0.9	31.4	1.7	44.1	2.3	57.4	2.3	58.4	3.1
22	93TII-V/7SF	Low	16.0	0.7	22.0	1.2	35.8	1.7	54.2	2.4	59.1	4.2	71.9	3.7
23	65TII-V/28F/7M	Low	16.5	1.6	24.6	1.0	32.9	1.3	41.3	2.0	58.5	2.0	97.8	4.7
24	75TII-V/20F/5SF	Very low	15.7	0.9	23.9	1.3	38.4	2.1	62.9	2.5	93.6	3.4	97.7	6.0
25	65TII-V/35G100S	Very low	16.1	0.5	25.3	0.7	39.4	1.5	52.5	2.8	56.6	1.8	50.2	1.9
26	50TII-V/35G100S/15C	Very low	19.9	2.3	29.4	2.8	41.3	3.9	47.5	3.3	57.7	4.4	59.1	5.5
27	58TII-V/35G120S/7M	Very low	15.6	0.4	26.8	0.8	43.8	1.4	58.0	1.5	69.4	2.2	80.8	3.3
28	60TII-V/35G100S/5SF	Very low	16.7	2.5	26.4	1.2	44.5	2.5	60.7	3.0	71.8	3.4	78.2	2.5
29	45TII-V/35G100S/20F	Very low	21.3	1.1	33.3	1.5	47.5	1.9	52.0	2.6	71.2	3.4	73.5	3.0
30	50TII-V/35G100S/15F	Very low	21.7	1.6	34.1	4.4	47.6	4.6	51.4	4.6	64.4	4.8	72.0	5.0
31	45TII-V/40G120S/15F	Very low	20.9	1.8	33.2	2.7	48.7	3.8	69.3	2.6	71.9	5.7	69.6	5.2
32	50TII-V/35G120S/15M	Very low	17.8	2.4	34.2	1.0	61.8	1.9	102.8	3.0	146.8	5.0	205.1	8.8
33	50TII-V/40G120S/10M	Very low	27.4	2.5	46.4	3.3	73.4	3.6	107.2	7.0	140.3	6.6	109.3	9.0

Table 3 Bulk resistivity of different mixtures at all ages.

No.	Mixture ID	Classification	Bulk resistivity (Kohm-cm)											
			7 days		14 days		28 days		56 days		91 days		161 days	
			Mean	Standard deviation	Mean	Standard deviation	Mean	Standard deviation	Mean	Standard deviation	Mean	Standard deviation	Mean	Standard deviation
1	100TII-V	Moderate	6.5	0.3	6.9	0.3	8.3	0.4	10.6	0.5	13.9	0.6	15.0	0.6
2	60TII-V/25C/15M	Moderate	3.5	0.1	4.5	0.2	5.7	0.1	8.6	0.2	9.5	0.2	16.5	0.2
3	60TII-V/30C/10M	Moderate	2.9	0.0	4.0	0.1	5.7	0.0	7.3	0.1	8.0	0.1	12.9	0.4
4	80TII-V/20C	Moderate	5.3	0.3	5.6	0.3	8.4	0.5	10.7	0.3	14.3	0.9	18.0	0.7
5	45TII-V/40G120S/15C	Moderate	5.8	0.3	5.9	0.3	6.8	0.4	9.9	0.5	9.4	0.6	11.7	0.6
6	65TII-V/35G120S	Moderate	5.9	0.1	6.5	0.1	6.5	0.1	7.4	0.2	8.5	0.1	11.2	0.2
7	65TII-V/5SF/30C	Moderate	4.5	0.2	6.0	0.2	9.0	0.2	16.0	0.7	18.5	1.1	18.9	0.3
8	60TII-V/20C/20F	Moderate	3.9	0.2	5.7	0.3	8.1	0.4	11.6	0.5	16.1	0.6	19.5	0.6
9	60TII-V/30F/10M	Low	4.1	0.0	6.8	0.0	8.5	0.1	10.5	0.1	13.1	0.0	27.5	0.2
10	65TII-V/28C/7M	Low	5.4	0.1	8.5	0.2	7.9	0.1	11.1	0.1	10.7	0.1	17.1	0.5
11	50TII-V/35G120S/15C	Low	5.3	0.2	4.8	0.2	6.8	0.2	10.1	0.1	14.6	0.4	12.2	0.1
12	55TII-V/5SF/40G120S	Low	5.3	0.2	5.8	0.2	8.3	0.2	14.8	0.3	15.8	0.2	21.8	0.3
13	60TII-V/25F/15M	Low	6.2	0.5	7.2	0.5	10.2	0.5	17.6	0.8	26.1	0.9	67.5	3.0
14	50TII-V/35G120S/15F	Low	5.6	0.2	5.3	0.1	7.3	0.2	12.9	0.2	18.9	0.7	17.7	0.6
15	60TII-V/30C/10F	Low	5.4	0.1	8.1	0.3	10.0	0.3	15.2	0.4	16.5	0.4	22.5	0.7
16	60TII-V/30F/10C	Low	5.8	0.2	9.5	0.4	10.7	0.3	16.6	0.6	18.3	0.6	24.2	0.8
17	65TII-V/5SF/30F	Low	5.1	0.4	7.6	0.3	12.0	0.5	19.9	0.5	26.8	2.3	28.6	0.6
18	80TII-V/20F	Low	4.8	0.2	6.7	0.2	9.5	0.3	15.3	0.6	21.5	0.5	26.8	1.3
19	60TII-V/35G120S/5SF	Low	6.0	0.0	7.9	0.0	9.9	0.2	11.8	0.3	15.4	0.2	21.7	0.2
20	75TII-V/20C/5SF	Low	5.4	0.1	8.5	0.2	12.1	0.3	21.1	0.4	27.7	0.5	26.1	0.6

Table 3 continued

No.	Mixture ID	Classification	Bulk resistivity (Kohm-cm)											
			7 days		14 days		28 days		56 days		91 days		161 days	
			Mean	Standard deviation	Mean	Standard deviation	Mean	Standard deviation	Mean	Standard deviation	Mean	Standard deviation	Mean	Standard deviation
21	95TII-V/5SF	Low	5.8	0.2	6.5	0.4	7.7	0.4	14.5	0.6	20.9	0.3	21.1	0.7
22	93TII-V/7SF	Low	5.8	0.3	6.9	0.2	9.7	0.2	18.3	0.4	26.3	0.9	27.1	0.6
23	65TII-V/28F/7M	Low	9.2	0.1	12.8	0.2	12.3	0.5	19.1	0.3	22.4	0.2	39.7	1.0
24	75TII-V/20F/5SF	Very low	7.0	0.2	9.4	2.3	15.6	0.5	27.9	0.9	37.9	1.2	36.7	1.1
25	65TII-V/35G100S	Very low	7.2	0.3	10.8	0.6	19.1	0.5	20.3	0.6	21.5	2.6	18.9	0.6
26	50TII-V/35G100S/15C	Very low	8.9	1.0	11.9	1.0	16.4	2.4	18.8	1.5	25.5	2.1	26.7	2.5
27	58TII-V/35G120S/7M	Very low	7.7	0.2	13.1	0.3	15.0	0.3	25.2	1.2	25.4	0.9	33.6	0.6
28	60TII-V/35G100S/5SF	Very low	6.8	0.4	11.9	0.6	22.2	1.0	25.2	1.6	28.3	0.7	28.5	1.2
29	45TII-V/35G100S/20F	Very low	10.4	0.5	14.4	0.6	18.6	0.6	22.2	0.8	29.8	0.9	35.8	1.5
30	50TII-V/35G100S/15F	Very low	10.1	0.7	13.6	1.0	17.9	1.5	21.4	1.6	28.7	2.2	33.1	2.4
31	45TII-V/40G120S/15F	Very low	9.6	0.4	13.8	0.7	24.9	1.1	26.2	0.5	29.6	1.1	26.6	1.0
32	50TII-V/35G120S/15M	Very low	9.2	0.5	16.7	0.7	21.2	0.4	43.6	1.1	53.5	0.7	79.9	0.8
33	50TII-V/40G120S/10M	Very low	10.7	0.4	21.1	0.5	24.2	1.0	51.4	1.0	56.7	0.7	71.1	0.6

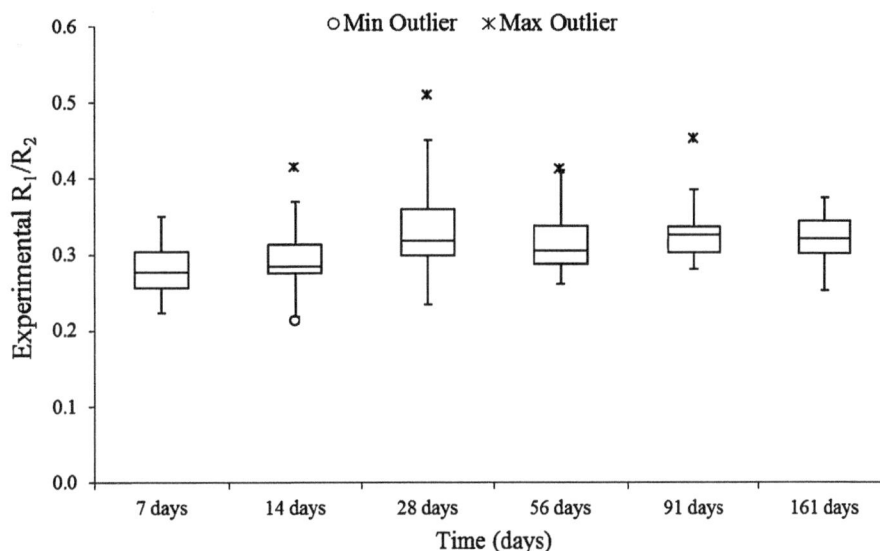

Fig. 6 Distribution of experimental surface vs. bulk resistance for all mixtures at all ages.

the specimen has to be totally water saturated and dry on the surface, which is sometimes hard to control in the laboratory environment. Additionally, the sponges soaked with water attached to both end of the specimen are sometimes stressed and water drains out into the surface of specimen and this causes variation of measurement of the bulk resistivity.

The IQR (IQR = Q3−Q1) for all days are small and varies from 0.03 to 0.06. However, it can be observed that at 7 days, Q1 and Q3 values are 0.26 and 0.3, respectively and it falls outside of the expected theoretical ratio value of 0.33. Similarly, at 14 days, the values are 0.28 and 0.31. Starting from 28 days, the expected value falls in the variation of Q1 and Q3 and the IQR decreases over time from 0.06 at 28 days to 0.03 at 91 days and 0.04 at 161 days. This is explained by the fact that the pozzolanic reactions of pozzolans or SCMs take longer time to complete.

Table 4 shows the ratio of surface vs. bulk resistivity at different test ages. It is evident from Table 4 that average ratio ranges in between 2.25 to 2.66 from 7 to 161 days. Overall, the ratio of surface vs. bulk resistivity is in good agreement with the theoretical geometric correction factor of 2.63 proposed by Morris et al.(1996), for a cylinder with a length of 200 mm (8 inches), diameter of 100 mm (4 inches) and probe spacing of 50 mm (2 inches). The average ratio is little smaller than 2.63 at early ages. It is also observed from Table 4 that theoretical geometric correction factor (2.63) closely matches with average ratios at 28, 56, 91 and 161 days for all cementitious mixtures. Recent study conducted by Spragg et al. (2012) also showed similar trend of results for comparison of surface and bulk resistivity at different ages. This relationship of ratio between surface and bulk resistivity actually represents an adjustment factor due to geometric size difference for measurement of electrical resistivity on different sizes of specimens. Sr measurement is an indirect way to obtain bulk resistivity using the ratio of surface vs. bulk resistivity namely the geometric shape factor. The numerical value of this ratio will help to obtain

the resistivity of in situ semi-infinite bridge deck slab from the laboratory SR measurement of a cylindrical specimen.

3.3 Correlation of Surface vs. Surface and Bulk vs. Bulk Resistivity at Different Days

The correlations of resistivity of surface vs surface and bulk vs bulk between 161 days and that obtained at 28, 56, 91 days are shown in Figs. 7, 8 and 9 respectively. Two types of concrete mixtures are chosen for investigation: one group is for permeable and control mixtures and the other group is for various ternary based HPC mixtures. Polynomial equation was used for regression analysis in most cases as best correlation except SR vs SR for 28 and 161 days and 91 and 161 days. Power equation was used for those two specific cases as it provided better correlation coefficient. It is observed that as the time progresses towards 161 days, the correlation coefficient increases significantly for both permeable and HPC mixtures. The effect is more prominent in case of HPC mixtures as pozzolanic effect of SCMs takes significantly longer time to provide beneficial effect to improve pore structure. It needs to be remembered that all HPC mixtures achieved significant gain in electrical resistivity values over longer period of time. Implementation of these mixtures in future bridge decks can be effective before active corrosion starts in bridge decks and will reduce the corrosion potential remarkably. For this reason, long term monitoring of the electrical resistivity of the bridges is the key strategy of the success of this philosophy of extending the life of the infrastructure. Some HPC mixtures gained the electrical resistivity at later ages exceptionally well (more than 80 Kohm-cm) compared to control and permeable mixtures. It is interesting to note that the correlation coefficient of bulk resistivity vs bulk resistivity was significantly higher than the correlation coefficient of SR vs SR at all ages for both permeable and HPC mixtures. This clarifies difference of precision levels between two experimental procedures. In addition, this correlation indicates chronological

Table 4 Ratio of experimental surface and bulk resistivity on different time periods.

No.	Mixture ID	7 days	14 days	28 days	56 days	91 days	161 days
		ρ1/ρ2	ρ1/ρ2	ρ1/ρ2	ρ1/ρ2	ρ1/ρ2	ρ1/ρ2
1	100TII-V	2.22	2.22	2.52	2.68	2.62	2.81
2	60TII-V/25C/ 15M	1.86	2.24	2.24	2.30	2.37	2.02
3	60TII-V/30C/ 10M	2.12	2.47	2.39	2.23	2.40	2.17
4	80TII-V/20C	2.11	2.24	2.16	2.55	2.62	2.53
5	45TII-V/ 40G120S/15C	2.29	2.59	2.68	2.48	2.69	2.25
6	65TII-V/ 35G120S	2.05	2.27	2.88	2.81	3.15	2.56
7	65TII-V/5SF/ 30C	2.43	2.23	2.21	2.23	2.34	2.41
8	60TII-V/20C/ 20F	2.23	2.24	2.51	2.76	2.63	2.78
9	60TII-V/30F/ 10M	2.17	2.49	2.48	2.31	2.58	2.40
10	65TII-V/28C/7M	1.82	1.99	2.71	2.15	2.63	2.22
11	50TII-V/ 35G120S/15C	2.56	3.31	3.17	2.70	2.51	3.03
12	55TII-V/5SF/ 40G120S	2.55	2.72	2.79	2.40	2.78	2.44
13	60TII-V/25F/ 15M	1.96	2.39	2.32	2.44	2.58	2.55
14	50TII-V/ 35G120S/15F	2.56	3.31	3.52	2.86	2.92	2.92
15	60TII-V/30C/ 10F	2.18	2.09	2.44	2.11	2.68	2.57
16	60TII-V/30F/ 10C	2.12	2.04	2.45	2.17	2.64	2.83
17	65TII-V/5SF/30F	2.43	2.21	2.21	2.45	2.35	2.40
18	80TII-V/20F	2.25	2.39	2.87	2.97	2.73	2.76
19	60TII-V/ 35G120S/5SF	2.29	2.55	3.15	3.29	3.62	3.34
20	75TII-V/20C/ 5SF	2.44	2.25	2.59	2.42	2.67	2.74
21	95TII-V/5SF	2.80	3.21	4.07	3.05	2.75	2.76
22	93TII-V/7SF	2.74	3.20	3.68	2.97	2.25	2.66
23	65TII-V/28F/7M	1.79	1.93	2.68	2.16	2.61	2.46
24	75TII-V/20F/5SF	2.24	2.55	2.46	2.25	2.47	2.66
25	65TII-V/ 35G100S	2.24	2.34	2.06	2.58	2.64	2.65
26	50TII-V/ 35G100S/15C	2.24	2.47	2.52	2.53	2.26	2.22
27	58TII-V/ 35G120S/7M	2.04	2.05	2.91	2.30	2.73	2.40

Table 4 continued

No.	Mixture ID	7 days	14 days	28 days	56 days	91 days	161 days
		$\rho1/\rho2$	$\rho1/\rho2$	$\rho1/\rho2$	$\rho1/\rho2$	$\rho1/\rho2$	$\rho1/\rho2$
28	60TII-V/ 35G100S/5SF	2.46	2.22	2.01	2.41	2.54	2.75
29	45TII-V/ 35G100S/20F	2.05	2.31	2.55	2.34	2.39	2.06
30	50TII-V/ 35G100S/15F	2.15	2.50	2.65	2.40	2.24	2.17
31	45TII-V/ 40G120S/15F	2.17	2.40	1.95	2.65	2.43	2.62
32	50TII-V/ 35G120S/15M	1.93	2.05	2.92	2.36	2.74	2.57
33	50TII-V/ 40G120S/10M	2.57	2.20	3.03	2.09	2.48	1.54
Average		2.24	2.41	2.66	2.50	2.61	2.52

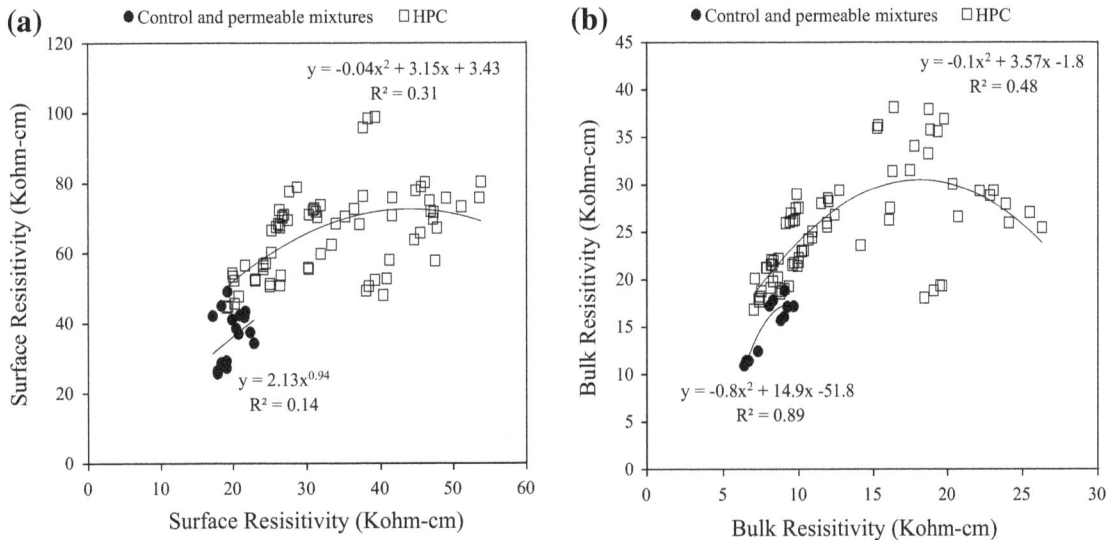

Fig. 7 Correlation of the resistivities between 28 and 161 days. **a** SR vs SR and **b** BR vs BR.

development of resistivity of HPC mixtures over an extended time period and differentiates from control and permeable mixtures in terms of accelerated rate of resistivity development at later ages. This is attributed to beneficial effect of high pozzolanic reactions of SCMs in ternary based HPC mixtures at later ages.

4. Conclusions

1. This comprehensive study presented here demonstrates the importance of variation and correlation of bulk and SR of different binary and ternary based HPC mixtures over longer period of time.
2. Both bulk and surface electrical resistivity are well correlated for different types of HPC and control mixtures over longer period of time (7 to 161 days).
3. Ratio of theoretical to experimental surface vs bulk resistance and ratio of surface and bulk resistivity are in

good agreement for the application of geometric correction factor provided by the previous research study.

4. Most of the ternary and binary mixtures studied here have substantial influence to increase the surface and bulk electrical resistivity and improve long term resistance against corrosion over an extended period of time. The key reason is probably the densification of the matrix brought about by the pozzolanic reactions of pozzolans which try to close the pores and result in reducing permeability. Another possible reason is that these mix designs have different pore solution chemistries which might increase the bulk and SR.

5. Combination of Class C pozzolan with Class F, silica fume, metakaolin and ordinary portland cement is not satisfactory in terms of development of surface and bulk resistivity. The possible reason is its incompatibility with other SCMs or chemical admixtures. It is also observed that metakaolin mixtures takes more than

(a)

(b)

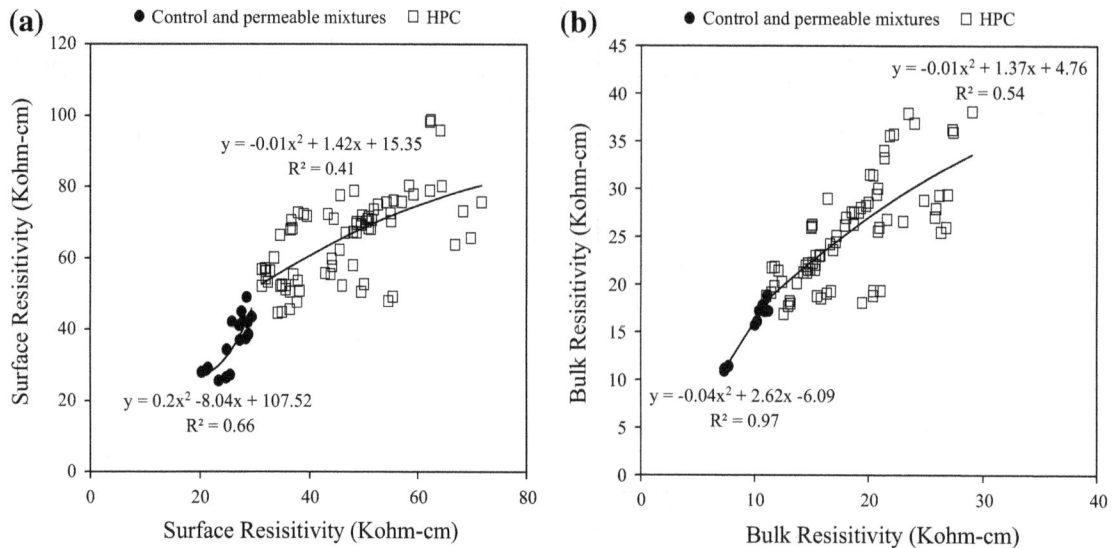

Fig. 8 Correlation of the resistivities between 56 and 161 days. **a** SR vs SR and **b** BR vs BR.

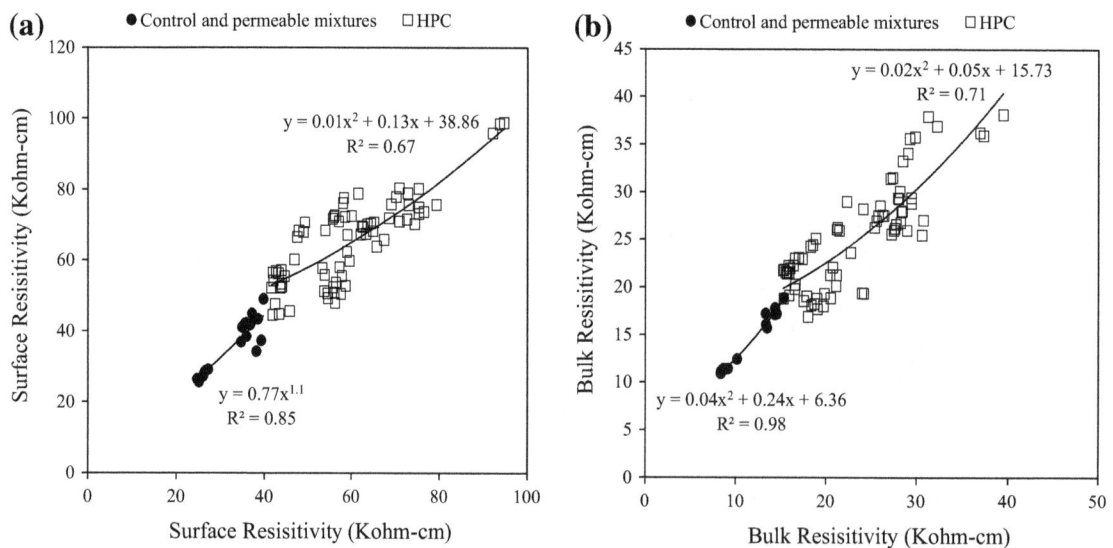

(a)

(b)

Fig. 9 Correlation of the resistivities between 91 and 161 days. **a** SR vs SR and **b** BR vs BR.

28 days to provide its superior beneficial effect by significant increment of bulk and surface electrical resistivity.

distribution, and reproduction in any medium, provided the original author(s) and the source are credited.

Acknowledgments

The authors would like to thank CSUF's grant office for the intramural grant to support this research. The authors also would like to thank BASF, Lehigh Cement Company, Mitsubishi Cement Corporation, Headwaters, and CMT Research Associates, LLC, Vulcan Materials Company, CalPortland Company for their generous donation of materials.

References

Christensen, B. J., Coverdale, R. T., Olson, R. A., Ford, S. J., Garboczi, E. J., Jennings, H. M., et al. (1994). Impedance spectroscopy of hydrating cement based materials: Measurement, interpretation, and application. *Journal of American Ceramic Society, 77*(11), 2789–2804.

Darren, T. Y., Lim, B., Sabet, D., Xu, D., & Susanto, T. (2011). Evaluation of High Performance Concrete Using Electrical Resistivity Technique. *In proceedings of 36th Conference on our World in Concrete & Structures*, Singapore, 14–16 August 2011.

FDOT Standard FM5-578. (2004). Florida method of test for concrete resistivity as an electrical indicator of its permeability. *Florida Department of Transportation*.

Icenogle, P. J., & Rupnow, T. D. (2012). Development of a precision statement for concrete surface resistivity. *92nd TRB Annual Meeting*, Paper No. 12-1078, Washngton D.C., 23–26 Jan 2012.

Julio-Betancourt, G. A., & Hooton, R. D. (2004). Study of the joule effect on rapid chloride permeability values and evaluation of related electrical properties of concretes. *Cement and Concrete Research, 34*(1), 1007–1015.

Katherine, K., Tinnea, J., Tinnea, R., Bellomio, S., Fanoni, M., Johnson, D., & Towns, J. (2010). High Electrical Resistivity Concrete Mixture Design Using Supplementary Cementitious Materials. *In Proceedings of Second International Conference on Sustainable Construction Materials and Technologies*, Università Politecnica delle Marche, Ancona, Italy, 28–30 June 2010.

Kessler, R. J., Powers, R. F. & Paredes, M. A. (2005). Resistivity measurements of water saturated concrete as an indicator of permeability, In *Proceedings of NACE International Corrosion Conference*. Houston, TX, Paper 5261, pp. 1–10.

Marriaga, J. L., Claisse, P., & Ganjian, E. (2010). Application of traditional techniques on chloride resistance assessment of GGBS concrete. *In Proceedings of Second International Conference on Sustainable Construction Materials and Technologies*, Università Politecnica delle Marche, Ancona, Italy, 28–30 June 2010.

Morris, W., Moreno, E. I., & Sagües, A. A. (1996). Practical evaluation of resistivity of concrete in test cylinders using a Wenner array probe. *Cement and Concrete Research, 26*(12), 1779–1787.

Newlands, M. D., Jones, M. R., Kandasami, S., & Harrison, T. A. (2008). Sensitivity of electrode contact solutions and contact pressure in assessing electrical resistivity of concrete. *Journal of Materials Structures, 41*(5), 621–632.

Paredes, M., Jackson, N. M., Safty, A. E., Dryden, J., Joson, J., Lerma, H., et al. (2012). Precision statements for the surface resistivity of water cured concrete cylinders in the laboratory. *Advances in Civil Engineering Materials, 1*(1), 1–23.

Polder, R. B., Andrade, C., Elsener, B., Vennesland, Ø., Gulikers, J., Weidert, R., et al. (2004). Test methods for on-site measurement of resistivity of concrete. *Materials and Structures, 33*(10), 603–611.

Rupnow, T. D., Schaefer, V. R., Wang, K., & Tikalsky, P. J. (2007). Effects of different air entraining agents (AEA), supplementary cementitious materials (SCM), and water reducing agent (WR) on the air void structure of fresh mortar, International Conference on Optimizing Paving Concrete Mixtures and Accelerated Concrete Pavement Construction and Rehabilitation, FHWA/ACI/ACPA, Nov 6–9, 2007.

Spragg, R. P., Castro, J., Nantung, T., Paredes, M., & Weiss, J. (2012). Variability analysis of the bulk resistivity measured using concrete cylinders. *Advances in Civil Engineering Materials, 1*(1), 1–17.

Tikalsky, P., Taylor, P., Hanson, S., & Ghosh, P. (2011). *Development of performance properties 1 of ternary mixtures: Laboratory study on concrete*. Ames, IA: Iowa State University.

4

Numerical Simulation of Prestressed Precast Concrete Bridge Deck Panels Using Damage Plasticity Model

Wei Ren[1),*], Lesley H. Sneed[2)], Yang Yang[2)], and Ruili He[2)]

Abstract: This paper describes a three-dimensional approach to modeling the nonlinear behavior of partial-depth precast prestressed concrete bridge decks under increasing static loading. Six full-size panels were analyzed with this approach where the damage plasticity constitutive model was used to model concrete. Numerical results were compared and validated with the experimental data and showed reasonable agreement. The discrepancy between numerical and experimental values of load capacities was within six while the discrepancy of mid-span displacement was within 10 %. Parametric study was also conducted to show that higher accuracy could be achieved with lower values of the viscosity parameter but with an increase in the calculation effort.

Keywords: bridge decks, concrete, concrete damage plasticity, cracking, finite element simulation.

1. Introduction

This paper presents the results of numerical simulations conducted using ABAQUS on hybrid partial-depth precast prestressed concrete (PPC) bridge deck panels using the concrete damage plasticity model to investigate the behavior and failure mechanism. The term "hybrid panel" in this paper describes a PPC panel that contains two different types of prestressing tendons: either epoxy-coated steel or carbon fiber reinforced polymer (CFRP) tendons at the panel edges, and uncoated steel tendons at the interior of the panel. Previous studies have shown that substitution of steel tendons with epoxy-coated steel could effectively reduce the occurrence of corrosion (Kobayashi and Takewaka 1984) and using FRP tendons as the addition of reinforced tendons, the ductility of the prestressed beams can be significantly improved (Saafi and Toutanji 1998).

2. Background

2.1 Bridge Deck Description

Partial-depth prestressed precast concrete deck panels span between girders and serve as stay-in-place (SIP) forms for a cast-in-place (CIP) concrete bridge deck. Typical panel geometries are 75–90 mm (3.0–3.5 in.) thick, 2.4 m (8 ft) long in the longitudinal direction of the bridge, and sufficiently wide to span between the girders in the bridge transverse direction. The panels are typically pretensioned with prestressing steel strands located at the panel mid-depth. Panels are placed adjacent to one another along the length of the bridge and typically are not connected to each other in the longitudinal bridge direction. After the panels are in place, the top layer of reinforcing steel is placed, and the CIP concrete portion of the deck [typically 125–140 mm (5.0–5.5 in) thick] is cast on top of the panels. At the bridge service state, the CIP concrete and SIP panels act as a composite deck slab.

2.2 Problem Statement

The most common problem reported with the use of partial-depth deck panels is reflective cracking on the top surface of the deck. Cracks in the transverse direction of the bridge may form at locations at which adjacent panels are placed (panel edges), while cracks in the longitudinal direction may form at the locations at which the panels are supported on the girders (panel ends).

The cause of the transverse reflective cracks is attributed primarily to the concentration of shrinkage and stress of CIP concrete at the joints between the precast panels (Hieber et al. 2005) (Fig. 1a). Transverse reflective cracks generally raise a deterioration concern because they permit the ingress of moisture and corrosion agents of steel reinforcement in the deck (Fig. 1b). When reflective cracks extend the full thickness of the CIP concrete layer, the ingress of moisture and corrosion agents can be concentrated at the panel edges (Fig. 1c), which has been observed to cause corrosion of steel prestressing tendons at the panel edges and spalling of

[1)]Key Laboratory of Bridge Inspection and Reinforcement Technology of China Ministry of Communications, Chang'an University, Xi'an 710064, Shaanxi, China.
*Corresponding Author; E-mail: rw20062@163.com
[2)]Department of Civil, Architectural & Environmental Engineering, Missouri University of Science and Technology, Rolla, MO 65409, USA.

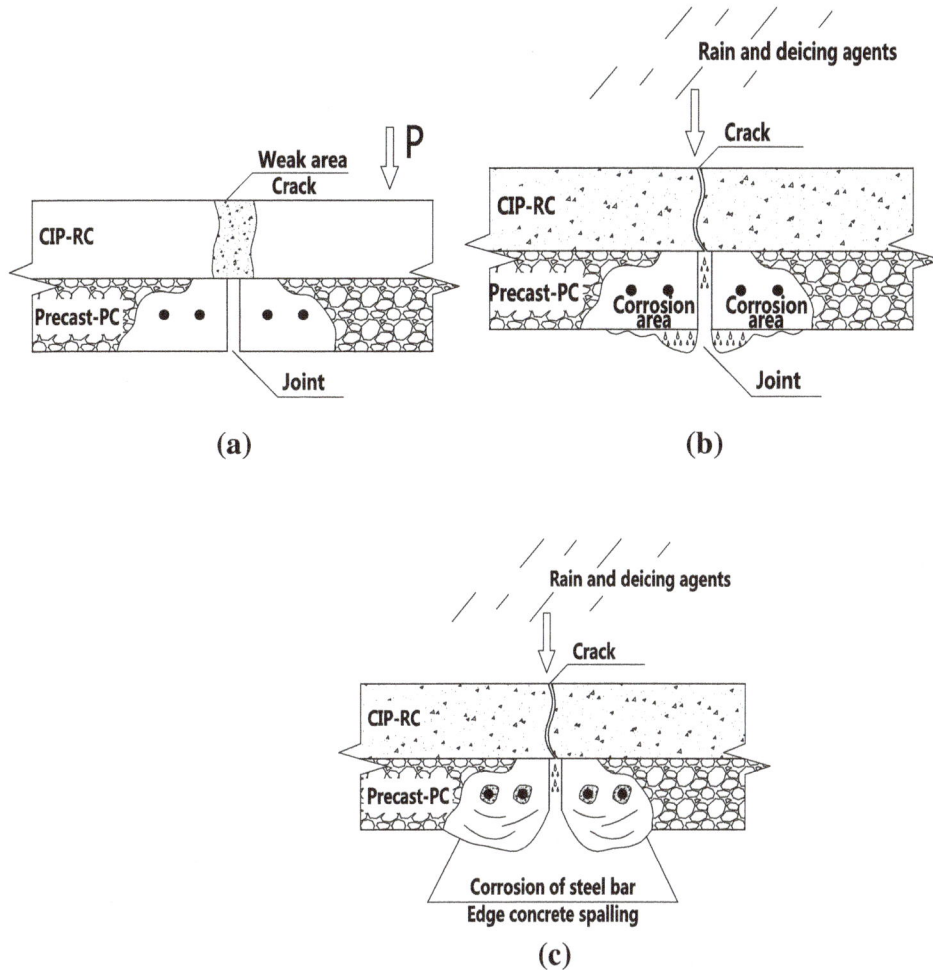

Fig. 1 Spalling mechanism observed in PPC panels.

Fig. 2 Cracking and spalling at bridge deck panel joints.

concrete along the panel edges (Fig. 2) (Wieberg 2010; Sneed et al. 2010). Critical combinations of panel geometry, material properties, and reinforcement details can lead to long-term serviceability problems (Young et al. 2012; Wenzlick 2008).

3. Experimental Program

Testing data of an experimental program (Sneed et al. 2010) on six full-size partial-depth precast concrete deck panels were used to verify the numerical model developed in

this study. The purpose of the experimental program was to investigate the structural behavior of the hybrid PPC panels as discussed previously. The dimension of the specimens was 75 mm thick, 2,440 mm length, and 2,440 mm width (Fig. 3 and Table 1). Three types of prestressed reinforcement were used in the experiments. Steel strand was 9.5 mm diameter, 7-wire, Grade 270 low-relaxation conforming to ASTM A 416 (2010). Epoxy-coated strand was 9.5 mm diameter, 7-wire, Grade 270 low-relaxation grit-impregnated conforming to ASTM A 882 (2010). Carbon fiber reinforced polymer (CFRP) tendons were No. 3 reinforcing bar. Table 2 summarizes the material properties of the prestressing reinforcement.

The specimens were testing under displacement control, where a 1.25 mm increment was used until the failure of specimens. All six panels failed with concrete crushing in the compression zone near the mid-span. The testing setup is shown in Fig. 4.

4. Modeling Approach

4.1 Material Models

The finite element models of the tested specimens were built and analyzed with software ABAQUS. For linear elastic materials, at least two material constants are required:

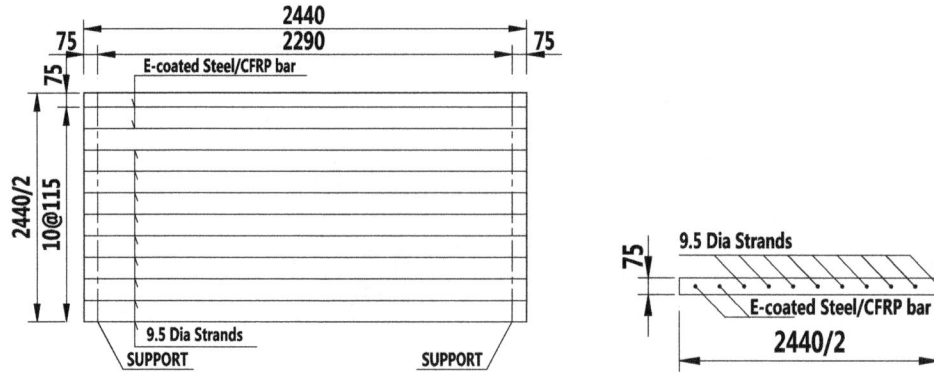

Fig. 3 Reinforcement details of specimens (mm).

Table 1 Test matrix.

Test specimen	Edge tendon type	Concrete type
ST-NC	Steel	Normal
ST-FRC	Steel	FRC
ECT-NC	Epoxy-coated steel	Normal
ECT-FRC	Epoxy-coated steel	FRC
CFRP-NC	CFRP	Normal
CFRP-FRC	CFRP	FRC

N normal concrete, *FRC* fiber reinforced concrete, *ST* steel strands, *ECT* epoxy-coated steel strands, *CFRP* CFRP tendons.

Table 2 Material parameters (MPa/Psi).

	Concrete compressive strength	Concrete tensile strength	Concrete modulus of rupture	Tendon f_y		Tendon f_u		
				ST	Epoxy-coated steel	ST	Epoxy-coated steel	CFRP
ST-NC	37.2/6,360	3.36	31,342/600	1,737/ 2.52×10^5	–	1,889/ 2.74×10^5	–	–
ST-FRC	32.6/5,580	3.08	29,357/855	1,737/ 2.52×10^5	–	1,889/ 2.74×10^5	–	–
ECT-NC	34.5/5,900	3.20	30,187/765	1,793/ 2.6×10^5	1,882/ 2.73×10^5	1,924/ 2.79×10^5	1,999/ 2.9×10^5	–
ECT-FRC	37.8/6,460	3.40	31,587/745	1,793/ 2.6×10^5	1,882/ 2.73×10^5	1,924/ 2.79×10^5	1,999/ 2.9×10^5	–
CFRP-NC	40.9/700	3.58	32,881/620	1,793/ 2.6×10^5	–	1,924/ 2.79×10^5	–	2,576.5
CFRP-FRC	37.4/6,390	3.37	31,416/585	1,793/ 2.6×10^5	–	1,924/ 2.79×10^5	–	2,576.5

Young's modulus (E) and Poisson's ratio (v). For nonlinear materials, the steel and concrete uniaxial behaviors beyond the elastic range must be defined to simulate their behavior at higher strains. ABAQUS provides different types of concrete constitutive models including, (1) a smeared crack model; (2) a discrete crack model; and (3) a damage plasticity model (ABAQUS Theory Manual 2010). The concrete damage plasticity model, which can be used for modeling concrete and other quasi-brittle materials, was used in this study. This model combines the concepts of isotropic damage elasticity with isotropic tensile and compressive plasticity to model the inelastic behavior of concrete. The model assumes scalar (isotropic) damage and can be used for both monotonic and cyclic loading conditions. Elastic stiffness degradation from plastic straining in tension and compression is accounted for (Lubliner et al. 1989; Lee and Fenves 1998). Cicekli et al. (2007) and Qin et al. (2007) proved that damage plasticity model provides an effective method for modeling the concrete behavior in tension and compression.

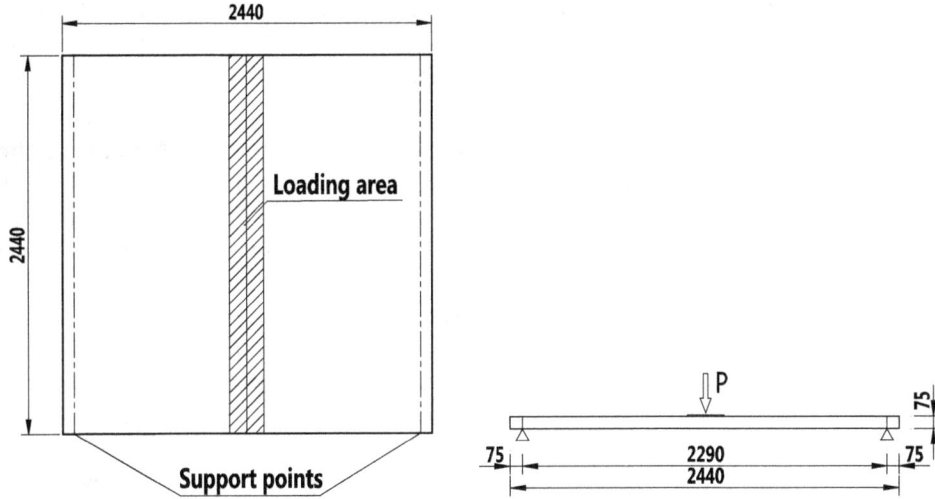

Fig. 4 Test setup (mm).

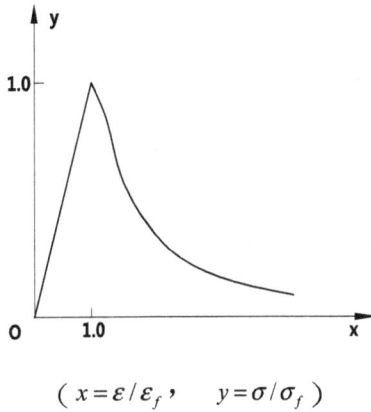

$$(x=\varepsilon/\varepsilon_f , \quad y=\sigma/\sigma_f)$$

Fig. 5 Concrete uniaxial tensile stress–strain curve.

4.1.1 Concrete Constitutive Model and Damage Indices

The concrete damage plasticity model requires input of parameters including the constitutive relationship of concrete, which can be customized by the user. This paper used the constitutive model of concrete developed by Zhenhai (2001) and Xue et al. (2010).

The stress–strain relationship as shown in Fig. 5 of concrete under uniaxial tension is described in Eq. (1). Damage is assumed to occur after the peak stress is reached.

$$y = \frac{x}{\alpha_t(x-1)^{1.7}+x} \quad x \geq 1, \tag{1a}$$

$$\alpha_t = 0.312f_t^2, \tag{1b}$$

where α_t is decline curve parameters of concrete under uniaxial tension (if $\alpha_t = 0$ the curve becomes a horizontal line corresponding to fully plastic behavior while in case of $\alpha_t = \infty$ the curve becomes a vertical line corresponding to the fully brittle behavior). f_t is concrete tensile strength.

The stress–strain relationship as shown in Fig. 6 for concrete under uniaxial compression is described in Eq. (2).

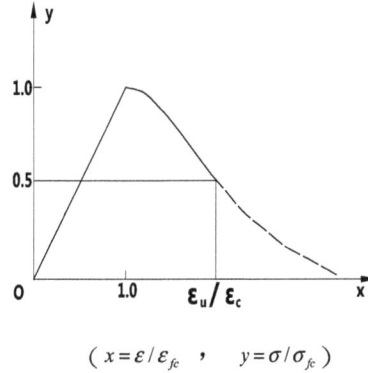

$$(x=\varepsilon/\varepsilon_{fc} , \quad y=\sigma/\sigma_{fc})$$

Fig. 6 Concrete uniaxial compressive stress–strain curve.

$$y = \frac{x}{\alpha_d(x-1)^2+x} \quad x \geq 1, \tag{2a}$$

$$\alpha_d = 0.157f_c^{0.785} - 0.905, \tag{2b}$$

where α_d is the declining parameter of concrete under uniaxial compression; f_c is concrete compressive strength.

4.1.2 Other Data of Concrete Models

(1) The dilation angle ψ is a ratio of vertical shear strain increment and strain increment, which is taken as 38 degrees.

(2) Flow potential eccentricity ε is a small positive number that defines the rate at which the hyperbolic flow potential approaches its asymptote. This paper takes a value of 0.1.

(3) National standard of the people's republic of China (2002) recommend,

$$-f_3/f_c^* = 1.2 + 33(\sigma_1/\sigma_3)^{1.8} \tag{3}$$

From FE model analysis, $\sigma_1 = -16.66$ Mpa, $\sigma_3 = -1.73$ Mpa, before the concrete cracks. So

$$-f_3/f_c^* = 1.2 + 33(16.66/1.73)^{1.8} = 1.7585$$

The ratio of initial equibiaxial compressive yield stress to initial uniaxial compressive yield stress σ_{bo}/σ_{co} was taken as 1.76.

(4) The ratio of the second stress invariant on the tensile meridian, $q(TM)$, to that on the compressive meridian K_c was taken as 2/3.
(5) The viscosity parameter μ used for the visco-plastic regularization of the concrete constitutive equations in Abaqus/Standard was taken as 0.0005.

4.1.3 Prestressing Tendons

CFRP tendons were modeled as linear elastic while steel strand and epoxy-coated steel strand were modeled as bilinear hardening model (Fig. 7).

4.2 Finite Element Model Description
4.2.1 Symmetry

Because the PC panels investigated had two axes of symmetry, it was possible to represent the full slab by modeling only one fourth of the panel (Fig. 8). This reduced

the analysis time (Wei et al. 2007). A linear elastic unit was also used to model the portion that stayed as elastic during testing.

4.2.2 Element Type and Meshing Scheme

The 3D eight-node solid element C3D8 (Tuo 2008) was used to model the concrete. The T3D2 element was used to represent the prestressing strands or tendons. The model contained 6,144 nonlinear concrete elements, 3,072 three-dimensional linear elastic solid elements, and 432 prestressing tendons elements. CFRP tendons (or epoxy coated steel tendons) were divided into 96 elements. Element sizes were 25.4 mm $\times$ 38.1 mm $\times$ 12.7 mm. A meshed model is shown in Fig. 9.

4.2.3 Bonding Between Reinforcement and Concrete

This is a technique used to place embedded nodes at desired locations with the constraints on translational degrees-of-freedom on the embedded element by the host element (Fig. 10). The rebar was modeled as embedded regions in the concrete in the interactive module, and making the concrete for the host. Thus, rebar elements can only had translations or rotations equal to those of the host elements surrounding them (Garg and Abolmaali 2009).

4.2.4 Boundary Conditions

Due to symmetry, only a quarter of the panel was modeled as shown in Fig. 8. The nodes on symmetry surfaces were constrained in X and Y directions, respectively. At the supports, nodes were constrained in the z direction.

4.2.5 Prestressing Effect

Prestressing effect is usually modeled through either (1) initial strain or (2) initial temperature load. This study used initial temperature load to apply the prestressing load. The applied temperature t (°C) can be obtained from Eq. 4.

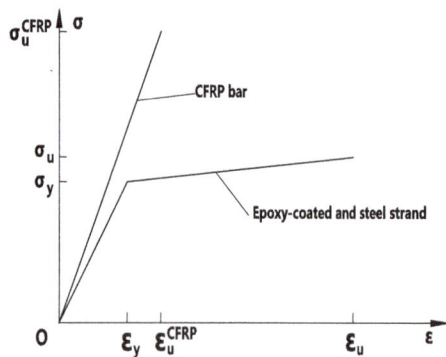

Fig. 7 Prestressed Reinforcement Material curve.

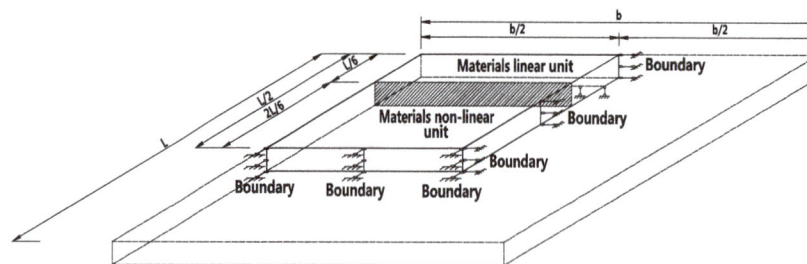

Fig. 8 Modeling scheme of one fourth of the panel.

Fig. 9 FE model of one fourth of the panel.

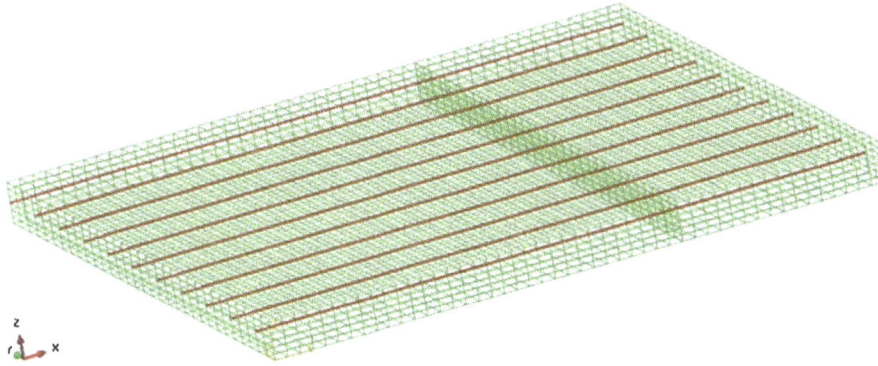

Fig. 10 Rebar was modeled as embedded regions.

$$C = -\frac{P}{c \cdot E \cdot A} \qquad (4)$$

C is coefficient of linear expansion taken as 1.0×10^{-5} (MPa/ °C); E is modulus elasticity of the tendon, in MPa; A is the cross-sectional area of the prestressing tendon in mm^2; P (in N) is prestressing force calculated based on the recorded force during pretension process and with consideration of loss of prestressing effect.

4.2.6 Convergence Considerations

Convergence issues were resolved with the following considerations:

(1) Loading steps were adjusted in consideration of the anticipated time of concrete cracking and the automatic time step was adopted.
(2) Constitutive relationship was modified by introducing the coefficient of viscosity. A higher viscosity coefficient would make the structure of "harder". Through extensive trials, a viscosity coefficient of 0.0005 was found to be helpful with convergence.
(3) In cases of computation time being more critical than accuracy (Jiang 2005), the force and displacement

convergence criteria were adjusted to reduce the computation time.

5. Validation of FE Model

5.1 Force–Displacement Relationships

The failure load for the numerical analysis was defined as peak load in the force –displacement relationship. Table 3 lists the failure loads and corresponding midspan displacements of the experimental work and the numerical analysis. The error of analysis was within 7 %. The difference in results can be a consequence of underestimation of the concrete's fracture energy. The error of analysis for midspan displacement ranged from 5 to 23 %.

Relatively large deviation between the analytical and experimental results was observed in panels ST-FRC, ECT-NC and CFRP-NC. This may be due to the fact that data of material concrete of these panels may be incorrect, such as concrete material parameters of panel CFRP-NC are much higher than others (Table 3), but its midspan displacement is significantly smaller. Panel ST-FRC and ECT-NC also have the same problem. So in this paper, on the basis of a large

Table 3 Failure loads and corresponding displacements.

	Failure load			Midspan displacement at failure load		
	P_{Test} (kN)	P_{FE} (kN)	P_{Test}/P_{FE}	δ_{Test} (mm)	δ_{FE} (mm)	$\delta_{Test}/\delta_{FE}$
ST-NC	98.21	101.4216	0.97	54.31	51.82	1.05
ST-FRC	90.55	88.1972	1.03	42.75	34.8868	1.23
		94.4248[a]	0.96[a]		42.5018[a]	1.01[a]
ECT-NC	82.51	88.3596	0.93	27.46	34.8623	0.79
		85.5892[a]	0.96[a]		29.4543[a]	0.93[a]
ECT-FRC	94.36	93.43	1.01	44.02	41.05	1.07
CFRP-NC	93.39	99.5872	0.94	36.45	45	0.81
		91.4712[a]	1.02[a]		33.4368[a]	1.09[a]
CFRP-FRC	92.85	97.0472	0.96	44.11	40.00	1.10

[a] Results of which the concrete material parameters were adjusted.

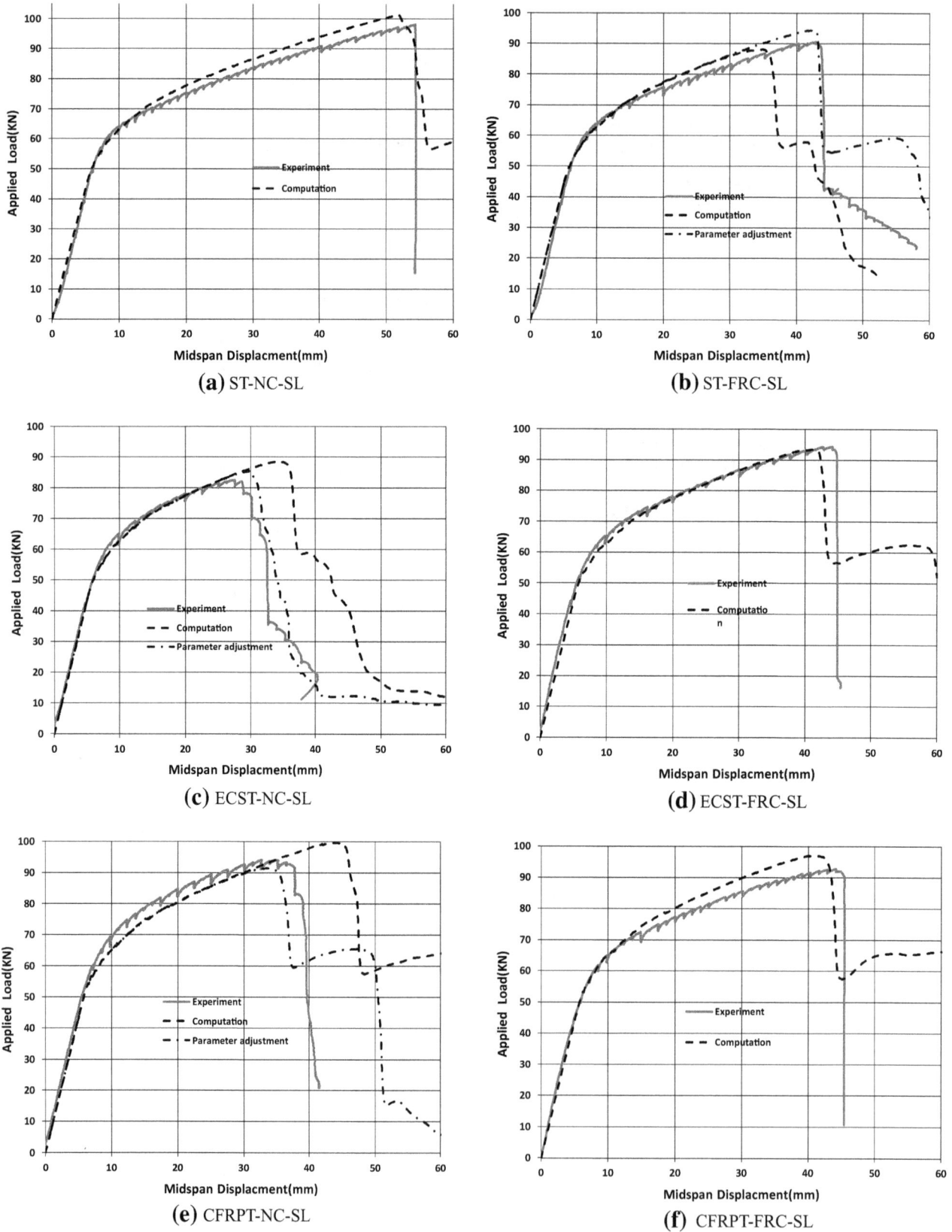

(a) ST-NC-SL

(b) ST-FRC-SL

(c) ECST-NC-SL

(d) ECST-FRC-SL

(e) CFRPT-NC-SL

(f) CFRPT-FRC-SL

Fig. 11 Experimental and numerical applied load—midspan displacement curves.

number of comparative analyses, these panels was analyzed again after adjusting the concrete material parameters, the results see Table 3 and Fig. 11.

The error of the failure loads was within 6 % and the error of the midspan displacement was within 10 % with the adjusted concrete model.

5.2 Failure Modes

Figures 12 and 13 compare the damage observed during the experiments with the damage predicted in the simulations. In the experiments, testing was terminated when brittle failure (concrete crushing) occurred. Figure 12 shows that at the peak load, concrete crushing occurred at the top of the panel at midspan, and flexural cracking (maximum principal plastic strain) was observed at the bottom surface of the panel near midspan. Figure 13 illustrates of the calculated compressive strain distribution of the panels. As shown in this figure, the failure due to the concrete crushing could be well predicted with the analysis.

6. Parametric Analysis

The influence of the concrete dilation angles, viscosity parameters, and prestressing effect on the analytical results was investigated through parametric analysis of panel ST-NC-SL.

6.1 Effect of Concrete Dilation Angle ψ

The dilation angle of a material is a measurement of the expansion of volume occurring when the material is under shear (as illustrated in Fig. 14) (Zhao and Cai 2010). For a Mohr–Coulomb material like concrete, the value of dilation angle generally varies in between zero (non-associative

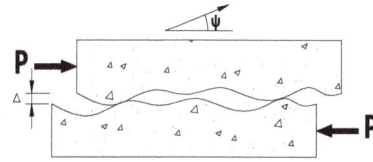

Fig. 14 Dilation angle.

flow rule) and the friction angle (associative flow rule). (Tuo 2008) recommended 30° for concrete material. However, in this paper, the analytical results with a value of 38 were the most closest to the experimental results.

From Fig. 15 it can be seen, as the dilation angle increased, the displacement capacity and the failure load of the panel was significantly increased while the required number of iterations to obtain a converged results decreased.

6.2 Effect of μ Viscosity Parameter

Material models exhibiting softening behavior and stiffness degradation often lead to severe convergence difficulties in implicit analysis programs, such as Abaqus/Standard. A common technique to overcome these convergence difficulties is the use of a visco-plastic regularization of the constitutive equations, which causes the consistent tangent stiffness of the softening material to become positive for sufficiently small time increments.

The lower value of the parameter would result in more accurate calculation and more computation time. The

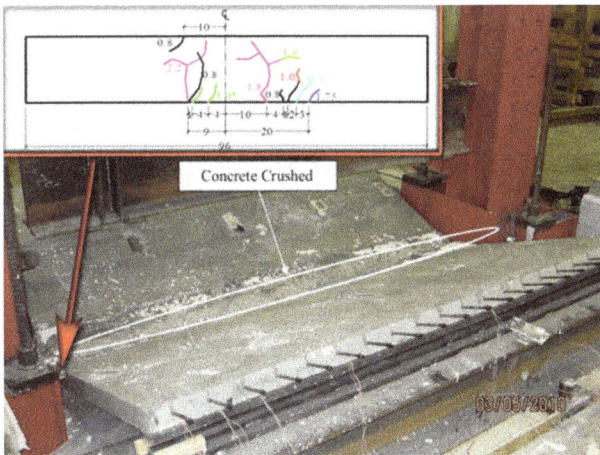

Fig. 12 Observed failure mode of test specimen (Panel ST-NC-SL shown).

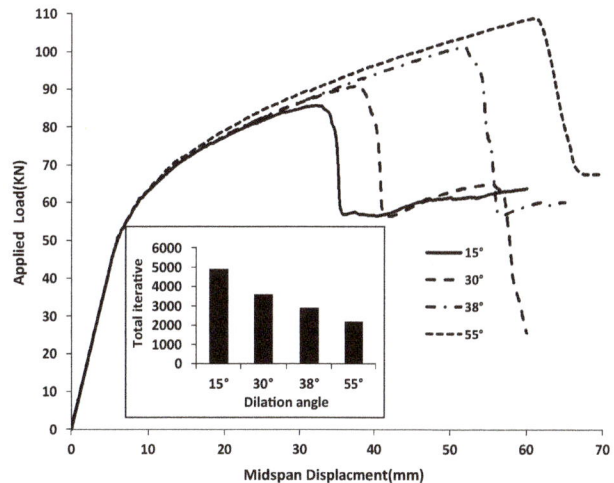

Fig. 15 Effect of concrete dilation angle.

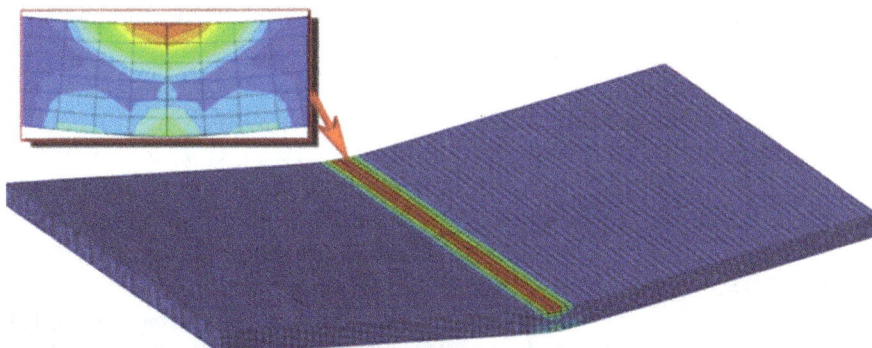

Fig. 13 Calculated maximum principal plastic strain distribution (Panel ST-NC-SL shown).

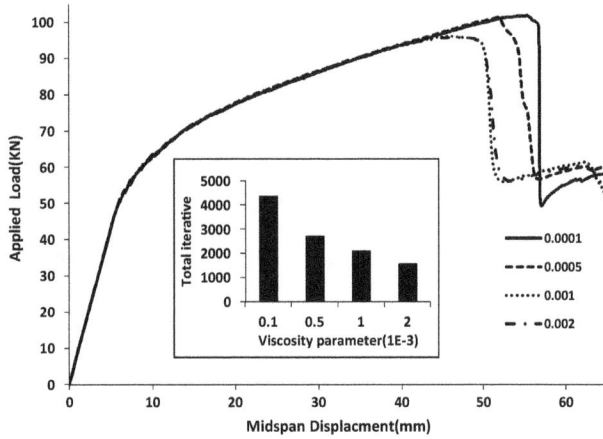

Fig. 16 Effect of viscosity parameter μ.

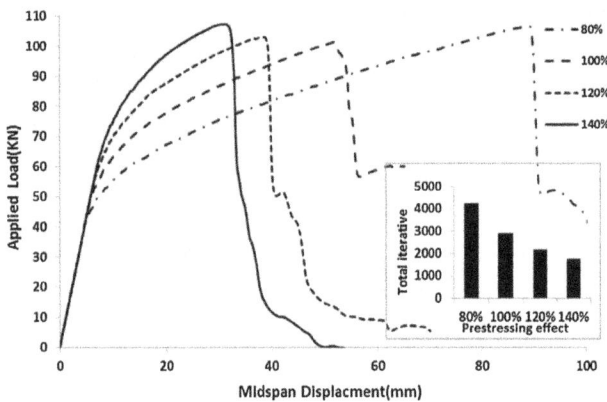

Fig. 17 Prestressed effect (Panel ST-NC-SL shown).

influence of the value of the viscosity parameter on the analytical results is shown in Fig. 16. As shown in this figure, with decreasing of the value of viscosity parameter, the displacement capacity and failure load increased and the required number of iterations to reach a converged solution increased. When the viscosity parameter was taken as 0.0005, the calculation results were close to the experimental results. On the premise of no apparent loss calculation accuracy and efficiency, proposals should be taken as far as possible to the lower value.

6.3 Effect of Prestressing Force

Figure 17 shows the influence of the prestressing effect. As shown in this figure, increasing the prestressing effect (from 60 to 140 %), resulted in increasing of the cracking load (from 37 to 69 KN and decreasing of displacement capacity.). However, the post-cracking stiffness and the peak load the panels was not sensitive to the prestress effect since the post-cracking curves was parallel to each other with varied prestressing forces.

7. Conclusions

In this paper, finite element analysis of PPC bridge deck panels was conducted using concrete damage plasticity

model. Based on the validation of finite element model against experimental results and parametric study with varied values of dilation angle, viscosity parameter, and prestressing force, following conclusions can be made:

(1) The concrete damage plasticity model in ABAQUS can predict the concrete crushing failure mode in PPC panels. The numerical error of the failure loads and mid-span displacement was within 6 % and 10 %, respectively.

(2) It was feasible and accurate enough to simulate the prestressing effect by applying temperature load to prestressing strands or tendons.

(3) Under tri-axial compression state in the case of this paper, the ratio of initial equibiaxial compressive yield stress to initial uniaxial compressive yield stress σ_{bo}/σ_{co} was taken as 1.76 and was shown accurate to predict the behavior of PPC panels.

(4) Increasing prestressing effect resulted in increasing of, the cracking load and decreasing of displacement capacity of the PPC panels as shown in the parametric study in this paper.

(5) Lower values of viscosity of parameter increased calculation accuracy and increased the calculation time.

Acknowledgments

The experimental program was funded by the Missouri Department of Transportation (MoDOT) and the National University Transportation Center (NUTC) at Missouri University of Science and Technology (Missouri S&T). Ministry of Transport of the People's Republic of China (2012 319 812 100) and the Shaanxi Department of Transportation (14–17 k) funded the other part of this study.

References

ABAQUS Theory Manual, version 6.9, Hibbitt Karlson & Sorensen, Inc 2010.

Cicekli, U., Voyiadjis, G. Z., & Abu Al-Rub, R. K. (2007). A plasticity and anisotropic damage model for plain concrete. *International Journal of Plasticity, 23*, 1874–1900.

Garg, A. K., & Abolmaali, A. (2009). Finite-element modeling and analysis of reinforced concrete box culverts. *Journal of Transportation Engineering, 135*(3), 121–128.

Hieber, D. G., Wacker, J. M., Eberhard, M. O., & Stanton, J. F. (2005). "State-of-the-art report on precast concrete systems for rapid construction of bridges." Olympia: Washington State Transportation Center (TRAC). (Report No. WA-RD 594.1).

Jiang, J., Lu, X., & Ye, L. (2005). *Finite element analysis of concrete structures*. Beijing, China: Tsinghua University Press.

Kobayashi, K., & Takewaka, K. (1984). Experimental studies on epoxycoated reinforcing steel for corrosion protection. *International Journal of Cement Composites and Lightweight Concrete, 6*(2), 99–116.

Lee, J., & Fenves, G. L. (1998). Plastic-damage model for cyclic loading of concrete structures. *Journal of engineering mechanics, 124*(8), 892–900.

Lubliner, J., Oliver, J., Oller, S., & Oñate, E. (1989). A plastic-damage model for concrete. *International Journal of Solids and Structures, 25*(3), 299–326.

National standard of the people's republic of China. (2002). *Code for design of concrete structures GB50010-2002.* Beijing, China: China architecture and building press.

Qin, F., Yi, H., Ya-dong, Z., & Li, C. (2007). Investigation into static properties of damaged plasticity model for concrete in ABAQUS. *Journal of PLA University of Science and Technology, 8*(3), 254–260.

Saafi, M., & Toutanji, H. (1998). Flexural capacity of prestressed concrete beams reinforced with aramid fiber reinforced polymer (AFRP) rectangular tendons. *Construction and Building Materials, 12*(5), 245–249.

Sneed, L., Belarbi, A., & You, Y-M. (2010). Spalling solution of precast-prestressed bridge deck panels. Jefferson, MO: Missouri Department of Transportation Report.

Tuo, L. (2008). Application of damaged plasticity model for concrete. *Structural Engineers, 24*(2), 22–27.

Wei, R., et al. (2007). Numerical method of bearing capacity for perloaded RC beam strengthened by bonding steel plates. *Journal of traffic and Transportation Engineering, 7*(6), 96–100.

Wenzlick, J. D. (2008). Inspection of deterioration of precast prestressed panels on bridges in Missouri. Jefferson, MO: Missouri Department of Transportation. (Report No. RI05-024B).

Wieberg, K. (2010). Investigation of spalling in bridge decks with partial-depth precast concrete panel systems. Master's thesis, Civil Engineering Department Missouri University of Science and Technology, Rolla, MO.

Xue, Z., et al. (2010). A damage model with subsection curve of concrete and its numerical verification based on ABAQUS. In *International Conference on Computer Design and Applications (ICCDA)* 2010 (Vol. 5, pp. 34–37).

You, Y. M., Sneed, L. H., & Belarbi, A. (2012). Numerical simulation of partial-depth precast concrete bridge deck spalling. *American Society of Civil Engineers, 17*(3), 528–536.

Zhao, X. G., & Cai, M. (2010). A mobilized dilation angle model for rocks. *International Journal of Rock Mechanics and Mining Sciences, 47*(3), 368–384.

Zhenhai, G. (2001). *Theory of reinforced concrete*. Beijing, China: Tsinghua University Press.

5

Effect of Morphology and Dispersibility of Silica Nanoparticles on the Mechanical Behaviour of Cement Mortar

Lok Pratap Singh[1],*, **Anjali Goel**[2], **Sriman Kumar Bhattachharyya**[1], **Saurabh Ahalawat**[1], **Usha Sharma**[1], **and Geetika Mishra**[1]

Abstract: The influence of powdered and colloidal nano-silica (NS) on the mechanical properties of cement mortar has been investigated. Powdered-NS (~ 40 nm) was synthesized by employing the sol–gel method and compared with commercially available colloidal NS (~ 20 nm). SEM and XRD studies revealed that the powdered-NS is non-agglomerated and amorphous, while colloidal-NS is agglomerated in nature. Further, these nanoparticles were incorporated into cement mortar for evaluating compressive strength, gel/space ratio, portlandite quantification, C–S–H quantification and chloride diffusion. Approximately, 27 and 37 % enhancement in compressive strength was observed using colloidal and powdered-NS, respectively, whereas the same was up to 19 % only when silica fume was used. Gel/space ratio was also determined on the basis of degree of hydration of cement mortar and it increases linearly with the compressive strength. Furthermore, DTG results revealed that lime consumption capacity of powdered-NS is significantly higher than colloidal-NS, which results in the formation of additional calcium-silicate-hydrate (C–S–H). Chloride penetration studies revealed that the powdered-NS significantly reduces the ingress of chloride ion as the microstructure is considerably improved by incorporating into cement mortar.

Keywords: cement mortar, nano-silica, strength, morphology, dispersibility.

1. Introduction

Use of nanomaterials in the construction sector is gaining widespread attention as significant improvements are expected to be achieved in the desired properties of construction materials. The most commonly used nanomaterials in the cement are nano-silica, nano-titania, nano-alumina, carbon nano-tubes (CNTs) etc. (Sanchez and Sobolev 2010). Among all, nano-silica has been proven an effective additive to cement matrix for accelerating cement hydration due to its high reactivity, ability to refine the microstructure and thus, leading to a reduced porosity (Toutanji et al. 2004). Various types of nano-silica (powder or in suspension) are available commercially, having specific density, specific surface area, pore structure and reactivity (Quercia et al. 2014). Several researchers (as reviewed by Singh et al. (2013)) reported that the mechanical properties and durability can be improved by adding nano-silica (powder or colloidal) in cement-based materials. The enhancement in compressive strength of cement mortar with 0.25 % powder nano-silica was achieved 63.9 and 95.9 MPa at the age of 1 and 28 days, respectively

(Flores 2010). The characteristics of cement mortar with powder nano-silica particles showed that nano-silica behaves not only as a filler to improve the microstructure but also as an activator to promote the pozzolanic reaction (Jo et al. 2007). The performance enhancing properties of nano-silica are achieved through two mechanisms: firstly, the ultrafine particles are able to fill the voids between the cement particles improving "packing" and creating a less permeable structure. Secondly, the nano-silica also reacts with the calcium hydroxide (CH) produced with the cement hydration to form additional C–S–H (Gaitero et al. 2008, 2009). The porosity and capillary pores decreased while the gel pores increased as a result of the inclusion of silica fume and fly ash in the cement-based composites (Lin et al. 2009). Several researchers (as reviewed by Shi et al. (2012)) have studied the role of mineral admixtures in concrete durability, methods of measuring chloride ingress into concrete, challenges in assessing concrete durability from its chloride diffusivity, and the service life modeling of reinforced concrete in chloride-laden environments. The ingress of gases, water or ions in aqueous solutions into concrete takes place through pore spaces in the cement paste matrix and paste-aggregate interfaces or microcracks. For the durability of concrete, permeability is believed to be the most important characteristic (Baykal 2000), related to its microstructural properties, such as the size, distribution, and interconnection of pores and microcracks (Savas 2000). The water permeability test shows that the nano-silica concrete has lower water permeability as compared to normal concrete (Ji 2005).

[1]CSIR-Central Building Research Institute, Roorkee 247 667, India.

*Corresponding Author; E-mail: lpsingh@cbri.res.in

[2]Gurukul Kangri University, Haridwar 249404, India.

The incorporation of nanoparticles (Fe_2O_3, Al_2O_3, TiO_2 and SiO_2) and nanoclays (montmorillonite) reduces the diffusion coefficient of the mortar as well as electrochemical impedance spectroscopy. The tests indicate that such effects are especially significant using nano-SiO_2 and nanoclays (He and Shi 2008).

In contrast, colloidal nano-silica (CNS) denotes small particles (1–100 nm) consisting of an amorphous silica core with a hydroxylated surface, which are insoluble in water (Coenen and Kruif 1988). The accelerating effects of colloidal silica on C_3S phase dissolution; C–S–H gel formation and silica polymerisation in cement paste hydration were studied (Bjornström et al. 2004). The surface treatment of colloidal nano-silica was found effective in decreasing water absorption of cement mortar at 50 °C, but a negligible effect at 20 °C and filled coarser pores (>50 nm) (Hou et al. 2014). The addition of 6 % CNS improves the compressive strength of mortar from 18.3 to 46.3 MPa, at 7 days (Jo et al. 2007). The improvements are attributed to three reasons: the acceleration effect of CNS on cement hydration, pozzolanic reaction of CNS and the improved particle packing of matrix. Cement with 2–4 % addition of CNS do not lead to an immediate mechanical strength gain due to the formation of agglomerates, later on hydration evolution takes place due to consumption of calcium hydroxide (Kontoleontos et al. 2012).

Nano-silica is extensively used in cement matrix, though their mixing is a challenge which needs to be addressed. When nanoparticles are added into the cement with water, they form agglomerates and may not reflect its original reactivity (Kong et al. 2013). In order to address this issue (i.e. homogeneous mixing of nanomaterials) dispersible silica nanoparticles were prepared and introduced into cement mortar. Further, the experiment comprises the comparison of powdered and colloidal NS with respect to their effect on gel/space ratio, compressive strength, pozzolanic reactivity of both nano-silica and silica fume and quantification of the C–S–H using thermogravimetric analysis were conducted. Moreover, chloride penetration of plain and nanomodified cement mortar was investigated.

2. Experimental Protocols

2.1 Materials

The present study was carried out with 43 grade OPC, type I cement, conforming to IS: 8112. The cement was analyzed for various proportions as per IS 4031-1988. Standard sand of grade (I, II and III) was chosen according to IS: 650. Grade I type sand consists course aggregate with particles size 1–2 mm, grade II consists fine aggregates with particles size 0.5–1 mm and grade III comprises very fine aggregates with particle size ranges between 0.09 to 0.5 mm. This sand attained a fineness modulus of 2.86 and a saturated surface dry specific gravity of 2.59. A high quality commercial grade silica fume (M/s Elkem) was used. The physical and chemical properties of the cement and silica fume are given in Table 1.

Further, two different type of nano-silica namely powder nano-silica ($\sim$40 nm) prepared in laboratory using sodium silicate as precursor and colloidal nano-silica (15–20 nm) was commercially available as an aqueous dispersion having the pH value of 9.2, and SiO_2 content (by weight) of 30 % corresponds to a density of 1.28 g/mL. The specific surface area of colloidal-NS is in the range of 170–200 m^2/g. For introducing nano-silica into cement mortar ultrasonic treatment was used to disperse the agglomerates. In this method nano-silica and water were sonicated for 30 min at 42 kHz frequency and 80 watt absorbed power using a 2.5 L capacity bath sonicator till the solution turned milky. Afterwards, this sonicated mixture was mixed for 1 min with cement and sand in the Hobart mixer having B-type blade. Mortar mixer as per the standard ASTM C144 was used for mortar preparation. The mixer was equipped with B- flat type peddles, having minimum and maximum revolution speed of 140 $\pm$ 5 rpm and 285 $\pm$ 10 rpm, respectively. The capacity of the bowl was $\sim$5 L.

2.2 Methodology
2.2.1 Preparation of Silica Nanoparticles: Sol Gel Method

For the cost-effective preparation of nano-silica, sol–gel technique was followed using sodium silicate as a precursor (Tan et al. 2005; Venkatathri and Nanjundan 2009). In this preparation cetrimonium bromide (CTABr) as dispersing agent and 1 N HCl as catalyst was used. CTABr, HCl and deionized water were mixed, stirred for 45 min, followed by dropwise addition of 1 M solution of sodium silicate with stirring at room temperature until the pH of reaction system reached $\sim$8.0. The resultant white suspension was filtered and washed with deionized water to remove all the sodium chloride formed (Fig. 1). The prepared powder was dried (50 °C) and then muffled at 700 °C for 4 h (Singh et al. 2012a, b). Finally, the white powder was characterized by scanning electron microscope (SEM), X-ray diffraction (XRD), BET techniques. The morphological attributes of two types of nano-silica and mortar samples containing nano-silica and silica fume were studied using SEM (LEO 438VP) at an accelerating voltage of 15–20 kV. The samples were analysed under variable pressure (VP) mode with gold coating so as to improve the surface conductivity. For SEM analysis, the slices of the mortar samples were cut directly and immersed in acetone to discontinue the hydration process. The dried samples were deposited on a sample holder with a double stick conducting carbon tape to develop their micrographs. For XRD studies, Rigaku make (DMax-2200) with a X-ray source of Cu Kα radiation ($\lambda = 1.54$ Å) was used. The powder sample of 75 micron sieved was used for analysis. The scan step size was 0.02°, in 2θ range from 5 to 80°. The X-ray tube voltage and current were fixed at 40 kV and 40 mA, respectively. The specific surface area of powder nano-silica was analyzed with BET (model Adsotrac DN-04, Microtrac SSA, USA).

Table 1 Chemical composition of OPC and silica fume used.

Parameters	OPC	Silica fume
SiO_2 (%)	22.1	96.03
Al_2O_3 (%)	6.8	0.43
Fe_2O_3 (%)	2.2	0.99
CaO (%)	61.6	0.25
MgO (%)	3.4	0.63
SO_3 (%)	2.4	0.30
Na_2O (%)	0.12	0.40
K_2O (%)	0.7	0.63
Average diameter (nm)	–	0.1–10 μm
Specific gravity	3.15	2.23
Specific surface area (m^2/g)	–	29
LOI	0.5	1.98

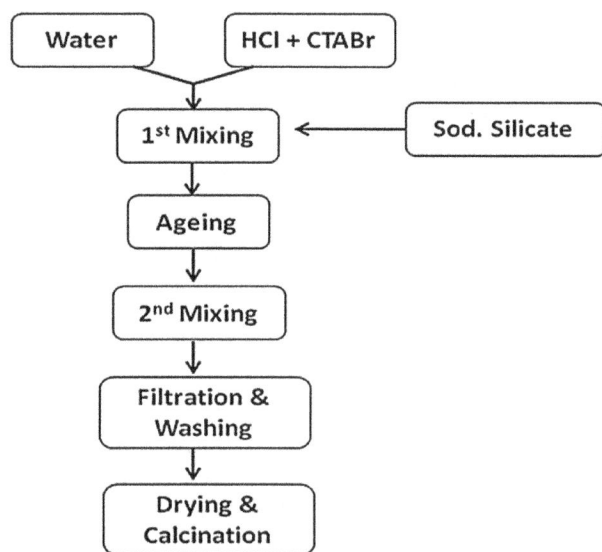

Fig. 1 Synthesis flow chart.

2.2.2 Compressive Strength

Standard cube molds (50 × 50 × 50 mm) were used to prepare mortar specimens for compressive strength determination. The mix proportions of cement and sand in mortar was 1:3 throughout. The water-to-cement ratio for all the samples was fixed at 0.4. The nano-silica addition was 1.0, 2.0 and 3.0 % by weight of cement. Water content in colloidal-NS was considered in the mix proportion calculation. The cubes were de-moulded after curing at 20 ± 1 °C and over 95 % of relative humidity for 24 h followed by immersing in water at 20 ± 1 °C till the testing time. Three specimens were used to determine the average compressive strength.

2.2.3 Quantification of Portlandite (CH) and C–S–H Through Thermo-gravimetric Analysis (TG/DTG)

CH content of mortars with and without NS addition was detected using thermo-gravimetric method. The procedure follows the heating of powdered sample from 50 to 1000 °C at the rate of 10 °C min^{-1} in a Perkin–Elmer thermogravimetric analyzer. Before the tests, at desired hydration time samples were crushed and dipped into acetone to stop hydration for 24 h. Further, the powdered samples were dried at 105 °C for 4 h. In the DTG curve, the weight loss between 400–500 °C was recorded and considered as the cause of CH decomposition (Singh et al. 2012a, b; Jain and Neithalath 2009). The temperature range of mass loss between 110 °C and the temperature at which CH loss begins (400 °C) is considered to indicate the loss of water from C–S–H gel (Ramachandran et al. 2003; Gallucci et al. 2013; Olsona and Jennings 2001). The amount of CH (%) and C–S–H (%) in the sample is calculated from the TG curves using the following Eqs. (1 & 2):

$$CH(\%) = WL\,\%(CH) \times \frac{MW(CH)}{MW(H)} \tag{1}$$

$$CSH(\%) = WL\,\%(CSH) \times \frac{MW(CSH)}{\text{Moles of water} \times MW(H)} \tag{2}$$

where WL %(CH) & WL %(CSH) correspond to the mass loss in percentage attributable to CH and C–S–H dehydration, and MW(CH), MW (CSH) and MW(H) are the molecular weights of CH, C–S–H and water, respectively.

2.2.4 Chloride Penetration Resistance

An accelerated electromigration test was performed on glass cell assembly for measuring chloride ion movement (Fig. 2). Plain and nanomodified cement mortar was shaped into disc of 7 mm thickness and 30 mm in diameter and stored for 24 h at 20 °C. The specimens were de-moulded after 24 h and then cured in distilled water at room temperature (20 ± 1 °C). Further, these disc shaped mortar specimens after 28 days of hydration were sandwiched

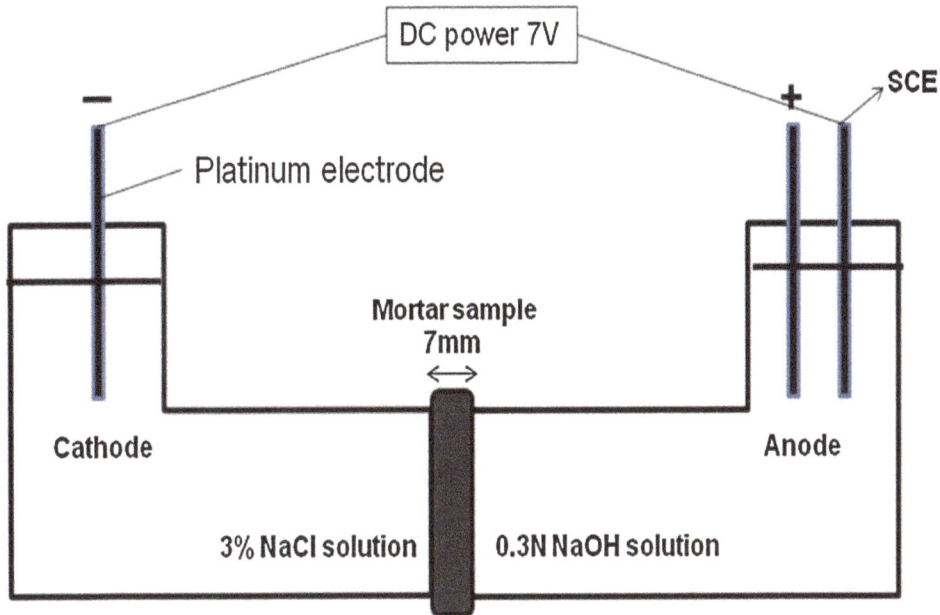

Fig. 2 Schematic diagram of experimental setup used for accelerated electromigration test.

between two glass cells. One of the cells was filled with 3 % NaCl solution and the other cell with 0.3 N NaOH solution as per ASTM C1202. Two platinum electrodes placed in both glass cells solution served as working electrodes whereas saturated calomel electrode (SCE) worked as reference electrode to monitor the potential applied. Once the mortar specimen disc, solutions and electrodes were in place, the potential voltage 7 V was applied and migration of chloride ions were measured periodically using UV–Vis spectrophotometer.

3. Results and Discussions

3.1 Characterization of Nano-Silica

Powdered silica nanoparticles were characterized using SEM, XRD and BET techniques. The SEM micrograph revealed that the average particle size of powdered-NS is ~40 nm (Fig. 3a) and the particles are spherical, non-

agglomerated and possess a smooth surface morphology. On the other hand the colloidal-NS appear in the form of agglomerates (Fig. 3b). From X-ray studies, the characteristic diffraction broad peak, centered on 22° (2θ), confirmed its amorphous nature (Fig. 4). BET results shows that the specific surface area of powered-NS is 116.23 m^2/g.

3.2 Effect of Nano-Silica on Gel/Space Ratio of Cement Mortar

The Gel/space ratio is defined as the volume of gel divided by the sum of the volumes of gel and capillary pores (Power et al. 1948; Acker 2001) whereby "gel" is a synonym for hydration products which include the gel pores of typically 0.5–2.5 nm. It is well-known that the compressive strength of cement based materials depends on the gel/space ratio (Pichler et al. 2013). For the Portland cement paste, it is assumed that 1 ml of hydrated cement occupies 2.06 mL of space, the gel/space ratio is given by (Neville 1981) Eq. (3):

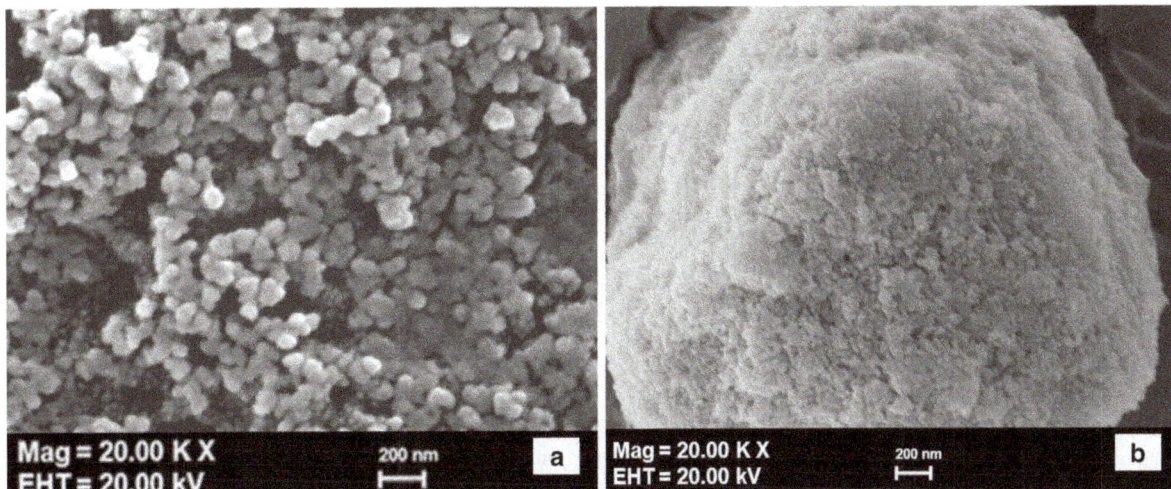

Fig. 3 SEM micrographs of **a** powdered-NS, **b** colloidal-NS.

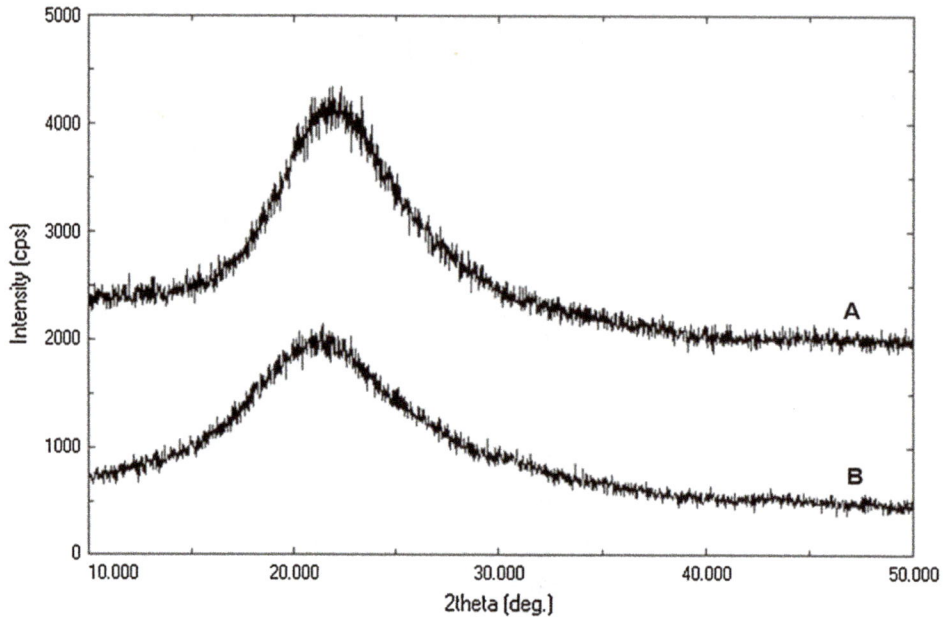

Fig. 4 XRD profile of *A* colloidal-NS, *B* powdered-NS.

Table 2 Calculated gel/space ratio of plain and blended cement mortar.

Curing time (days)	Cement mortar	3 %SF	3 % powdered-NS	3 % colloidal-NS
1	0.37	0.40	0.42	0.43
3	0.47	0.49	0.63	0.53
7	0.55	0.58	0.74	0.60
28	0.71	0.74	0.94	0.76

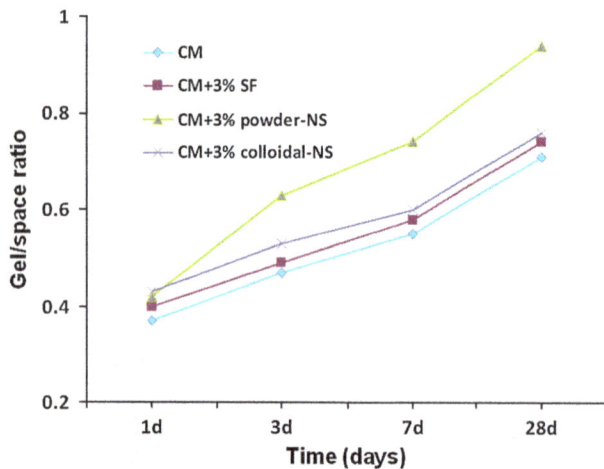

Fig. 5 Gel/space ratio with hydration time.

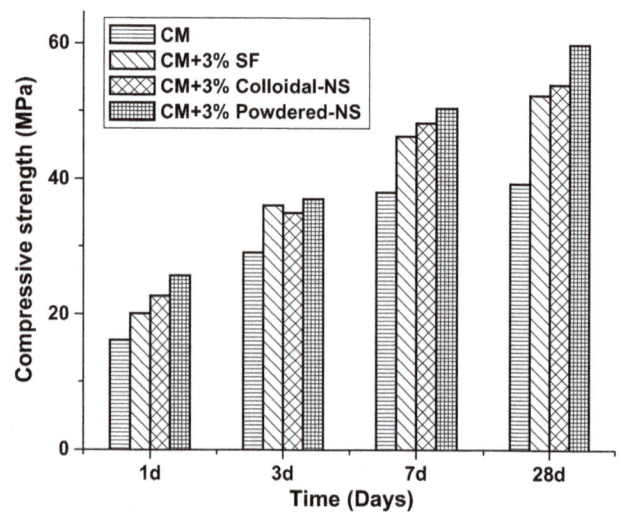

Fig. 6 Compressive strength of cement mortar with additives.

$$X_c = \frac{2.06 V_c \, \alpha_c}{V_c \, \alpha_c + {}^w\!/_c} \tag{3}$$

where X_c is gel/space ratio of plain cement, V_c is specific volume of anhydrous cement, α_c is degree of hydration of cement and w/c is water-to-cement ratio. For supplementary materials (SF and NS) pozzolanic reaction, 1 mL of supplementary materials reacted is considered to occupy 2.52 mL of space (Lam et al. 2000), therefore, the gel/space ratio of blended cement is given by Eq. (4):

$$X_{bc} = \frac{2.06 V_c \, \alpha_c \, C + 2.5 V_s \, \alpha_s \, S}{V_c \, \alpha_c \, C + V_s \, \alpha_s \, S + {}^w\!/_c} \tag{4}$$

where X_{bc} is gel/space ratio of blended cement; V_c and V_s are specific volumes; α_c and α_s are the degrees of hydration of cement and with the addition of supplementary materials (SF and NS), respectively, C and S are the original fractions of cement and SF, NS in blend.

Fig. 7 SEM micrographs at 1 & 28 of hydration of **a**, **b** plain cement mortar, **c**, **d** with SF, **e**, **f** with colloidal-NS, and **g**, **h** with powered-NS.

With the addition of powdered-NS (3 %), gel space ratio increases from 0.43 to 0.94, while with colloidal-NS (3 %) 0.42–0.74 from 1 to 28 days (Table 2). This result reveals that powdered-NS is more reactive and homogeneously mixed with cement matrix than that colloidal-NS (Fig. 5).

The continuous increasing trend of gel/space ratio signifies that the powder nano-silica act as a centre of nucleation for cement hydrates, which accelerate the hydration. The mechanism is related to the non-agglomerated nature (well-dispersed particles) and high surface area of powdered-NS,

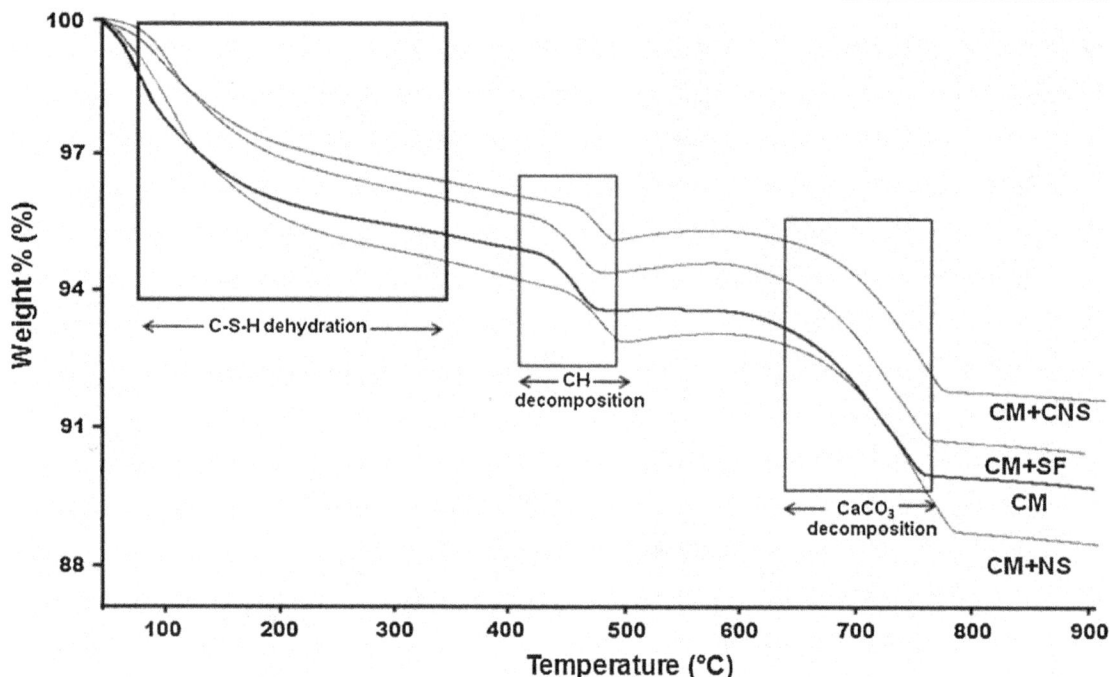

Fig. 8 TGA curve of cement mortar with NS and SF.

which works as a nucleation site for the precipitation of additional C–S–H gel. The formation of additional C–S–H occupies the available space leading to denser structure.

3.3 Effect of Nano-Silica on Compressive Strength of Cement Mortar

The modification on the cement hydration caused by nanoparticles can be reflected by their effect on the mechanical properties of cementitious materials. The addition of silica fume and nano-silica significantly improves the compressive strength with the increase of doses due to pozzolanic reaction. Figure 6 illustrates that the strength of the cement mortar with 1 % powdered-NS increases by 19 and 17 % at 1 and 3 day, respectively, whereas those with 1 % colloidal-NS were increased 15 % compared to those of the plain cement mortar. However, the strength gain of 1 % powdered-NS at 28 days is 26 % more than the control.

In addition, with 3 % powdered-NS the strength goes to 37 % higher than that the plain cement mortar at 1 day and 34 % at 28 days. In contrast, with the colloidal-NS (3 %) strength gain is 28 % at 1 day and 27 % at 28 days. On the other hand, SF (3 %) is able to increase compressive strength 19 % at 1 day and 25 % at 28 days. From these results it may be inferred that SF shows its reactivity at later stage of hydration (Fig. 6). The development of compressive strength with and without NS and SF were expressed as a function of gel/space ratio. The result signifies that the gel/space ratio increases with curing time. A significant increase in strength compared with the control was also observed. This may be attributed to higher content of calcium silicate hydrate (C–S–H) in NS and SF blended specimens, due to the pozzolanic reaction of CH produced from cement hydration with nano-silica and silica fume. These results are further supported by microstructure studies of cement mortar at 28 days of hydration (Fig. 7). The SEM micrographs

revealed that with the addition of nano-silica, more C–S–H is appearing at the early stage of hydration, so that later on the microstructure at 28 days become more compacted, uniform and denser. This mechanism emphasise that the addition of powdered-NS increases the strength at the early stage, mainly because of packing effect. It actually acted as filler material, which filled the interstitial spaces and pores, inside the matrix of hardened cement mortar, resulting into increase in density as well as its strength.

3.4 Portlandite Quantification

The TG/DTG curves showed that the typical reactions occurring in cement matrix when subjected to a progressive temperature rise from room temperature to 1000 °C in 200 mL/min Nitrogen gas flow. The first change was observed between 60 and 105 °C, can be attributed to the departure of weekly bound water (Fig. 8). The second significant loss between ~120–400 °C corresponding to dehydration of some hydrates like C–S–H and ettringite (Gabrovsek et al. 2006; Ramachandran et al. 2003). The third reduction at ~400–500 °C, causes a loss in mass corresponding to de-hydration of CH; hence, the portlandite decomposes into free lime (dehydroxylation) (Gaitero et al. 2008; Jain and Neithalath 2009). Subsequently, the final weight loss area was observed at ~650–800 °C, occurs due to the decomposition of calcium carbonate (Alonso and Fernandez 2004; Gabrovsek et al. 2006). Figure 9a showed a strong decrease in the CH peak, appears considerably smaller with 3 % powdered NS than the same peak of other samples at 1 day (Table 3). As well as the hydration proceeds up to 28 days, the reduction in this region (~400–500) is observed significantly more with powdered NS (~68 %). Similarly, colloidal-NS and SF reduce ~57 and 31 % at 28 days, respectively (Fig. 9b). The pozzolanic reaction with CH is proportional to the amount of surface

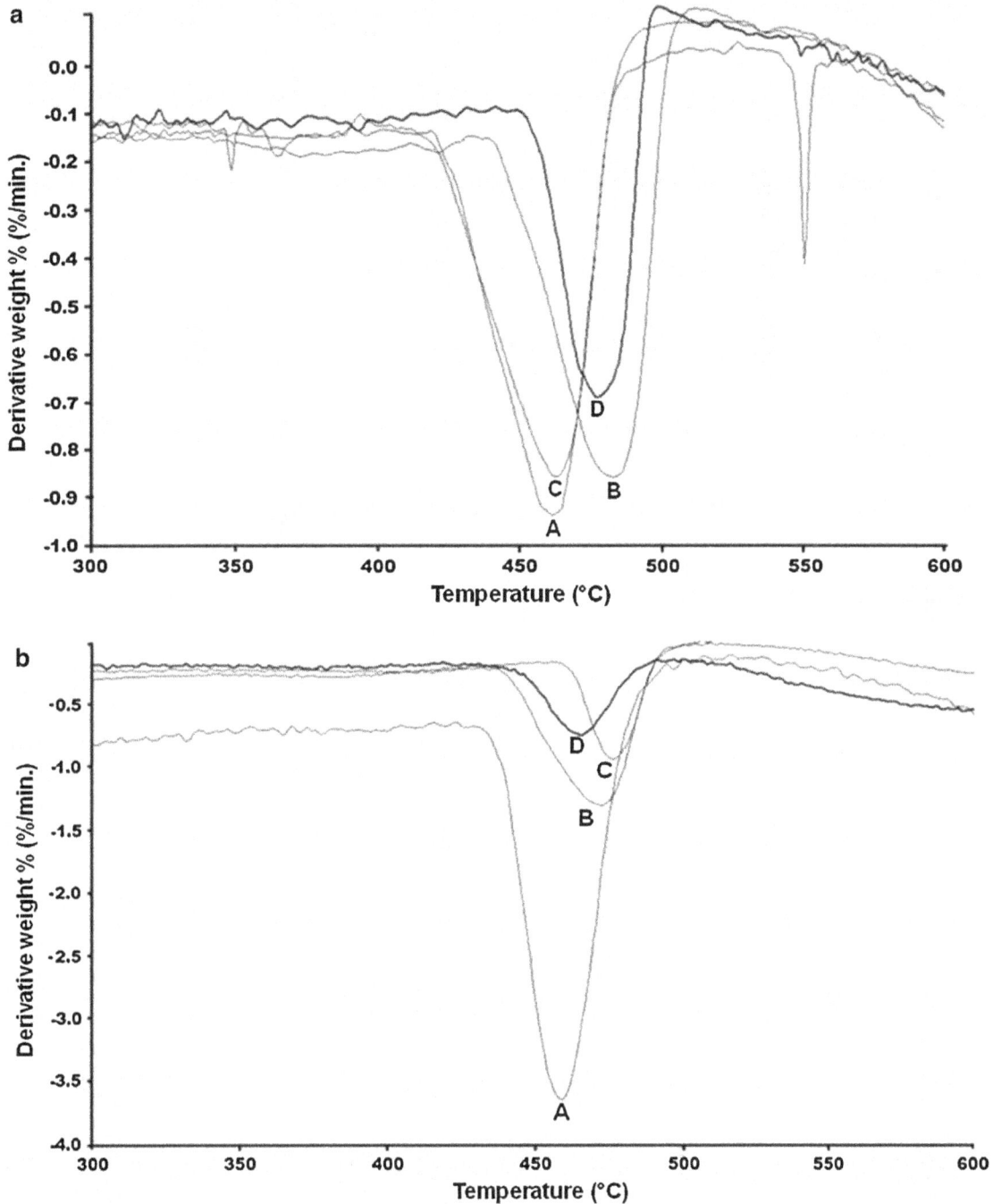

Fig. 9 a DTG of samples at 1 day. *A* plain cement mortar, *B* CM + 3 %SF, *C* CM + 3 %CNS, and *D* CM + 3 %NS. **b** DTG of samples at 28 day. *A* plain cement mortar, *B* CM + 3 %SF, *C* CM + 3 %CNS, and *D* CM + 3 %NS.

Table 3 Portlandite content in cement mortar incorporated nanosilica.

Sample	Portlandite content (%)			
	1 days	3 days	7 days	28 days
Cement mortar	5.85	9.25	11.6	15.32
CM + 3 %SF	5.38	6.9	7.55	10.5
CM + 3 % colloidal-NS	4.2	6.3	6.7	6.5
CM + 3 % powder-NS	4.0	5.8	5.4	4.8

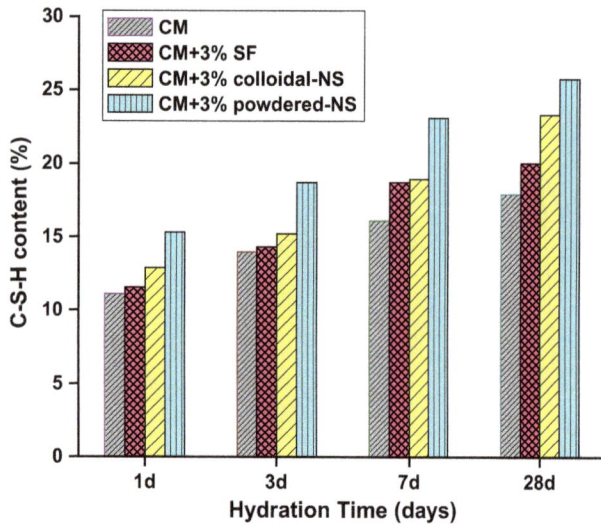

Fig. 10 C–S–H content in plain and nano-silica incorporated cement mortar.

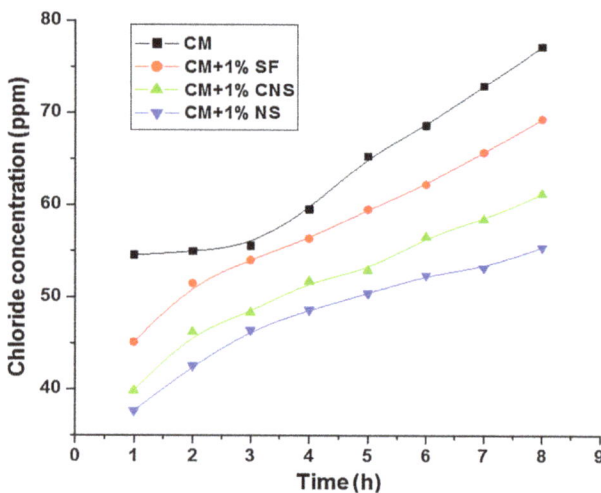

Fig. 11 Concentration of chloride ion with 1 % additives.

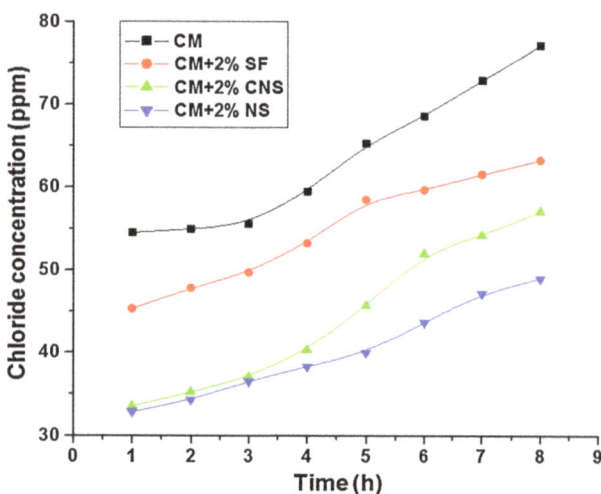

Fig. 12 Concentration of chloride ion with 2 % additives.

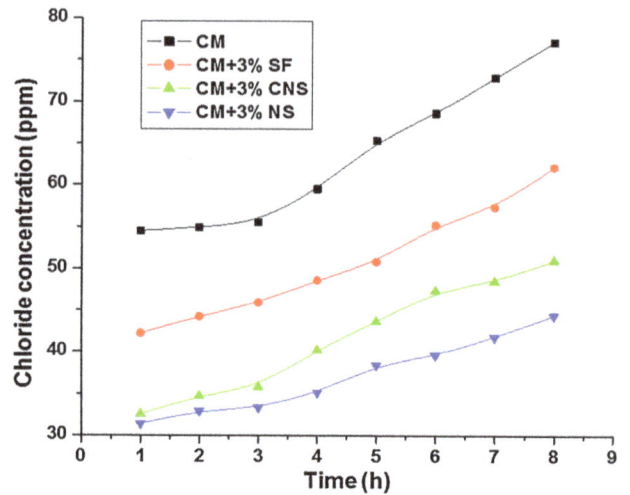

Fig. 13 Concentration of chloride ion with 3 % additives.

consuming CH. Hence, the pozzolanic reactivity of powdered-NS is observed to be higher than colloidal-NS and SF.

3.5 C–S–H Quantification

A detailed approach is given in this section to quantify the C–S–H through TG/DTG. The composition of C–S–H gel approximately $1.7CaO \cdot SiO_2 \cdot 4H_2O$ for saturated C_3S pastes as well as cement, where 4 mol of water including adsorbed water on the surface of hydration products, will be removed at temperature below 110 °C (Taylor 1997). Therefore, the equilibrium composition of C–S–H becomes $1.7CaO \cdot SiO_2 \cdot 2.1H_2O$. Similarly in the present study, the formation of C–S–H with chemical composition $11.7CaO \cdot SiO_2 \cdot 2.1H_2O$ was considered (Young and Hansen 1987). The results indicate that the C–S–H formation at 1 day with plain cement mortar, colloidal nano-silica and powder nano-silica is 11.3, 12.9 and 15.3 %, respectively.

Further, at 28 days, powdered-NS enhances the C–S–H content approx 30 % higher than that plain cement mortar, whereas the colloidal-NS increased approx 21 % (Fig. 10). The mechanism of significant increase in C–S–H content with powdered-NS, colloidal-NS, SF as compared to cement mortar is due to high surface area, resulting to the deposition of cement hydration products on them. The pozzolanic reactivity of powdered-NS is observed to be higher causing continuous consumption of CH to form additional C–S–H from the early stage of hydration. The results are consistent with CH reduction. The non-agglomerated nature of NS, allow it to mix with cement grains homogeneously and develop more compact and denser C–S–H. This agrees with the remarkable increase in the compressive strength when powdered NS particles are incorporated into cement matrix.

3.6 Chloride Penetration Resistance

The electromigration test shows that the cement mortar incorporating nano-silica resist the chloride ingress of mortar, as indicated by the reduced chloride ion concentration with increasing doses of nano-silica and SF. Such improvements are significant with powdered-NS. At 7 V the chloride ion concentration drops to ∼28 % with 1 % nano-

area available for the reaction. The results reveal that the powered-NS has non-agglomerated particles which allow the formation of additional C–S–H on their surface by silica, whereas with colloidal-NS ~ 20 % only at 8 h, results are illustrated in Fig. 11. As the percentage addition of NS and SF increases up to 3 %, the ingress of chloride ions in destination solution (0.3 N NaOH) through mortar specimen reduces by 43, 34 and 19 % in powdered-NS, colloidal-NS and SF, respectively. It was observed that the penetration of mortar specimen decreases (Figs. 12 and 13). The mechanism interpreted that the presence of powdered-NS contribute more in the reduction of chloride ions ingress as compared to colloidal-NS, because powdered-NS is non-agglomerated in nature, which increases the packing density of solid materials by occupying space between cement grains. These physical effects of powered-NS may contribute to refine the pore system and reduce the chloride penetration of microstructure of hydration products. These findings are also consistent with gel/space ratio. The colloidal-NS is agglomerated, does not mix evenly with cement mortar and develop weak zone, which are favourable for migration of ionic species. The electromigration test reveals that the small amount of nano-silica markedly improves the chloride penetration resistance of the cement mortars. Finally, it can be concluded that non-agglomerated powdered-NS can be used to improve cement and concrete durability.

4. Conclusions

Based on the experimental results presented in this study, the following conclusions can be drawn:

(1) The dispersibility of nanomaterials plays a key role, therefore spherical and non-agglomerated nano-silica was prepared to address the issue of mixing of nanomaterials into cement matrix. It is found that non-agglomerated powdered-NS is effective in improving the mechanical properties of cement mortar.

(2) The improvement in compressive strength and gel/space ratio of hardened mortar adding powered-NS is greater than that colloidal-NS which was associated with the higher content of C–S–H (30 %), leading to more compacted and denser microstructure.

(3) Powdered-NS due to higher pozzolanic reactivity reduced CH content up to 68 %, while colloidal-NS reduced 57 % at 28 days of hydration.

(4) SEM micrographs revealed that the powdered-NS behaved as a filler to improve cement microstructure leading to denser morphology.

(5) For durability assessment, chloride ingress was also monitored in cement mortar with powered-NS, colloidal-NS and SF. Among these Powdered-NS reduced the chloride ion concentration up to ~ 43 % as compared to plain cement mortar.

References

Acker P. (2001). Micromechanical analysis of creep and shrinkage mechanisms. In F. J. Ulm, Z. P. Bazant, F. H. Wittmann (Eds.), *Creep, shrinkage and durability mechanics of concrete and other quasi-brittle materials, 6th international conference* (pp. 15–26) Amsterdam, Netherlands: CONCREEP@MIT Elsevier.

Alonso, C., & Fernandez, L. (2004). Dehydration and rehydration processes of cement paste exposed to high temperature environments. *Journal of Materials Science, 39*(9), 3015–3024.

Baykal, M. (2000). *Implementation of durability models for portland cement concrete into performance-based specifications*. Austin, TX: University of Texas at Austin.

Bjornström, J., Martinelli, A., Matic, A., Borjesson, L., & Panas, I. (2004). Accelerating effects of colloidal nano-silica for beneficial calcium-silicate-hydrate formation in cement. *Chemical Physics Letters, 392*(1–3), 242–248.

Coenen, S., & Kruif, C. G. (1988). Synthesis and growth of colloidal silica particles. *Journal of Colloid and Interface Science, 124*(1), 104–110.

Flores, I., Sobolev K., Torres-Martinez L. M., Cuellar E. L., Valdez P. L., Zarazua E. (2010). Performances of cement systems with nano-SiO$_2$ particles produced by using the sol-gel method. In *Transportation Research Record: Journal of the Transportation Research Board, No. 2141* (pp. 10–14). Washington, DC: Transportation Research Board of the National Academies.

Gabrovsek, R., Vuk, T., & Kaucic, V. (2006). Evaluation of the hydration of Portland cement containing various carbonates by means of thermal analysis. *Acta Chimica Slovenica, 53*(2), 159–165.

Gaitero, J. J., Campillo, I., & Guerrero, A. (2008). Reduction of the calcium leaching rate of cement paste by addition of silica nanoparticles. *Cement and Concrete Research, 38*(8–9), 1112–1118.

Gaitero, J. J., Zhu, W., & Campillo, I. (2009). Multi-scale study of calcium leaching in cement pastes with silica nanoparticles. *Nanotechnology in construction 3, Berlin* (pp. 193–198). Heidelberg, Germany: Springer.

Gallucci, E., Zhang, X., & Scrivener, K. L. (2013). Effect of temperature on the microstructure of calcium silicate hydrate (C-S-H). *Cement and Concrete Research, 53*, 185–195.

He, X., & Shi, X. (2008). Chloride permeability and microstructure of Portland cement mortars incorporating nanomaterials. *Transportation Research Record: Journal of the Transportation Research Board, 2070*, 13–21.

Hou, P., Cheng, X., Qian, J., & Shah, S. P. (2014). Effects and mechanisms of surface treatment of hardened cement-based materials with colloidal nano-SiO$_2$ and its precursor. *Construction and Building Materials, 53*, 66–73.

Jain, J., & Neithalath, N. (2009). Analysis of calcium leaching behavior of plain and modified cement pastes in pure water. *Cement and Concrete Composite, 31*(3), 176–185.

Ji, T. (2005). Preliminary study on the water permeability and microstructure of concrete incorporating nano-SiO_2. *Cement and Concrete Research, 35*(10), 1943–1947.

Jo, B., Kim, C., & Lim, J. (2007). Characteristics of cement mortar with nano-SiO_2 particles. *Construction and Building Materials, 21*(6), 1351–1355.

Kong, D., Su, Y., Xi, D., Yang, Y., Wei, S., & Shah, S. P. (2013). Influence of nano-silica agglomeration on fresh properties of cement pastes. *Construction and Building Materials, 43*, 557–562.

Kontoleontos, F., Tsakiridis, P. E., Marinos, A., Kaloidas, V., & Katsioti, M. (2012). Influence of colloidal nano-silica on ultrafine cement hydration: Physicochemical and microstructural characterization. *Construction and Building Materials, 35*, 347–360.

Lam, L., Wong, Y. L., & Poon, C. S. (2000). Degree of hydration and gel/space ratio of high-volume fly ash/cement systems. *Cement and Concrete Research, 30*(5), 747–756.

Lin, W. T., Huang, R., Chang, J. J., & Lee, C. L. (2009). Effect of silica fume on the permeability of fiber cement composites. *Journal of the Chinese Institute of Engineers, 32*(4), 531–541.

Neville, A. M. (1981). *Properties of concrete* (3rd ed., pp. 257–279). London, UK: ELBS with Longman.

Olsona, R. A., & Jennings, H. M. (2001). Estimation of C-S-H content in a blended cement paste using water adsorption. *Cement and Concrete Research, 31*(3), 351–356.

Pichler, B., Hellmich, C., Eberhardsteiner, J., Wasserbauer, J., Termkhajornkit, P., Barbarulo, R., & Chanvillard, G. (2013). Effect of gel–space ratio and microstructure on strength of hydrating cementitious materials: An engineering micromechanics approach. *Cement and Concrete Research, 45*, 55–68.

Powers, T. C., & Brownyard, T. L. (1948). Studies of the physical properties of hardened Portland cement paste. *Research Laboratories of the Portland Cement Association Bulletin, 22*, 101–992.

Quercia, G., Spiesz, P., Hüsken, G., & Brouwers, H. J. H. (2014). SCC modification by use of amorphous nano-silica. *Cement & Concrete Composites, 45*, 69–81.

Ramachandran, V. S., Paroli, R. M., Beaudoin, J. J., & Delgado, A. H. (Eds.). (2003). *Handbook of thermal analysis of construction materials*. Norwich: Noyes Publications.

Sanchez, F., & Sobolev, K. (2010). Nanotechnology in concrete—A review. *Construction and Building Materials, 24*(11), 2060–2071.

Savas B. Z. (2000). Effects of microstructure on durability of concrete, Ph.D. thesis. Raleigh: North Carolina State University.

Shi, X., Xie, N., Fortune, K., & Gong, J. (2012). Durability of steel reinforced concrete in chloride environments: An overview. *Construction and Building Materials, 30*, 125–138.

Singh, L. P., Bhattacharyya, S. K., & Ahalawat, S. (2012a). Preparation of size controlled silica nano particles and its functional role in cementitious system. *Journal of Advanced Concrete Technology, 10*(11), 345–352.

Singh, L. P., Bhattacharyya, S. K., Mishra, G., & Ahalawat, S. (2012b). Reduction of calcium leaching in cement hydration process using nanomaterials. *Materials Technology, 27*(3), 233–238.

Singh, L. P., Karade, S. R., Bhattacharyya, S. K., & Ahalawat, S. (2013). Beneficial role of nano-silica in cement based materials—a review. *Construction and Building Materials, 47*, 1069–1077.

Tan, B., Lehmler, H. J., Vyas, S. M., Knuston, B. L., & Rankin, S. E. (2005). Controlling nanopore size and shape by fluorosurfactant templating of silica. *Chemistry of Materials, 17*(4), 916–925.

Taylor, H. F. W. (1997). *Cement chemistry*. London, UK: Thomas Telford.

Toutanji, H., Delatte, N., & Aggoun, S. (2004). Effect of supplementary cementitious materials on the compressive strength and durability of short-term cured concrete. *Cement and Concrete Research, 34*(2), 311–319.

Venkatathri, N., & Nanjundan, S. (2009). Synthesis and characterization of a mesoporous silica microsphere from polystyrene. *Materials Chemistry and Physics, 113*(2–3), 933–936.

Young, J. F., & Hansen, W. (1987). Volume relationships for C-S-H formation based on hydration stoichiometries. *Materials Research Society, 85*, 313.

6

Residual Strength of Corroded Reinforced Concrete Beams Using an Adaptive Model Based on ANN

Ashhad Imam[1],*, Fatai Anifowose[2], and Abul Kalam Azad[3]

Abstract: Estimation of the residual strength of corroded reinforced concrete beams has been studied from experimental and theoretical perspectives. The former is arduous as it involves casting beams of various sizes, which are then subjected to various degrees of corrosion damage. The latter are static; hence cannot be generalized as new coefficients need to be re-generated for new cases. This calls for dynamic models that are adaptive to new cases and offer efficient generalization capability. Computational intelligence techniques have been applied in Construction Engineering modeling problems. However, these techniques have not been adequately applied to the problem addressed in this paper. This study extends the empirical model proposed by Azad et al. (Mag Concr Res 62(6):405–414, 2010), which considered all the adverse effects of corrosion on steel. We proposed four artificial neural networks (ANN) models to predict the residual flexural strength of corroded RC beams using the same data from Azad et al. (2010). We employed two modes of prediction: through the correction factor (C_f) and through the residual strength (M_{res}). For each mode, we studied the effect of fixed and random data stratification on the performance of the models. The results of the ANN models were found to be in good agreement with experimental values. When compared with the results of Azad et al. (2010), the ANN model with randomized data stratification gave a C_f-based prediction with up to 49 % improvement in correlation coefficient and 92 % error reduction. This confirms the reliability of ANN over the empirical models.

Keywords: corrosion, reinforced concrete beam, flexural strength, artificial neural networks.

1. Introduction

Corrosion of reinforcement steel has been proved to be a major cause of deterioration of reinforced concrete (RC) structures, resulting in the reduction of the service life of concrete structures. A substantial amount of research related to reinforcement corrosion has been carried out in the past, addressing various issues related to the corrosion process, its initiation and damaging effects. Assessment of the flexural strength of corrosion-damaged RC members has been studied (Azad et al. 2010; Cabrera 1996; Huang and Yang 1997; Rodriguez et al. 1997; Uomoto and Misra 1988). A number of studies have also been conducted on the prediction of residual flexural strength of corroding concrete beams (Azad et al. 2007; Mangat and Elgarf 1999; Nokhasteh and Eyre

1992; Ravindrarajah and Ong 1987; Tachibana et al. 1990; Wang and Liu 2008; Jin and Zhao 2001). Some of these studies had been conducted in the laboratory. They involve the casting of concrete beam specimens sometimes in large scale, in the order of meters in dimension (Ou et al. 2012), and sometimes in small scale, in the order of millimeters (Azad et al. 2007; Mangat and Elgarf 1999; Nokhasteh and Eyre 1992; Revathy et al. 2009; Tachibana et al. 1990; Wang and Liu 2008; Jin and Zhao 2001). The specimens are then subjected to various degrees of corrosion damage after which the samples are tested for their bending or flexural performances. These procedures take a lot of time as some of the specimens need to be left for several days to attain their required degree of corrosion. They also require the use of expensive and specialized laboratory equipment, exorbitant man-hours and concerted effort. An average experiment can take up to 6 months to complete. Though, experiments are the best sources of real data but the associated costs often make them prohibitive.

In order to reduce the completion time and avoid the cost associated with such studies without compromising on accuracy, some attempts have been made on the use of numerical modeling methods (Azad et al. 2007, 2010; Cabrera 1996; Coronelli and Gambarova 2004; Ou et al. 2012). These methods are however static and cannot be generalized well on datasets outside those for which they were designed. Most of them do not consider the non-linearity of the attributes of the natural phenomena involved in the corrosion

[1]Department of Civil Engineering, King Fahd University of Petroleum & Minerals, Dhahran 31261, Saudi Arabia.
*Corresponding Author; E-mail: ashhad.ce@gmail.com

[2]Center for Petroleum and Minerals, Research Institute, King Fahd University of Petroleum & Minerals, Dhahran 31261, Saudi Arabia.

[3]Department of Civil & Environmental Engineering, College of Engineering Sciences, King Fahd University of Petroleum & Minerals, Dhahran 31261, Saudi Arabia.

process. Since corrosion is a natural process, it is expected that its attributes be non-linearly related to the corrosion property being studied. Hence, modeling the process with linear relations is inadequate. To make such models more generalized, they need to be recalibrated with new sets of data. Doing this will result in re-generating new sets of co-efficients to evolve a new model, which requires considerable time and effort.

With the limitations in the experimental and theoretical methods, the quest for cost-effective, easy to use and adaptive models that offer scalability and efficient generalization capability to new cases continues. With the huge amount of data generated from various experiments over the years, robust data mining techniques that are based on computational intelligence (CI) and machine learning paradigms are hypothesized to be capable of overcoming the limitations of the conventional methods. The prediction and generalization capability of artificial neural networks (ANN) had been investigated in this paper. With the capability of ANN to handle the non-linearity in natural phenomena such as corrosion, coupled with its capability to adaptively learn from hidden patterns in experimental data, we presented two optimized models of ANN to efficiently predict the flexural strength of corrosion-damaged RC beams.

The motivations for choosing the proposed ANN models are:

- ANN is the most commonly used of the CI techniques in various application areas (Abdalla et al. 2007; VanLuchene and Sun 1990; Waszczyszyn and Ziemiański 2001; Wu et al. 1992).
- Though ANN has been applied in modeling other civil engineering problems, they have not been adequately applied to the problem of estimating the residual flexural strength of corrosion-damaged RC beams, which is the focus of this paper.
- ANN is easy to use and understand by researchers outside the Computer Science field.
- Following the principle of Occam's Razor (Jefferys and Berger 1991), starting an investigation with a simple model like ANN is preferred to using more complex and state-of-the-art techniques.
- Since the performance of a model is determined by the nature of the problem (represented by data), there is no guarantee that using a more sophisticated algorithm will perform better. This agrees with the No Free Lunch theorem (Wolpert and Macready 1997).

Further to ensuring simplicity in design and implementation, proposing optimization- and feature selection-based hybrid models is considered undesirable. Optimization algorithms such as Particle Swam, Genetic Algorithm, Bee Colony, Ant Colony, etc. are based on heuristic and exhaustive search paradigms (Bies et al. 2006). Due to this, they take much time to converge, require considerable memory resources and are complex. Also, since this study does not involve high-dimensional dataset, proposing feature selection-based hybrid models will be of no use.

The rest of this paper is organized as follows: Sect. 2 presents a literature survey on the proposed study. Section 3 gives a brief background on the proposed technique as well as the previously published empirical equation. Section 4 describes the datasets used for this study and the details of the proposed methodology. Results are presented and discussed in Sect. 5 while conclusions, highlighting the contributions of this study as well as its limitations, are presented in Sect. 6.

2. Literature Survey

The estimation of the flexural strength of corroded RC beams has been a focus of keen research for almost three decades (Cabrera 1996; Huang and Yang 1997; Ravindrarajah and Ong 1987; Rodriguez et al. 1997; Uomoto and Misra 1988). That shows the importance of this phenomenon in the construction engineering field. One of the earliest studies on this subject include the experiment carried out by Ravindrarajah and Ong (1987) to study the effect of corrosion on steel bars in mortar with the use of an accelerated corrosion technique. This was followed by Uomoto and Misra (1988) who studied the behavior of concrete beams and the changes in columns as corrosion of reinforcing bars increases. They presented an idea on when to repair the structures in marine environment. Another interesting research was carried out by Rodriguez et al. (1997) who induced corrosion to some RC beams. The data extracted from the experiment was used to develop some numerical models for the assessment of concrete structures affected by steel corrosion and other deterioration mechanisms. A similar work was carried out by Huang and Yang (1997) who tested thirty-two concrete beams in order to assess their structural behavior due to corrosion. The afore-mentioned studies were based on experimental methods whose limitations have been highlighted in Sect. 1.

Numerical methods of estimating the effect of corrosion have been applied since the experimental studies started. One of the earliest of such studies is that of Cabrera (1996) who used laboratory data to derive numerical models to relate the rate of corrosion to cracking and loss of bond strength. Rodriguez et al. (1997) used one of the Euro Code 2 conventional models to predict the ultimate bending moment and shear force. Later, Coronelli and Gambarova (2004) used a finite-element-based numerical procedure to estimate the bond deterioration index of concrete beams under corrosive conditions. Azad et al. (2007) and Al-Gohi (2008) employed a regression analysis method on the data obtained from experiments to predict the residual flexural strength of RC beams. They formulated a correction factor that can be used to calculate the flexural strength of a corroded beam using the reduced area of corroded bars. They later improved this correction factor in Azad et al. (2010) using new sets of data. When compared with Azad et al. (2007), they concluded that the improved model yielded values that are in good agreement with the test data, lending confidence to the proposed method to serve as a reliable analytical tool to predict the flexural capacity of a corroded concrete beam. The most recent study is the empirical equation proposed by Ou et al. (2012) who concluded that the results closely corresponds to experimental values.

It is well known that all these regression analysis and analytical modeling techniques, though performing well in their respective applications, do not handle the non-linearity between independent variables and their target values. They are rather based on the assumption that the dependent variables are linearly related to one or more of the independent variables. They do not consider the hidden and the non-linear relationships that exist in such natural phenomena. CI techniques, on the other hand, have the capability to extract hidden patterns and "learn" from historical knowledge to make predictions about unknown future cases (Eskandari et al. 2004). Applications of CI techniques have demonstrated superior performance over regression analysis and analytical modeling techniques. ANN has featured in a number of engineering problems with excellent performance (Castillo et al. 2001; Mohaghegh 1995; Rafiq et al. 2001; Tsai and Hsu 2002).

A set of pragmatic guidelines for designing ANN for engineering applications were proposed by Rafiq et al. (2001). Hsu and Chung (2002) evolved a model of damage diagnosing for RC structures using the ANN technique. The network learning procedure showed that the rate and the accuracy of the convergence is acceptable while the test results showed that the technique is efficient for the problem. ANN has been used to predict the shear strength (Cabrera 1996), deformation capacity (Inel 2007) and shear resistance (Abdalla et al. 2007) of various shapes of RC beams. The prediction of residual flexural strength has not been addressed with CI techniques. We intend to fill this research gap by proposing ANN models for this problem.

The success of the few attempts at utilizing the learning and predictive power of ANN in construction engineering as discussed above is the major motivation for this study. The main objective of this study is to use the ANN technique in the prediction of residual flexural strength of corroded RC beams. This technique utilizes the capability of the supervised machine learning concept to model the complex non-linear relationship between the properties of RC beams and their independent variables.

The next section gives a brief theoretical background of ANN and the necessary details of the empirical model proposed by Azad et al. (2010).

3. Theoretical Background

3.1 Artificial Neural Networks

Artificial Neural Networks (ANN) was inspired by the functioning of the human brain. It is an emulation of the biological nervous system that is made up of several layers of neurons (nodes) interconnected by links. Each node is assigned a weight. ANN can be thought of as a "computational system" that accepts inputs and produces outputs (Baughman 1995). Figure 1 shows how the human nervous system is mapped to evolve a typical ANN structure. The dendrites and synapses that serve as receptacles of excitatory signals in Fig. 1a is equivalent to the input layers of ANN in Fig. 1b from where input variables are admitted into the system. As

the nucleus gathers all input signals and prepares them for processing (as shown in Fig. 1a), the input signals in ANN are multiplied by the weights and biases, and aggregated in the summation layer (in Fig. 1b) where they are "excited" with an activation function. Finally, while the gathered signals are sent through the axon (Fig. 1a) to the brain for processing, the optimized results produced by ANN are sent to the output layer for interpretation and decision-making processes.

The individual inputs: $x_1, x_2, \ldots, x_k$ are multiplied by weights and the weighted values are fed to the summing junction ($\sum$). Their sum is simply wx, which is the dot product of the matrix w and the vector x. The neuron has a bias μ_i, which is summed up with the weighted inputs to form the predicted output y_i subject to the transfer function f. The input vector enters the network through the weight matrix w expressed as:

$$w = \begin{bmatrix} w_{11}, w_{12}, \ldots, w_{1k} \\ w_{21}, w_{22}, \ldots, w_{2k} \\ w_{i1}, w_{i2}, \ldots, w_{ik} \end{bmatrix} \qquad (1)$$

where i is the number of neurons and k is the number of inputs.

The ANN processes are mathematically represented and generalized as:

$$y_i = f\left(\sum_k w_{ik}x_k + \mu_i\right) \qquad (2)$$

where x_k are inputs to the neuron, w_{ik} are weights attached to the inputs to the neuron, μ_i is a threshold, offset or bias, $f(\cdot)$ is a transfer function and y_i is the output of the neuron. The transfer function $f(\cdot)$ can be any of: linear, non-linear, piecewise linear, sigmoidal, tangent hyperbolic and polynomial functions.

As each input x_k is applied to the network, the network output is compared to the target. The error is calculated as the difference between the target output and the network output. The goal is to minimize the average of the sum of these errors. The least mean square (LMS) algorithm adjusts the weights and biases of the linear network so as to minimize this mean square error:

$$mse = \frac{1}{Q}\sum_{k=1}^{Q} e(k)^2 = \frac{1}{Q}\sum_{k=1}^{Q}(t(k) - a(k))^2 \qquad (3)$$

where Q is the number of data samples, e is the error criterion, t is the original target values and a is the model predicted values.

The mean square error performance index for the linear network is a quadratic function. Thus, the performance index will either have one global minimum, a weak minimum, or no minimum, depending on the characteristics of the input vectors. Specifically, the characteristics of the input vectors determine whether or not a unique solution exists.

The number of neurons in the input layer corresponds to the number of input variables fed into the neural network.

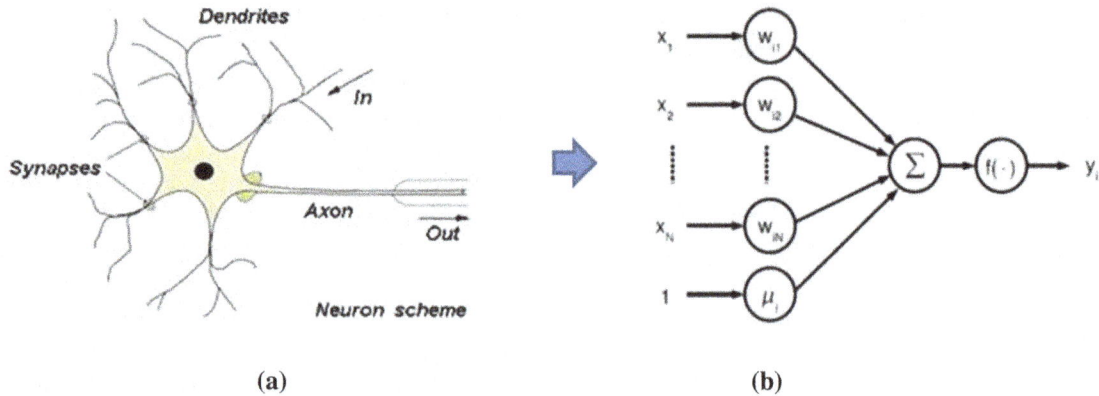

(a)   **(b)**

Fig. 1 The Architecture of artificial neural networks.

The number of hidden layer(s) and the corresponding number of neurons in each determine the "power" of the network. This is determined through the training process. The complexity of a problem determines the "power" it requires. Using more power than necessary will lead to the model being overfitted. In this case, the models perform well on the training data but poorly on the validation data (Hastie et al. 2009). On the other hand, using less power than required will lead to underfitting (Hastie et al. 2009). In this case, the model performs poorly on both training and validation datasets.

In order to concentrate more on the subject of this study, readers are referred to (Petrus et al. 1995) for more details of the architecture, structure and mathematical bases of ANN. This technique has caught the interest of most researchers and has today become an essential part of the technology industry, providing a good ground for solving many of the most difficult prediction problems in various areas of engineering applications (Baughman 1995; Guler 2005; Inan et al. 2006; Li and Jiao 2002; Moghadassi et al. 2009; Mohaghegh 1995; Nascimento et al. 2000; Phung and Bouzerdoum 2007; Übeyli 2009). ANN has also gained vast popularity in solving various Civil Engineering problems (Baughman 1995; Beale and Demuth 2013; Chen et al. 1995; Flood and Kartam 1994; Hasancebi and Dumlupınar 2013; Kang and Yoon 1994; Kirkegaard and Rytter 1994; Neaupane and Adhikari 2006; Pandey and Barai 1995; Rafiq et al. 2001).

3.2 The Previously Published Empirical Model

Azad et al. (2010) proposed the following two-step procedure to predict the residual flexural strength of corroded beams for which the cross-sectional details, material strengths, corrosion activity index $I_{corr}T$, and diameter of rebar, D were known. The procedure used is:

- First, the moment capacity $(M_{th,c})$ was calculated using reduced cross-sectional area of tensile reinforcement, A'_s, in the conventional manner.
- The computed value of $M_{th,c}$ was then multiplied by a correction factor (C_f) to obtain the predicted residual flexural strength of the beam (M_{res}) using this relation:

$$M_{res} = C_f M_{th,c} \tag{4}$$

where C_f was assumed to represent the combined effect of bond loss and factors pertaining to loss of flexural strength other than the reduction of the metal area.

- The value of C_f was taken as a function of the two important variables, namely $I_{corr}T$ and D. Finally, using regression analysis of the data and a gravimetric analysis of the weight loss of steel (Beale and Demuth 2013), the following empirical equation for the correction factor C_f was derived:

$$C_f = \frac{5.0}{D^{0.54}(I_{corr}T)^{0.19}}; \quad C_f \leq 1.0 \tag{5}$$

where D is the diameter of rebar in mm, I_{corr} is the corrosion current density in mA/cm^2 and T is the duration of corrosion in days.

The residual flexural strength of a rectangular beam could be determined using Eq. (4) by substituting the value of the calculated C_f from Eq. (5).

As stated earlier, the limitation of the above equations is that they were based on the assumption that $I_{corr}T$ and D are linearly related to the residual flexural strength. In such a natural phenomenon as corrosion, this will not give an optimal solution to the estimation problem. Basing construction projects on the results of this equation may result in suboptimal life spans and increase maintenance costs. In order to overcome this limitation, ANN models, with their capability to utilize the non-linear relationship of the variables and their ability to extract hidden patterns from datasets, are proposed.

4. Research Methodology

4.1 Description of Data

For effective validation and comparison with the results of previous work, we designed and implemented our ANN models by using the same experimental data obtained and used by Azad et al. (2010) to predict the residual strength of corroding concrete beams. The data was obtained from an experiment consisting of 48 RC beams of different cross-sections and reinforcements. The beams were made of three different depths viz. 215, 265 and 315 mm, two different diameters of tension bars viz. 16 and 18 mm, and different durations of corrosion. The corrosion was induced by applying a direct current at a constant rate of 1.78 mA/cm². Out of the 48 beams, 36 were subjected to accelerated corrosion. Both the corroded and un-corroded beams were tested in a four point bend to find their load carrying capacity using a span length of 900 mm and a flexure span of 200 mm. Statistically, the values of both C_f and M_{res} follow normal distribution as shown in the histograms in Figs. 2 and 3 respectively.

The results of the basic descriptive statistics (maximum, minimum, range, mean, variance, standard deviation, skewness, and kurtosis) that further describe the dataset are presented in Table 1. The maximum statistic is the largest or the greatest value in a set of data. The minimum is the smallest or the least value in a given set of data. The range, the arithmetical difference between the maximum and the minimum of a set of data, is a statistical measure of the spread of a dataset. The most common expression for the mean of a statistical distribution with a discrete random variable is the mathematical average of all the terms. The variance, a measure of dispersion in a data, is the average squared distance between the mean and each item in the population or sample.

The standard deviation, also a measure of dispersion, is the positive square root of the variance. An advantage of the standard deviation (compared to the variance) is that it expresses dispersion in the same units as the original values in the sample or population. Skewness is a measure of the extent to which a probability distribution of a real-valued random variable "leans" to either side of the mean. The skewness value can be positive or negative, or even undefined. When the skewness is positive, then it is said that the distribution is skewed to the right of the mean. Similarly, when it is negative, the distribution is skewed to the left. A zero skewness is said to be undefined. Kurtosis is a measure of the "peakedness" or the "flatness" of the probability distribution of a real-valued random variable. In a similar way to the concept of skewness, kurtosis is a descriptor of the shape of a probability distribution.

Table 1 shows that the predictor variables are diameter (D) and corrosion activity index ($I_{corr}T$) while the target properties are correction factor (C_f) and residual strength (M_{res}). The variable D has a range of 2 with the maximum and minimum values being 18 and 16 respectively and an average value of 17.03. The variable $I_{corr}T$ has a higher range of 30.16 with the maximum and minimum values being 32.06 and 1.90 respectively and an average value of 15.36. The maximum, minimum, range and average values of the C_f are much less than those of M_{res}. The values of C_f range between 0.5 and 1.08 (Table 1) with most of the values concentrated around 0.73 (Fig. 2) while M_{res} has its range between 16.1 and 66 (Table 1) and those with the highest frequency around 36 (Fig. 3). The variances and standard deviations of D and C_f are much less than those of $I_{corr}T$ and M_{res} respectively. The distribution of D is skewed to the left with a negative value while that of others is skewed to the right with positive values. The values of the skewness, together with the statistical trends shown in Figs. 2 and 3, are an indication that the data is not symmetrical. The negative kurtosis of all the predictor and target variables shows that the distribution of the data is flat rather than being peaked or Gaussian. All these are an indication that the data is typical of real-life experimental data and not simulated.

In order to simulate the practical application of the ANN models, the data was divided into training and testing subsets. The training subset represented the available experimental data comprising both the cross-sectional details and their corresponding flexural strength estimations while the testing subset represented only the experimental measurements without their equivalent flexural strength estimations. The models used the hidden non-linear relationship between the experimental measurements and the target values to train the models. The trained models were then used to predict the required flexural strength estimations given the desired experimental measurements. This was intended to save time, cost and man-hours while improving the accuracy.

To achieve the above implementation objective, the data was divided in the 70:30 ratio in which 70 % was used for training

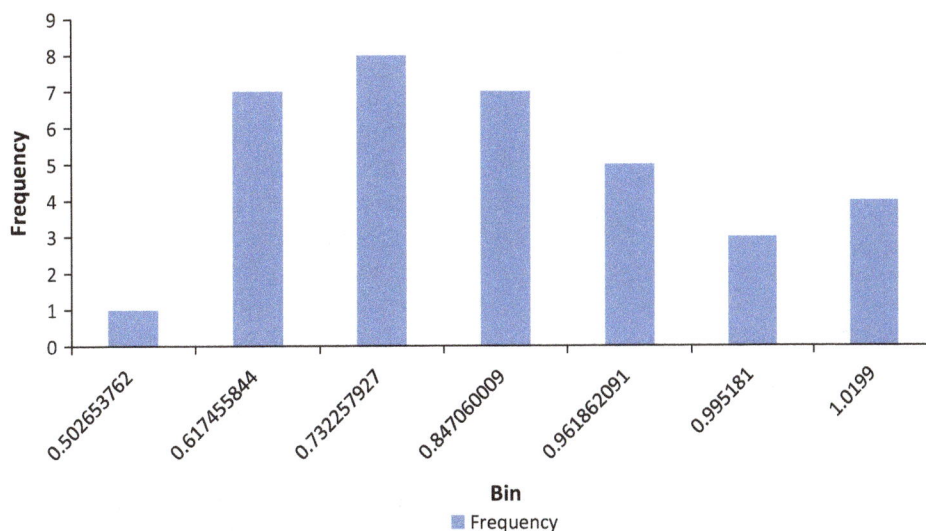

Fig. 2 The histogram of C_f values.

Histogram

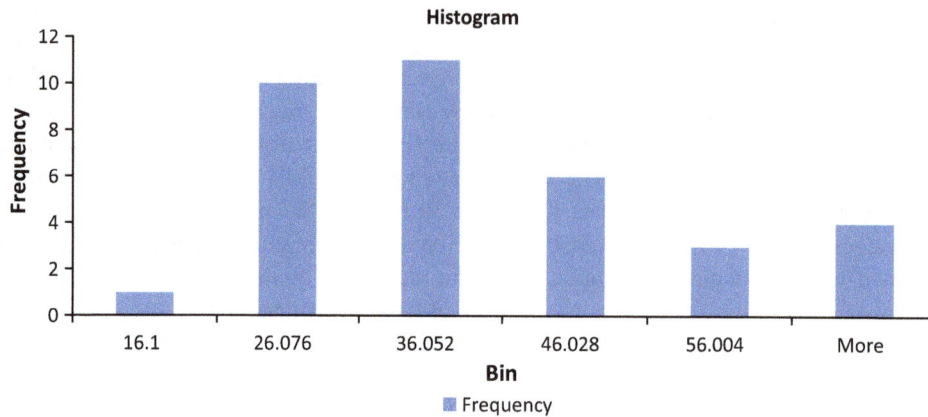

Fig. 3 The histogram of M_{res} values.

Table 1 Basic descriptive statistics of the data.

Statistics	Predictor variables		Target properties	
	Diameter (D)	Corrosion activity index (I_{corr} T)	Correction factor (C_f)	Residual strength (M_{res})
Maximum	18.00	32.06	1.0767	65.98
Minimum	16.00	1.90	0.5026	16.10
Mean	17.03	15.36	0.7749	34.60
Range	2.00	30.16	0.5740	49.88
Variance	0.99	76.38	0.0275	167.11
Standard deviation	1.01	8.87	0.1683	13.12
Skewness	−1.77	0.70	0.4308	0.40
Kurtosis	−2.12	−0.93	−1.2438	−0.17

and the remaining 30 % for testing. This is similar to the k-fold cross-validation technique and follows the standard machine learning paradigm that requires that the training subset should be much more than the testing. This is partly meant to avoid the overfitting and underfitting problems (Hastie et al. 2009). The authors experimented with two data stratification strategies: fixed and random. In the former, we used the first 70 % for training and the remaining for testing. However, in the latter, the authors took 70 % of the randomized sample of the data for training and the remaining for testing. The authors opined that the latter option is more representative of the experimental measurements and has the potential to avoid the bias that is usually associated with the choice of the training data. Hence, it was hypothesized that the results from the randomized stratification will be of more confidence. The outcome of this hypothesis is revealed in Sect. 5.

4.2 Criteria for Models Evaluation

For ease of comparison and due to their common use in predictive modeling literature, we used the comparative coefficient of determination (R^2) and root mean square error ($RMSE$) to evaluate the comparative performance of the ANN models and the empirical equation of Azad et al. (2010). The R^2 measures the statistical correlation between the predicted (y) and actual values (x). It is expressed as:

$$R^2 = \frac{n \sum xy - (\sum x)(\sum y)}{\sqrt{n(\sum x^2) - (\sum x)^2} \sqrt{n(\sum y^2) - (\sum y)^2}} \quad (6)$$

The $RMSE$ is a measure of the spread of the actual x values around the average of the predicted y values. It computes the average of the squared differences between each predicted value and its corresponding actual value. It is expressed as:

$$RMSE = \sqrt{\frac{\sum_{i=1}^{n} (x_i - y_i)^2}{n}} \quad (7)$$

4.3 Details of Design and Implementation of the ANN Models

Four different ANN models were developed in this study: A model each to directly predict the flexural strength of corroded RC beams (M_{res}) and indirectly through the prediction of the correlation factor (C_f). The implementation of each of these models was carried out with fixed and random data stratifications as discussed in Sect. 4.1. Each of the models is basically made up of a two-hidden-layer configuration with two neurons in the input layer and one in the output layer. Our choice of this configuration is based on the report of Beale and Demuth (2013) that a 2-layer neural network is enough to solve most problems. The two neurons in the input layer represent the diameter of reinforced steel

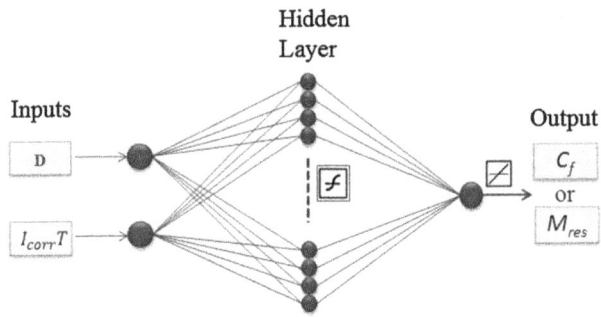

Fig. 4 Basic ANN architecture used in this work.

(D) and corrosion activity index $(I_{corr}T)$ while the output neuron represents each of the target properties, C_f and M_{res}. Figure 4 is a sketch of the basic architecture used in this study showing the input layer, hidden layer, output layer and the transfer functions in the hidden and output layers.

Following the assumption that no problem would require an ANN model with more than 50 neurons in each hidden layer (Beale and Demuth 2013), the optimal number of hidden neurons was investigated from 1 through 50 while trying different learning algorithms and activation functions for the hidden and output layers. This procedure to search for optimal parameters is necessary since each problem requires different values of these parameters for optimal performance. Choosing the appropriate number of hidden neurons is essential for the ultimate performance of ANN models. If the number is too small, the model will not adequately capture the pattern that is hidden in the data. This leads to underfitting, which is characterized by a generally poor performance in both training and testing (Hastie et al. 2009). If the number is too large, the model will exert too much energy than necessary to solve the problem. This leads to overfitting characterized by a model performing excellently well in training but poor in generalization on new cases (Hastie et al. 2009).

In order to avoid cases of underfitting and overfitting, we followed the standard training procedure of ANN as presented in the following algorithm:

1.	Divide data into training and testing subsets.
2.	Initialize the weights and biases to some random initial values.
3.	Find the dot product of weights and input vector.
4.	Feed forward the results of each hidden layer to the output layer.
5.	Compare the network outputs with the original target values.
6.	If the error is equal or less than the pre-set error goal then
7.	Use the network to predict the target for the testing subset.
8.	Present prediction results.
9.	Go to 13.
10.	Otherwise
11.	Back-propagate results to the end of the input layer.
12.	Go to 2.
13.	End.

While the optimal number of hidden neurons obtained for each model of ANN is presented under the description of the respective models in the following sections, the other parameters used in the design of the ANN models are:

- Number of training epochs = 100
- Error goal = 0.001
- Training algorithm = Levenberg–Marquardt
- Error criterion = Mean squared error
- Transfer function in the hidden layer = Sigmoidal
- Transfer function in the output layer = Purelin

The first two parameters above were used to determine when to stop the training process: either the attainment of the error goal or using the most optimal configuration obtained on reaching the maximum number of epochs. Following the standard machine learning paradigm, the feed-forward back-propagation ANN models were trained by feeding a matrix of training data complete with the input and target values. The main objective of the training process is to optimize the connection weights that contribute toward reducing the prediction errors between the predicted and actual target values to a satisfactory level (the preset error goal). This process is carried out through the minimization of the defined error function by updating the connection weights.

The feed-forward back-propagation algorithm operates in two passes: the feed-forward pass and the back-propagation pass. During the feed-forward pass, the initial weights (either fixed or randomized) are applied on the input matrix and propagated through the hidden layer(s) to the output layer. At the output layer, the model is evaluated with a part of the training data (called validation data) and the error between the prediction results and the actual values of the target variable are compared with the error goal. If the error obtained is higher than the goal, the back-propagation pass is launched. This is where the connection weights are propagated back to the input layer where a new set of weights are computed (or re-initialized) and the feed-forward process continues. This loop continues until either the error goal is achieved or the pre-defined number of epochs is attained. The optimized parameters obtained at this point were then used. After the errors are minimized, the trained models with all the optimized parameters can be saved as virtual models and used for the prediction of the target values given any set of hypothetical or real-life input values to be experimented with.

The following sections explain the optimization details of each of the four models.

4.3.1 Prediction of C_f with Fixed Data Stratification

In this model (called Model 1-1), the data was divided by taking the first 70 % and the remainder of the samples for training and testing respectively. The model was used to predict C_f, which would be multiplied with the computed moment capacity $(M_{th,c})$ using Eq. (2). The result of the search for the optimal number of neurons in the hidden layer is shown in Fig. 5. From the plot, the optimal number of hidden neurons for this model is 13. This corresponds to the minimum number (x-axis) that gave the highest testing accuracy (y-axis) with the least overfitting. The least overfitting is determined by the least separation between the training and testing points. Although, several points qualify for this criteria but the most optimal was chosen.

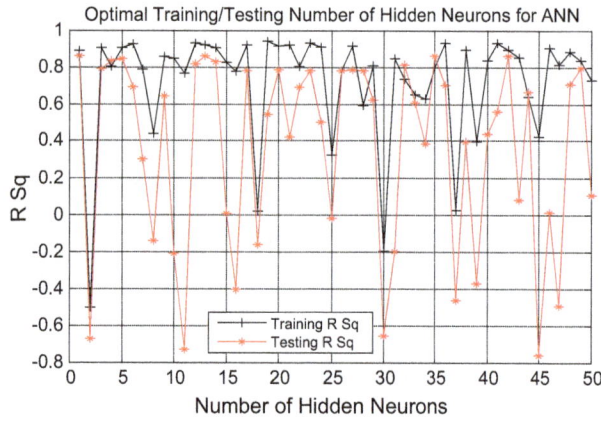

Fig. 5 Optimal number of hidden neurons for Model 1-1.

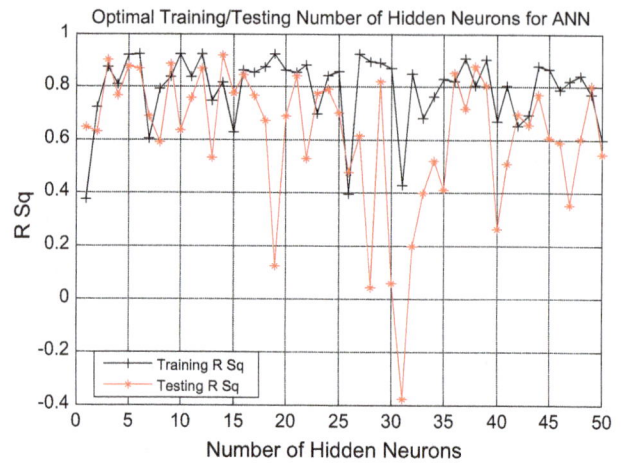

Fig. 7 Optimal number of hidden neurons for Model 1-2.

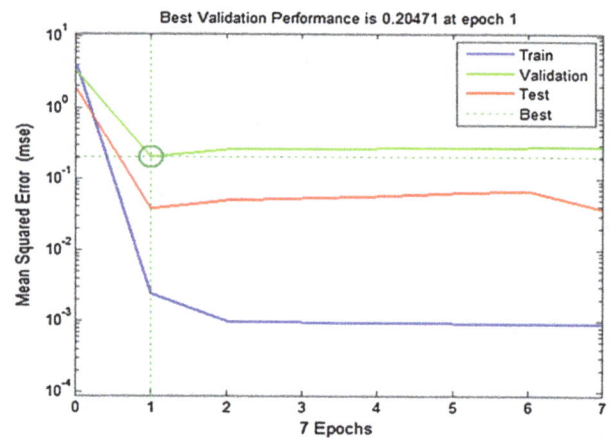

Fig. 6 Results of the training and validation for Model 1-1.

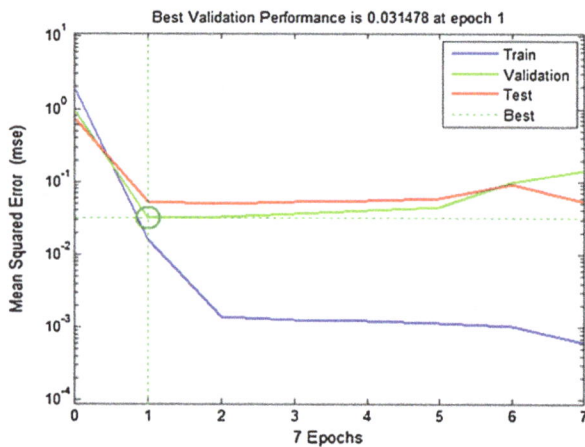

Fig. 8 Results of the training and validation for Model 1-2.

Figure 6 shows how the minimum error (hence the optimal model) was attained with respect to the validation and testing during the training process. Out of the 100 pre-defined for the model, the minimum error was attained within 7 epochs after which the validation error continued to increase. Hence, the best validation error attained so far at the first epoch was chosen. It could be seen that the errors decreased sharply at the beginning. However, after the first epoch, while training error continues to decrease, the test and validation errors could not converge in the same manner. Hence, the model decided to stop the training process and selected the best validation error attained so far.

4.3.2 Prediction of C_f with Random Data Stratification

This model (called Model 1-2) uses the random stratification of the data. A randomly selected 70 % and the remainder of the samples were taken for training and testing respectively. The randomization process ensured that each sample has equal chance of being selected for training or testing. This also ensures that there is a good mix of the data and all experimental cases are represented in each subset. Unlike the case of Model 1-1 (explained in Sect. 4.3.1), the randomization procedure avoids bias and skewness in the data given to the model for training and testing.

The optimal number of hidden neurons found and used for this model was 3 (as shown in Fig. 7). As defined in Sect. 4.2.1, this corresponds to the minimum number (x-axis) that gave the highest testing accuracy (y-axis) with the least overfitting. Similar to Fig. 4, the attainment of the optimal validation error is shown in Fig. 8. Seeing that the validation error could not converge after 7 epochs, the learning algorithm stopped the training process. Hence, the training process that gave the best error attained so far at epoch 1 was chosen.

4.3.3 Prediction of M_{res} with Fixed Data Stratification

Unlike the previous two, this model (named model 2-1) was developed to directly predict the residual flexural strength (M_{res}) without using C_f but with the same set of input variables. Like in model 1-1, the first 70 % of the data were used for training an ANN model while the last 30 % was used for testing and validation. The optimal number of hidden neurons found to be optimal for this model was 11 (as shown in Fig. 9). Details of how this value was obtained have been explained in Sects. 4.3.1 and 4.3.2. Hence, unnecessary repetitions will be avoided here. Figure 10 also shows the attainment of the optimal model parameters with

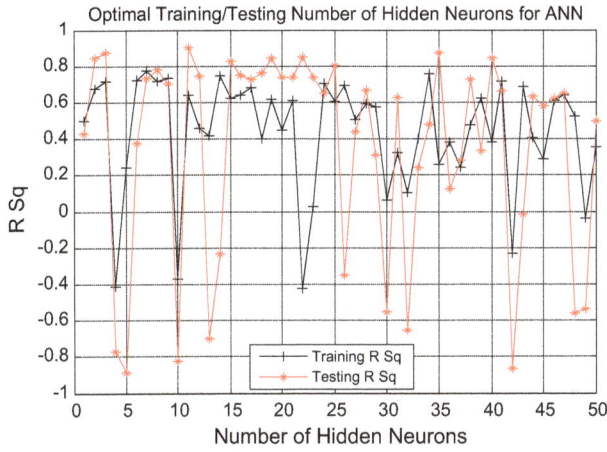

Fig. 9 Optimal number of hidden neurons for Model 2-1.

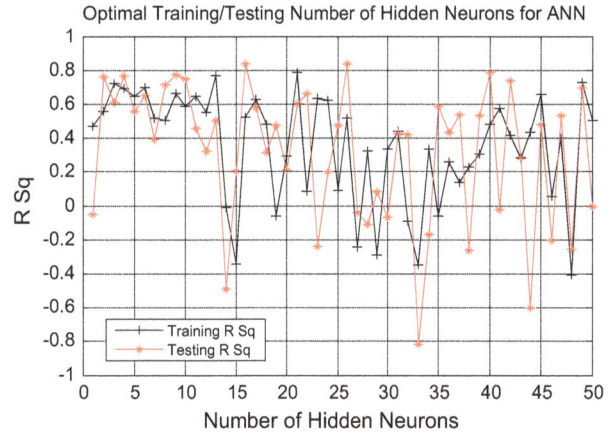

Fig. 11 Optimal number of hidden neurons for Model 2-2.

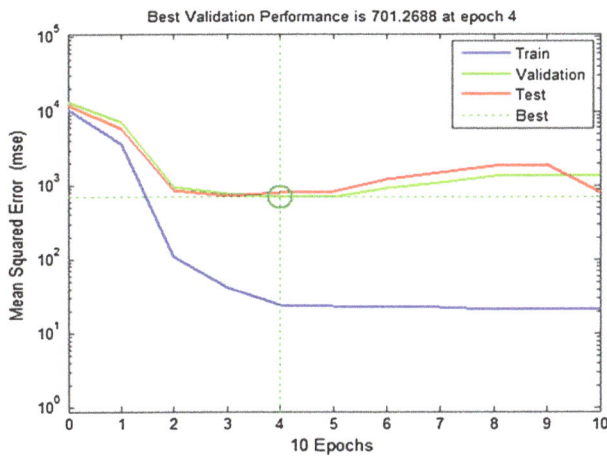

Fig. 10 Results of the training and validation for Model 2-1.

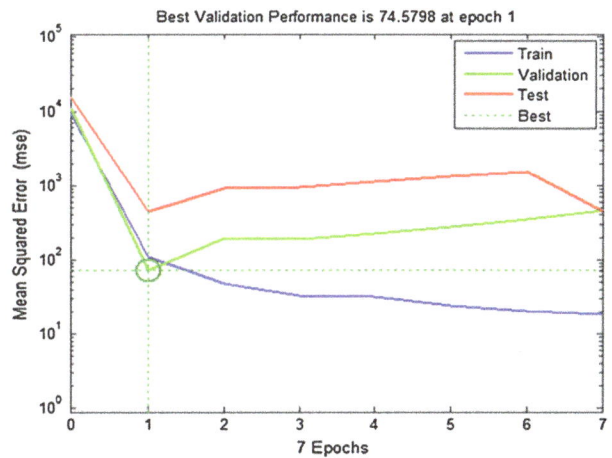

Fig. 12 Results of the training and validation for Model 2-2.

respect to the least validation error. According to the plot, since the validation error could not converge after 10 epochs, the least validation error attained at epoch 1 was chosen by the learning algorithm.

4.3.4 Prediction of M_{res} with Randomized Data Stratification

Similar to model 2-1, this model (called model 2-2) directly predicts the residual flexural strength (M_{res}) without using C_f but with the randomized stratification of the dataset. Like in model 1-2, 70 % of the data was randomly selected for training while the remaining percentage was used to test the generalization capability of the model. The randomized stratification procedure resulted in data subsets that are representative of the experimental cases. More of such details have been given in Sects. 4.3.1. and 4.3.2.

The optimal number of hidden neurons for this model was found to be 26 (as shown in Fig. 11) while the least validation error was attained at epoch 1. The model was chosen when the validation error did not improve up to 7 epochs. This is shown in Fig. 12.

After the implementation of the ANN models, the results are presented and discussed in Sect. 5.0.

5. Results and Discussion

Since the testing process simulates the capability of the models to generalize on new and never-seen-before cases, we focus on the testing performance of the models in the presentation and analysis of the results. Based on the methodology described in Sect. 4, the results of the comparative R^2 and RMSE that were obtained from the indirect prediction of flexural strength through C_f and the direct prediction of M_{res} for the testing data subset (representing new measurements) are shown in Figs. 13–16.

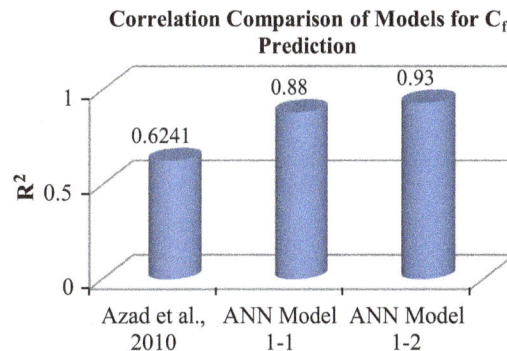

Fig. 13 R^2 Comparison of empirical model and ANN models for C_f prediction.

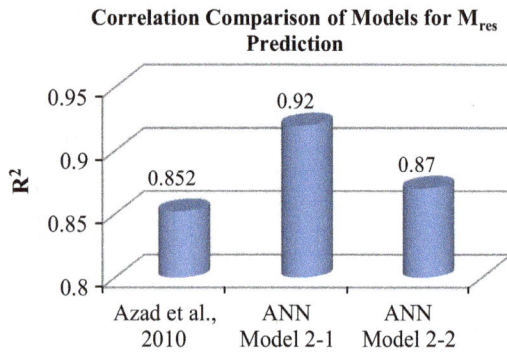

Fig. 14 R^2 of empirical model and ANN models for M_{res} prediction.

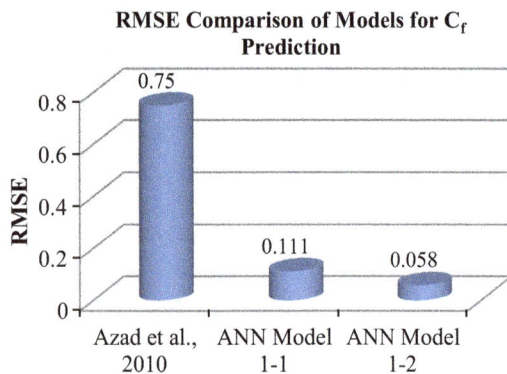

Fig. 15 RMSE Comparison of empirical model and ANN models for C_f prediction.

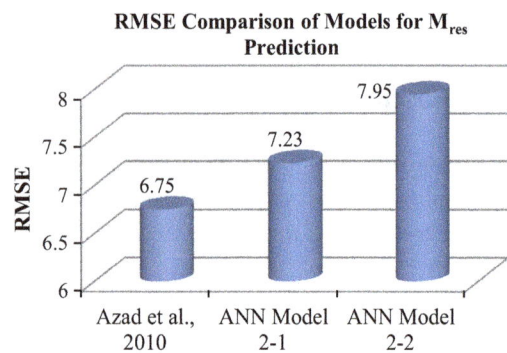

Fig. 16 RMSE Comparison of empirical model and ANN models for M_{res} prediction.

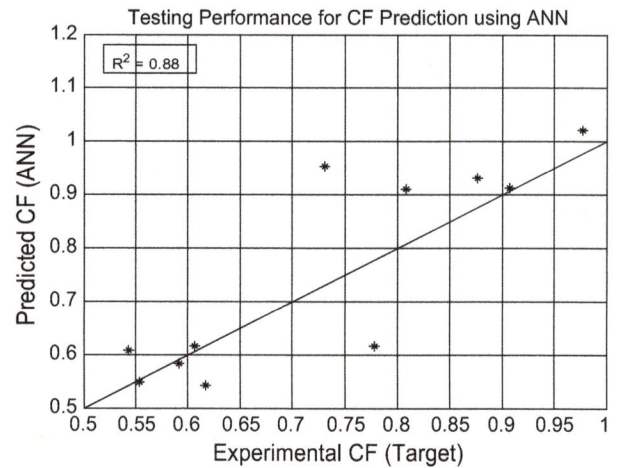

Fig. 17 Experimental and ANN prediction of M_{res} for testing data with fixed stratification.

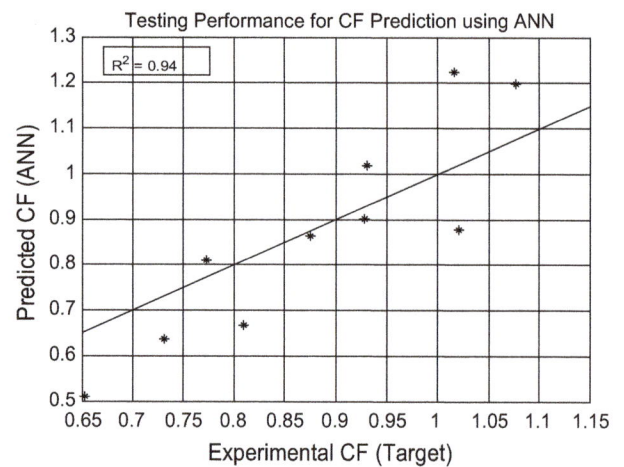

Fig. 18 Experimental and ANN prediction of M_{res} for testing data with random stratification.

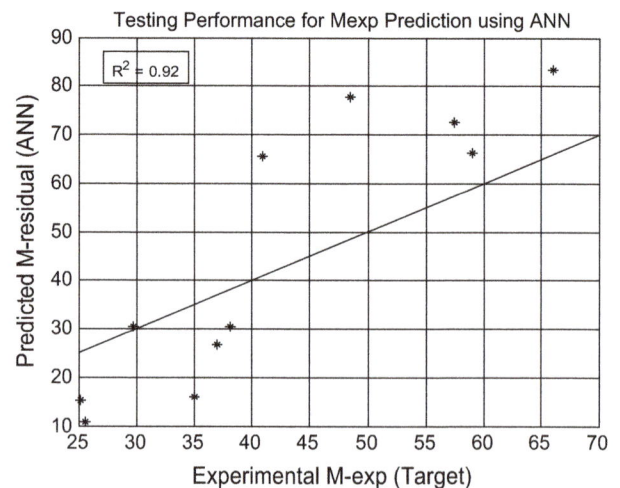

Fig. 19 Experimental and ANN prediction of M_{res} for testing data with fixed stratification.

Cross-plots showing the degree of correlation between the ANN predicted results and the experimental values of C_f and M_{res} are presented in Figs. 17–20.

Figure 13 compares the R^2 of the empirical model by Azad et al. (2010) with those of the ANN models in the prediction of C_f. The figure showed that the two ANN models performed better than the Azad's empirical equation. The R^2 of model 1-1 had a 41 % improvement while Model 1-2 had a 49 % improvement over the empirical equation. This also confirmed that the ANN model with a randomized data stratification (Model 1-2) performed better than the other one with fixed stratification (Model 1-1). For the prediction of M_{res}, Fig. 14 also showed that the ANN models

predicted better than the Azad's equation with Model 2-1 having a 10 % improvement and Model 2-2 having a 2 % improvement over the empirical equation. However,

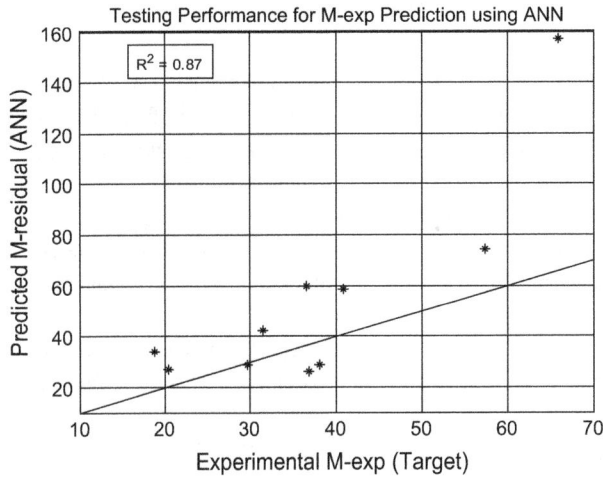

Fig. 20 Experimental and ANN prediction of M_{res} for testing data with random stratification.

between the ANN models, the one with the fixed data stratification (Model 2-1) showed better correlation with experimental data than the other one with randomized stratification (Model 2-2).

In terms of RMSE for the prediction of C_f, Fig. 15 agreed with the R^2 results by having the RMSE of the ANN models lower than that of the empirical equation. Model 2-1 and 2-2 respectively had an 85 and a 92 % reduction in error over the empirical equation. This also agrees with the R^2 results in Fig. 13 that the ANN model with randomized data stratification performed better than that with fixed stratification. For the prediction of M_{res}, Fig. 16 showed that the ANN models, despite their better performance in terms of higher R^2, had higher errors than the empirical equation. This implies that the ANN models had higher errors associated with their better predictions.

In the overall, the comparative results showed that the ANN models that were used to predict C_f exhibited better predictive capabilities than the empirical equation. However, for optimal results, the ANN model with randomized data stratification is preferred. Also, we recommend that it is better to estimate the flexural strength of corroded RC beams indirectly through the prediction of C_f. This implies that, according to our study, going through the prediction of C_f is a better way to predict the residual flexural strength of corroded RC beams.

With the emergence of the ANN model with randomized data stratification as a better tool to predict the flexural strength of RC beams, we further analyze the degree of correlation of the prediction results of the ANN models with experimental values. Figure 17 showed that the prediction results of Model 1-1 were in better agreement with the experimental values falling in the extremes than those in the middle. C_f values that are less than 0.7 and more than 0.8 were predicted more accurately than those that fall between the two. However, Fig. 18 showed the opposite as C_f values between 0.85 and 0.95 were predicted more accurately than those in the extremes. For the prediction of M_{res}, Fig. 19 showed that desptite the higher correlation, the prediction errors are high but evenly distributed over the values.

However, Fig. 20 shows that despite having the least R^2 among the ANN models, the prediction errors are low and consistent over the values except at the highest extreme. Since few samples fall in the highest extremes, the overall prediction error became low. This further confirmed that the models with random data stratification perform better than those with fixed stratification.

Giving more credence to the preference of the random data stratification over the fixed version is the visual comparison of the results of the search for the optimal number of hidden neurons for the prediction of C_f (Figs. 5 and 7) and M_{res} (Figs. 9 and 11). It would be observed that Fig. 5 (fixed stratification) is more haphazard in its fluctuation while Fig. 7 (random stratification) is smoother and more consistent. Similarly, Fig. 9 (fixed stratification) has sharper fluctuations than Fig. 11 (random stratification), which is smoother and more consistent. This behavior is probably due to the bias created by using fixed stratification instead of the fairer and unbiased randomized stratification. In the latter, there is fairness as each sample in the data has equal chance of being selected for training or testing. This perfectly agrees with the theory of randomization (Arora and Barak 2009; Cormen et al. 2001) and an existing study related to the subject (Helmy et al. 2010).

This fair analysis is based on the results obtained from the ANN models using the data obtained from a previous experimental work that was used to develop the empirical equation. Since datasets used by other authors are not available in the public domain, they could not be obtained for further testing of our ANN models. However, we hypothesize that, with the successful performance of the proposed ANN models on the available dataset, a similar successful performance is expected to be recorded on a wide array of other experimental datasets. Effort will be made in our future work to obtain such datasets for the purpose of testing the proposed models.

6. Conclusions, Limitation and Future Work

In order to reduce the cost, effort and time associated with persistent laboratory experiments, with the aim of benefiting from the high volume of experimental data gathered over the years and to increase the degree of accuracy of predictive models, we presented four optimized ANN models to predict the residual flexure strength of corroded RC beams with the diameter of reinforced steel (D) and corrosion activity index ($I_{corr}T$) as input variables. Computational techniques, especially ANN, have the capability to handle the non-linear relationship between predictor variables and their target values. A number of empirical equations have been proposed in literature. However, since they are based on assumed linear relationships among various predictor variables, they could not adequately handle such a natural phenomenon as corrosion.

This study was conducted in two parts: direct prediction of the residual strength, M_{res} and indirect prediction through the correction factor, C_f. Each part was further experimented with fixed and randomized stratification of the data into training and testing subsets. To ensure fairness in comparison and to increase the confidence in the outcome of the study, the same data that was used by Azad et al. (2010) was used to train and evaluate the performance of the ANN models. From the rigorous analysis of the ANN results and a comparison of the results with those of Azad's empirical equation, the ANN models demonstrated superior performance.

The outcome of this study can be highlighted as follows:

- With their higher coefficients of determination, the ANN models could be alternative modeling tools to the empirical equation of Azad et al. (2010) in the prediction of flexural strength of corroded RC beams. This is due to the excellent learning capability and the dynamic nature of ANN. Empirical correlations are static and have no learning capability.

- The ANN models with randomized data stratification gave better predictions than those with fixed stratification. This is due to the fairness of the randomization process.

- The ANN models with C_f as the target variable gave higher prediction accuracies and reduced errors than those that direct predict M_{res}. Hence, we recommend to Construction Engineers to estimate the residual flexural strength of RC beams by multiplying the correction factor, C_f, predicted by ANN, with the theoretical moment capacity, $M_{th,c}$.

- Our study demonstrated that ANN models are simpler, adaptive and more reliable tools for the prediction of flexural strength of corroded RC beams.

Since the main objective of this work is to investigate the capability of ANN models to predict the flexural strength of reinforced concrete, the focus is to keep the algorithm design and implementation simple. This follows the Occam's Razor principle (Jefferys and Berger 1991). Investigating the effect of various parameters (such as the effect of data normalization, activation function, different learning algorithms and different layers) on the performance of the ANN models will be carried out in our future work. More complex learning paradigms such as hybrid and ensemble concepts will also be considered.

The authors plan to confirm the consistency of the results of this study in the continued and future work by using more experimental data from various types of concrete samples and published datasets from previous studies. In the continued search for better predictive tools, it is planned to implement other types of ANN such as Generalized Regression Neural Networks, Radial Basis Functional Networks and Functional Networks. More advanced computational intelligence techniques such as Extreme Learning Machines, Support Vector Machine and Type-2 Fuzzy Logic Systems will also be considered.

Acknowledgement

The authors would like to acknowledge the support of King Fahd University of Petroleum and Minerals (KFUPM) for providing the resources used in the conduct of this research.

References

Abdalla, J. A., Elsanosi, A., & Abdelwahab, A. (2007). Modeling and simulation of shear resistance of R/C beams using artificial neural network. *Journal of the Franklin Institute, 344*(5), 741–756.

Al-Gohi, B. H. A. (2008). *Time-dependent modeling of loss of flexural strength of corroding RC beams.* Master Thesis, King Fahd University of Petroleum and Minerals, Dhahran, Saudi Arabia.

Arora, S., & Barak, B. (2009). *Computational complexity: a modern approach* (1st ed.). Cambridge, UK: Cambridge University Press.

Azad, A., Ahmad, S., & Al-Gohi, B. (2010). Flexural strength of corroded reinforced concrete beams. *Magazine of Concrete Research, 62*(6), 405–414.

Azad, A., Ahmad, S., & Azher, S. A. (2007). Residual strength of corrosion-damaged reinforced concrete beams. *ACI Material Journal, 104*(1), 40–47.

Baughman, D. R. (1995). *Neural networks in bioprocessing and chemical engineering.* PhD Dissertation, Virginia Tech, Blacksburg, VA.

Beale, M., & Demuth, H. (2013). *Neural network toolbox user's guide.* Natick, MA: The Mathworks Inc.

Bies, R. R., Muldoon, M. F., Pollock, B. G., Manuck, S., Smith, G., & Sale, M. E. (2006). A genetic algorithm-based hybrid machine learning approach to model selection. *Journal of Pharmacokinetics and Pharmacodynamics, 33*(2), 195–221.

Cabrera, J. (1996). Deterioration of concrete due to reinforcement steel corrosion. *Cement & Concrete Composites, 18*(1), 47–59.

Castillo, E., Gutiérrez, J. M., Hadi, A. S., & Lacruz, B. (2001). Some applications of functional networks in statistics and engineering. *Technometrics, 43*, 10–24.

Chen, H., Tsai, K., Qi, G., Yang, J., & Amini, F. (1995). Neural network for structure control. *Journal of Computing in Civil Engineering, 9*(2), 168–176.

Cormen, T. H., Leiserson, C. E., Rivest, R. L., & Stein, C. (2001). *Introduction to algorithms.* Cambridge, MA: MIT press.

Coronelli, D., & Gambarova, P. (2004). Structural assessment of corroded reinforced concrete beams: modeling guidelines. *Journal of Structural Engineering, 130*(8), 1214–1224.

Eskandari, H., Rezaee, M. R., & Mohammadnia, M. (2004). Application of multiple regression and artificial neural network techniques to predict shear wave velocity from wireline log data fora carbonate reservoir, South-West Iran. *CSEG Recorder, 42*, 48.

Flood, I., & Kartam, N. (1994). Neural networks in civil engineering. II: Systems and application. *Journal of Computing in Civil Engineering, 8*(2), 149–162.

Guler, I. (2005). ECG beat classifier designed by combined neural network model. *Pattern Recognition, 38*(2), 199–208.

Hasancebi, O., & Dumlupınar, T. (2013). Linear and nonlinear model updating of reinforced concrete T-beam bridges using artificial neural networks. *Computers & Structures, 119*, 1–11.

Hastie, T., Tibshirani, R., & Friedman, J. (2009). *The elements of statistical learning: data mining, inference, and prediction* (2nd ed.). Berlin, Germany: Springer.

Helmy, T., Anifowose, F. A., & Sallam, E. S. (2010). An efficient randomized algorithm for real-time process scheduling in PicOS operating system. In K. Elleithy (Ed.), *Advanced techniques in computing sciences and software engineering* (pp. 117–122). New York, NY: Springer.

Hsu, D. S., & Chung, H. T. (2002). Diagnosis of reinforced concrete structural damage base on displacement time history using the back-propagation neural network technique. *Journal of Computing in civil engineering, 16*(1), 49–58.

Huang, R., & Yang, C. (1997). Condition assessment of reinforced concrete beams relative to reinforcement corrosion. *Cement & Concrete Composites, 19*(2), 131–137.

Inan, O. T., Giovangrandi, L., & Kovacs, G. T. (2006). Robust neural-network-based classification of premature ventricular contractions using wavelet transform and timing interval features. *IEEE Transactions on Biomedical Engineering, 53*(12), 2507–2515.

Inel, M. (2007). Modeling ultimate deformation capacity of RC columns using artificial neural networks. *Engineering Structures, 29*(3), 329–335.

Jefferys, W. H. & Berger, J. O. (1991). *Sharpening Ockham's Razor on a Bayesian Strop*. Technical Report #91-44C, Department of Statistics, Purdue University, West Lafayette, IN.

Jin, W.-L., & Zhao, Y.-X. (2001). Effect of corrosion on bond behavior and bending strength of reinforced concrete beams. *Journal of Zhejiang University (Science), 2*(3), 298–308.

Kang, H. T., & Yoon, C. J. (1994). Neural network approaches to aid simple truss design problems. *Computer-Aided Civil and Infrastructure Engineering, 9*(3), 211–218.

Kirkegaard, P. H. & Rytter, A. (1994). Use of neural networks for damage assessment in a steel mast. In *Proceedings of the 12th International Modal Analysis Conference of the Society for Experimental Mechanics*. Honolulu, HI.

Li, L., & Jiao, L. (2002). Prediction of the oilfield output under the effects of nonlinear factors by artificial neural network. *Journal of Xi'an Petroleum Institute, 17*(4), 42–44.

Mangat, P. S., & Elgarf, M. S. (1999). Flexural strength of concrete beams with corroding reinforcement. *ACI Structural Journal, 96*(1), 149–158.

Moghadassi, A., Parvizian, F., Hosseini, S. M., & Fazlali, A. (2009). A new approach for estimation of PVT properties of pure gases based on artificial neural network model. *Brazilian Journal of Chemical Engineering, 26*(1), 199–206.

Mohaghegh, S. (1995). Neural network: What it can do for petroleum engineers. *Journal of Petroleum Technology, 47*(1), 42–42.

Nascimento, C. A. O., Giudici, R., & Guardani, R. (2000). Neural network based approach for optimization of industrial chemical processes. *Computers & Chemical Engineering, 24*(9), 2303–2314.

Neaupane, K. M., & Adhikari, N. (2006). Prediction of tunneling-induced ground movement with the multi-layer perceptron. *Tunnelling and Underground Space Technology, 21*(2), 151–159.

Nokhasteh, M. A., & Eyre, J. R. (1992) The effect of reinforcement corrosion on the strength of reinforced concrete members. In *Proceedings of Structural integrity assessment*. London, UK: Elsevier Applied Science.

Ou, Y. C., Tsai, L. L., & Chen, H. H. (2012). Cyclic performance of large-scale corroded reinforced concrete beams. *Earthquake Engineering and Structural Dynamics, 41*(4), 593–604.

Pandey, P., & Barai, S. (1995). Multilayer perceptron in damage detection of bridge structures. *Computers & Structures, 54*(4), 597–608.

Petrus, J. B., Thuijsman, F., & Weijters, A. J. (1995). *Artificial neural networks: An introduction to ANN theory and practice*. Berlin, Germany: Springer.

Phung, S. L., & Bouzerdoum, A. (2007). A pyramidal neural network for visual pattern recognition. *IEEE Transactions on Neural Networks, 18*(2), 329–343.

Rafiq, M., Bugmann, G., & Easterbrook, D. (2001). Neural network design for engineering applications. *Computers & Structures, 79*(17), 1541–1552.

Ravindrarajah, R. S., & Ong, K. (1987). Corrosion of steel in concrete in relation to bar diameter and cover thickness. *ACI Special Publication, 100*, 1667–1678.

Revathy, J., Suguna, K., & Raghunath, P. N. (2009). Effect of corrosion damage on the ductility performance of concrete columns. *American Journal of Engineering and Applied Sciences, 2*(2), 324–327.

Rodriguez, J., Ortega, L., & Casal, J. (1997). Load carrying capacity of concrete structures with corroded reinforcement. *Construction and Building Materials, 11*(4), 239–248.

Tachibana, Y., Maeda, K.-I., Kajikawa, Y., & Kawamura, M. (1990). Mechanical behavior of RC beams damaged by corrosion of reinforcement. *Elsevier Applied Science*, 178–187.

Tsai, C.-H., & Hsu, D.-S. (2002). Diagnosis of reinforced concrete structural damage base on displacement time history using the back-propagation neural network technique. *Journal of Computing in Civil Engineering, 16*(1), 49–58.

Übeyli, E. D. (2009). Combined neural network model employing wavelet coefficients for EEG signals classification. *Digital Signal Processing, 19*(2), 297–308.

Uomoto, T., & Misra, S. (1988). Behavior of concrete beams and columns in marine environment when corrosion of reinforcing bars takes place. *ACI Special Publication, 109*, 127–146.

VanLuchene, R., & Sun, R. (1990). Neural networks in structural engineering. *Computer-Aided Civil and Infrastructure Engineering, 5*(3), 207–215.

Wang, X. H., & Liu, X. L. (2008). Modeling the flexural carrying capacity of corroded RC beam. *Journal of Shanghai Jiaotong University (Science), 13*(2), 129–135.

Waszczyszyn, Z., & Ziemiański, L. (2001). Neural networks in mechanics of structures and materials—new results and prospects of applications. *Computers & Structures, 79*(22), 2261–2276.

Wolpert, D. H., & Macready, W. G. (1997). No free lunch theorems for optimization. *IEEE Transactions on Evolutionary Computation, 1*(1), 67–82.

Wu, X., Ghaboussi, J., & Garrett, J. H. (1992). Use of neural networks in detection of structural damage. *Computers & Structures, 42*(4), 649–659.

7

Lightweight Self-consolidating Concrete with Expanded Shale Aggregates: Modelling and Optimization

Abdurrahmaan Lotfy[1], Khandaker M. A. Hossain[2],*, and Mohamed Lachemi[2]

Abstract: This paper presents statistical models developed to study the influence of key mix design parameters on the properties of lightweight self-consolidating concrete (LWSCC) with expanded shale (ESH) aggregates. Twenty LWSCC mixtures are designed and tested, where responses (properties) are evaluated to analyze influence of mix design parameters and develop the models. Such responses included slump flow diameter, V-funnel flow time, J-ring flow diameter, J-ring height difference, L-box ratio, filling capacity, sieve segregation, unit weight and compressive strength. The developed models are valid for mixes with 0.30–0.40 water-to-binder ratio, high range water reducing admixture of 0.3–1.2 % (by total content of binder) and total binder content of 410–550 kg/m³. The models are able to identify the influential mix design parameters and their interactions which can be useful to reduce the test protocol needed for proportioning of LWSCCs. Three industrial class ESH–LWSCC mixtures are developed using statistical models and their performance is validated through test results with good agreement. The developed ESH–LWSCC mixtures are able to satisfy the European EFNARC criteria for self-consolidating concrete.

Keywords: expanded shale aggregates, lightweight self-consolidating concrete, multi-objective optimization, water to binder ratio, high range water reducing admixture, total binder content, statistical model.

1. Introduction

Lightweight self-consolidating concrete (LWSCC) is expected to provide high workability without segregation and high durability with reduced weight. The success to production of high quality LWSCC lies in the use of aggregates. Expanded shale (ESH) is a ceramic material produced by expanding and vitrifying select shale's, in a rotary kiln. The process produces a high quality ceramic aggregate that is non-toxic, absorptive, dimensionally stable, structurally strong, durable, environmentally inert and light in weight. The use of expanded shale aggregate with other quality supplementary cementing materials (such as fly ash and silica fume) can provide highly workable and durable LWSCCs. ESH and other lightweight aggregates such as: clayey diatomite, pumice, slate, perlite, bottom ash etc. have been successfully used in the production of lightweight concretes (LWCs) over the decades (Stamatakis et al. 2011; Wu et al. 2009; Hwang and Hung 2005; Hossain 2004; Fragoulis et al. 2003, 2004). Use of these aggregates has contributed to the sustainable development by conserving energy, maximizing structural efficiency and increasing the service life of structural lightweight concrete (LWC). These benefits add to those of LWSCC to further support sustainable development and contribute to projects becoming Leadership in Energy and Environmental Design (LEED) certified (ESCSI 2004).

LWSCC is capable of filling up the formwork and encapsulate reinforcement by its self-weight without the need for additional compaction or external vibration. It has excellent segregation resistance, high flowability and passing ability at fresh state as well as better mechanical and durability properties in the hardened state. LWSCC has more continuous aggregate-paste contact zone and more moisture in the pores of aggregates for continued internal curing—these improvements lead to reduced concrete cracking and improved hardened properties (Holm 1994).

Although numerous investigations have been made on SCC and LWC, to the authors' best knowledge little research has been conducted on the design procedures and statistical modeling of LWSCC (Hwang et al. 2012; Bogas et al. 2012; Topçu and Uygunoğlu 2010; Andiç-Çakır and Hızal 2012).

Wu et al. (2009) investigated workability of LWSCC and its mix design using expanded shale as both fine and coarse aggregates. The study demonstrated that fixed aggregate contents can be used effectively in volumetric method to design LWSCC mixtures. An increase in the paste content of the mix increased the flow velocity but reduced resistance to segregation. Lachemi et al. (2009) developed three different classes of LWSCC mixtures using combination of blast furnace slag and expanded shale aggregates. Hwang and

---

[1]Lafarge Canada Inc., Toronto, ON, Canada.

[2]Department of Civil Engineering, Ryerson University, Toronto, ON, Canada.
*Corresponding Author; E-mail: ahossain@ryerson.ca

Hung (2005) evaluated the performance of LWSCC mixtures containing bottom ash, for varying water to cement ratio (w/c) and cement paste content. Kim et al. (2010) studied the semi-lightweight SCC characteristics using two types of coarse aggregates with different densities. Nine mixes were evaluated in terms of flowability, segregation resistance and filling capacity of fresh concrete. The mechanical properties of hardened LWSCC, such as compressive strength, splitting tensile strength, elastic modulus and density were assessed. Müller and Haist (2002) proposed three mix proportions for LWSCC and assessed their self-compacting properties. No significant difference in the mix proportion design was found compared with SCC except for the aggregate used.

Design procedures and statistical models for normal weight SCC have been developed in previous research studies (Khayat et al. 1998; Patel et al. 2004; Sonebi 2004a, b). However, lack of research studies on LWSCC technology warrants investigations. Authors' research based on statistical design approach to identify primary mix design parameters and their effects on relevant properties of ESH lightweight SCC (ESH–LWSCC) is a timely initiative. The knowledge of influence of mixture variables on fresh state and hardened characteristics (which is the objectives of the current study) is essential for successful development of ESH–LWSCCs.

This paper presents the development and validation of statistical models for the design of ESH–LWSCC mixtures with desired fresh and hardened properties. The developed statistical models can be used as tools for practical production of ESH–LWSCCs. The recommendations of this research will be useful for engineers, designers and manufacturers involving in the development, production and use of ESH–LWSCCs.

2. Research Program

This research was conducted in three phases. The phase I focused on the experimental study of the fresh and hardened properties of mathematically derived ESH–LWSCC mixes. Twenty concrete mixtures were designed. Three key mix design parameters namely water (w) to binder (b) ratio (w/b) (0.30–0.40), dosage of high range water reducing admixtures (HRWRA) (0.3–1.2 % by total content of binder) and total binder content (B) (410–550 kg/m^3) were selected to derive mathematical models for the design of ESH–LWSCC mixtures. The tested ESH–LWSCC properties were, slump flow, V-funnel flow time, J-ring flow diameter/height difference, L-box ratio, filling capacity, segregation resistance, unit weight and compressive strength.

Phase II focused on the model development. Based on the test results, the influences of various parameters (w/b, HRWRA% and binder content) on ESH–LWSCC fresh and hardened properties were analyzed. The relative significance of these primary mixture design parameters and their coupled effects on relevant properties of ESH–LWSCCs were established. Afterward, statistical models were developed for prediction of these properties.

In phase III, the developed statistical models were used to derive optimized industrial class ESH–LWSCCs. ESH–LWSCC mixtures were mathematically optimized to satisfy three classes of EFNARC industrial classifications and their performance was experimentally validated through fresh and hardened properties. In addition, the relationship between theoretical and experimental results was further investigated, where validation of the statistical models were performed.

3. Phase-I Investigation

3.1 Materials
ASTM Type I cement, Class F fly ash (FA) and silica fume (SF) were used. The physical and chemical properties of cement, FA and SF are presented in Table 1. FA and SF were incorporated into the mixture at a fixed percentage by mass of total binder at 12.5 and 7.5 %, respectively. Nominal sizes of 4.75 and 12 mm lightweight expanded shale were used as fine and coarse aggregates, respectively. Expanded shale produced by TXI aggregate company, Colorado, USA, was used. In manufacturing process, natural shale is expanded in an oil fired rotary kiln maintained between 1,900 and 1,200 °C. At this temperature, the shale is in a semi-plastic state at which entrapped gases are formed and expansion results creating individual non-connecting air cells. After discharged from the kiln, it is cooled and stored. Table 1 presents the chemical properties of expanded shale aggregates, and Table 2 presents their grading and physical properties. Mineralogical composition of silica fume consists of an amorphous silica structure with very little crystalline particles. No undesirable trace elements were recorded in the manufacturer's material analysis sheet for all the materials.

The proposed ESH–LWSCC mixtures contained no viscosity-modifying admixture (VMA). The use of VMA is associated with reduction in paste volume which is believed to be detrimental to the LWSCC mixture stability, passing ability, filling ability and segregation resistance. Further, many successful LWSCC mixtures were developed without the use of VMA (Lachemi et al. 2009; Kim et al. 2010; Karahan et al. 2012). The silica fume is used to enhance the fresh properties as it helps to improve the cohesiveness and homogeneity of the LWSCCs; holding the lightweight coarse aggregates in place, and preventing them from floating. Further, fly ash and silica fume also enhance the durability characteristics of the mixture. A polycarboxylate ether type HRWRA with a specific gravity of 1.05 and total solid content of 26 % was used as superplasticizer (SP).

3.2 Mix Design Methodology and Mixture Proportions (Phase I)
Twenty concrete mixtures were designed using the Box–Wilson central composite design (CCD) method (Schmidt and Launsby 1994). Three input factors were used in the test program: X_1 (water to binder ratio: w/b), X_2 (percentage of HRWRA as a percentage of mass of total binder content),

Table 1 Characteristics of cement fly ash, silica fume and expanded shale.

	Cement	Fly ash	Silica fume	Expanded shale
Chemical				
SiO_2 (%)	19.6	46.7	95.21	67.6
Al_2O_3 (%)	4.9	22.8	0.21	15.1
Fe_2O_3 (%)	3.1	15.5	0.13	4.1
TiO_2 (%)	–	–	–	0.6
CaO (%)	61.4	5.8	0.23	2.2
MgO (%)	3	–	–	3.5
SO_3 (%)	3.6	0.5	0.33	0.24
Alkalis as Na_2O (%)	0.7	0.7	0.85	3.7
LOI (%)	2.3	2.2	1.97	3.06
Physical				
Blaine (cm^2/g)	3,870	3,060	21,000	–
+45 μm (%)	3.00	17	2.85	–
Density (g/cm^3)	3.15	2.48	2.20	–

Table 2 Grading and physical properties of aggregates.

Sieve size (mm)	% Passing			
	ASTM C-330 specification		E-shale	
	Fine	Coarse	Fine	Coarse
13.20	100	100	100	100
9.50	80–100	100	100	91
4.75	5–40	85–100	100	18.8
2.36	0–20	–	95	2.5
1.18	0–10	40-80	65	1.6
0.60	–	–	41	0.6
0.30	–	10–35	23.5	0.1
0.15	–	5–25	14.7	0
Bulk specific gravity (dry)	–	–	1.40	1.33
Bulk specific gravity (SSD)	–	–	1.81	1.71
Dry loose bulk density (kg/m^3)	1,120 (max)	880 (max)	1,070	862
Absorption (%)	–	–	13	14

and X_3 (total binder content: B). The ranges of the input factors were set at 0.30–0.40 for X_1, 0.3–1.2 % for X_2, and 410–550 kg/m^3 for X_3. Table 3 presents the coded value and limits of each factor.

The CCD method consists of three portions: the fraction factorial portion, the center portion, and the axial portion (Table 3). The mix design and statistical evaluation of the test results were performed at a 0.05 level of significance. Table 4 presents the mixture proportions for ESH–LWSCCs developed by the software.

3.3 Casting of Test Specimens

All concrete mixtures were prepared in 35 batches in a drum rotating mixer. Due to the high water absorption capacity, the expanded shale lightweight aggregates were pre-soaked for a minimum of 72 h. The saturated surface dry expanded shale aggregates were mixed for 5 min with 75 % of the mixing water then added to the cementitious materials and mixed for an additional minute. Finally, the remaining water and HRWRA were added to the mixture, and mixed for another 15 min. Just after mixing, the slump flow, L-box,

Table 3 Limit andcoded value of factors.

Factor	Range	Coded value				
		−1.414	−1	0	+1	+1.414
X1 = (w/b)	0.30–0.40	0.28	0.30	0.35	0.40	0.42
X2 = (% of HRWRA)	0.3–1.2 %	0.11	0.30	0.75	1.2	1.39
X3 = (B) kg/m³	410–550	380	410	480	550	580

	Factors			
CCD portion	Mixture	X1	X2	X3
Fractional factorial	1–8	±1	±1	±1
Center point	15–20	0	0	0
Axial	9–14	0, ±1.414	0, ±1.414	0, ±1.414

Table 4 Mixture proportions for ESH–LWSCC (Phase I).

Mix no.	X1 (w/b)	X2 (HRWRA)	X3 (B)	Cement (kg/m³)	FA (kg/m³)	SF (kg/m³)	HRWRA (l/m³)	Water (l/m³)	E-shale aggregate	
									Coarse	Fine
1	0.40	1.2	550	440	69	41	6.6	220	385	613
2	0.40	1.2	410	328	51	31	4.9	164	456	726
3	0.40	0.3	550	440	69	41	1.6	220	388	618
4	0.40	0.3	410	328	51	31	1.2	164	459	730
5	0.30	1.2	550	440	69	41	6.6	165	422	672
6	0.30	1.2	410	328	51	31	4.9	123	484	771
7	0.30	0.3	550	440	69	41	1.6	165	426	678
8	0.30	0.3	410	328	51	31	1.2	123	487	775
9	0.42	0.75	480	384	60	36	3.6	201	415	661
10	0.28	0.75	480	384	60	36	3.6	134	461	734
11	0.35	1.39	480	384	60	36	6.7	168	436	695
12	0.35	0.11	480	384	60	36	0.5	168	440	701
13	0.35	0.75	580	464	73	44	4.3	203	391	622
14	0.35	0.75	380	304	48	29	2.9	133	486	773
15	0.35	0.75	480	384	60	36	3.6	168	438	698
16	0.35	0.75	480	384	60	36	3.6	168	438	698
17	0.35	0.75	480	384	60	36	3.6	168	438	698
18	0.35	0.75	480	384	60	36	3.6	168	438	698
19	0.35	0.75	480	384	60	36	3.6	168	438	698
20	0.35	0.75	480	384	60	36	3.6	168	438	698

V-funnel, J-ring flow, filling capacity, sieve segregation, and unit weight tests were conducted. Ten 100 × 200 mm cylinders from each batch were cast for compressive strength determination. All ESH–LWSCC specimens were cast without any compaction or mechanical vibration. After casting, all the specimens were covered with plastic sheets and water-saturated burlap and left at room temperature for 24 h. They were then demolded and transferred to the moist curing room, and maintained at 23 ± 2 °C and 100 % relative humidity until testing. The cylinders for the oven dry unit weight test were stored in lime-saturated water for 28 days prior to transfer to the oven at 100 °C. The cylinders for the air dry unit weight test were stored in room temperature for 28 days.

3.4 Testing Procedures

All fresh tests were conducted as per EFNARC Self-Compacting Concrete Committee test procedures (EFNARC 2005). The slump flow test was conducted to assess the workability of concrete without obstructions to determine flow diameter. The deformability of ESH–LWSCC was measured using the V-funnel test, where flow time under gravity was determined. The filling capacity, J-ring and L-box tests determined the passing ability of concrete. The sieve segregation resistance (SSR) test was conducted according to EFNARC test procedures: 5 kg of fresh concrete was poured over 5 mm mesh, and the mass of the mortar passing through the sieve was recorded. The fresh unit weight was tested according to per ASTM C 138 (2010) and both air dry and oven dry densities were determined according to ASTM C 567 (2011). The compressive strength of ESH–LWSCC mixtures was determined by using 100 × 200 mm cylinders, as per ASTM C 39 (2011).

3.5 Phase I: Test Results, Analysis and Discussion

3.5.1 Fresh and Hardened Properties of ESH–LWSCC Mixtures

The fresh and hardened properties of ESH–LWSCC mixtures are summarized in Table 5. Ranges of the test values for ESH–LWSCC mixtures were between 365 and 850 mm for slump flow, 1.2 and 24 s for V-funnel flow time, 360 and 850 mm for J-ring flow, 0 and 14 mm for J-ring height difference, 0.28 and 1 for L-box ratio, 29 and 100 % for filling capacity and 4 and 38 %, for SSR. The compressive strength ranged from 20 to 40 and 28 to 53 MPa at 7 and 28 days, respectively. The fresh unit weight ranged from 1,742 to 1,892 kg/m^3 and the 28-day air dry density values were less than 1,840 kg/m^3 which classified all ESH–LWSCC mixtures as lightweight concrete. It is understood that the long-term strength of LWSCC mixes is very important since FA is used. This will be subject matter of future research studies in association with long-term durability properties of LWSCC mixes.

In order to qualify as SCC, the mixes should satisfy EFNARC industrial classifications, with 550–850 mm

Table 5 Test results on fresh and hardened properties.

Mix no.	Slump flow (mm)	V-funnel (s)	J-ring flow (mm)	J-ring height diff (mm)	L-box ratio	Filling capacity (%)	SSR (%)	Compressive strength (MPa)		28-day Unit weight (kg/m^3)		
								7-days	28-days	Fresh	Air dry	Oven dry
1	850	1.6	850	0	1.00	100	14	27	36	1,800	1,688	1,650
2[a]	810	1.2	770	0	1.00	100	38	21	28	1,826	1,700	1,645
3[a]	530	1.8	540	2	0.55	58	6	29	40	1,840	1,728	1,672
4[a]	535	5.6	510	5	0.53	58	24	23	31	1,850	1,740	1,690
5[a]	640	11.1	650	2	0.77	76	10	34	48	1,859	1,747	1,690
6[a]	625	11.9	590	4	0.63	67	20	31	43	1,866	1,754	1,707
7[a]	365	19.7	370	9	0.31	29	4	38	51	1,873	1,761	1,704
8[a]	380	18.5	360	14	0.37	31	6	34	46	1,751	1,611	1,566
9[a]	810	1.4	805	0	1.00	100	30	20	28	1,770	1,658	1,573
10[a]	395	24.0	415	5	0.28	31	7	40	53	1,807	1,667	1,623
11[a]	820	3.2	795	0	1.00	100	24	26	40	1,817	1,684	1,630
12[a]	390	6.0	390	8	0.33	29	7	32	46	1,779	1,652	1,603
13[a]	595	6.5	630	0	0.72	73	6	36	51	1,892	1,765	1,729
14[a]	755	1.9	715	0	1.00	100	34	22	31	1,742	1,630	1,601
15	675	3.6	680	2	1.00	98	13	31	44	1,807	1,695	1,635
16	705	3.7	710	2	0.98	100	11	34	48	1,789	1,676	1,604
17	685	4.0	680	1	1.00	99	12	32	44	1,779	1,667	1,611
18	700	3.7	700	1	0.97	97	13	31	45	1,782	1,670	1,614
19	685	3.5	680	1	1.00	97	10	33	46	1,787	1,662	1,597
20	705	4.1	700	2	0.99	99	12	31	43	1,800	1,675	1,625

[a] Mixture disqualified as LWSCC

slump flow (Nagataki and Fujiwara 1995), less than 8 s of V-funnel time, 80–100 % of filling capacity, greater than 0.8 of L-box ratio (Sonebi et al. 2000), and less than 20 % of segregation resistance (EFNARC 2005). To be classified as LWSCC, a mix should satisfy EFNARC-SCC industrial classifications as well as it should develop a minimum 28-day compressive strength of 17.2 MPa and attain an air dry unit weight of less than 1,840 kg/m^3 (ACI Committee 213R 2003).

Using basic knowledge of concrete technology, it is expected that fresh and hardened properties of LWSCC mixtures will be influenced by the same parameters and in same way as normal weight SCC mixtures, with exception to the V-funnel time. Theoretically speaking, when reducing the unit weight to less than 1,840 kg/m^3, it might be expected that the velocity of flow can be affected; leading to lower V-funnel time values than the ones reported for normal weight SCC.

The filling capacity test is more relevant for assessing the deformability of SCC among closely spaced obstacles. A filling capacity between 50 and 95 % indicates moderate to excellent flowability among closely spaced obstacles (Khayat et al. 2002). For a desirable SCC mixture performance, different range of V-funnel time is suggested by researchers: between 3 and 7 s, between 2.2 and 5.4 s and between 2.1 and 4.2 s (Khayat et al. 2002; Bouzoubaa and Lachemi, 2001; Ghezal and Khayat 2002).

It is reported that the SCC with L-box ratio greater than 0.8 exhibited good performance without blocking, hence 0.8 is considered as the lower critical limit for a mix to be SCC ratio (Sonebi et al. 2000). According to several studies, the L-box and the filling capacity test results should be simultaneously considered to evaluate the concrete passing ability through heavily reinforced sections without the need of vibration. One of the most important requirements for any SCC is that the aggregates should not be segregated from the paste and the mix should remain homogeneous during the production and placement. It is also equally important that the particles move with the matrix as a cohesive fluid during the flow of SCC. A stable SCC should exhibit a segregation index less than 10 % (Khayat et al. 1998). However it is expected that the allowable segregation index for LWSCC should be higher than normal weight SCC. Therefore, the limits for fresh state properties of LWSCC mixtures should be changed. For LWSCC mixtures, the criteria can be as follows: slump flow diameter (550–850 mm), V-funnel time (0–25 s), L-box ratio (≥0.80), sieve segregation resistance (0–20 %), 28-day air dry unit weight (<1,840 kg/m^3) and 28-day compressive strength (>17.2 MPa).

From the results of the present study (Table 5), mixes 3–8 and 10–13 exhibited low flowability, poor workability and passing ability as the slump flow diameter, V-funnel time and L-box ratio were below the acceptable EFNARC performance criteria for SCC (EFNARC 2005). On the other hand, mixes 2, 4, 9, 11 and 14 are considered segregated mixes due to high segregation index beyond the prescribe limits. Mixes 1, 6, 15, 16, 17, 18, 19 and 20 met all SCC fresh performance with no sign of segregation (Table 5). Out of 20 tested mixtures, only 8 mixtures satisfied the outlined criteria for structural LWSCC. This demonstrates the significant challenges associated with the development of LWSCC mixtures.

4. Phase II: Influence of Mix Design Parameters and Development of Statistical Models

The fresh and hardened properties of twenty ESH–LWSCC mixtures obtained in Phase I were used to analyze the influence of mix design parameters and development of statistical models.

4.1 Influence of Mix Design Parameters on Fresh and Hardened Properties
4.1.1 Influence on the Slump Flow
Figure 1 presents contour diagrams of the slump flow diameter changes of ESH–LWSCC mixtures depending on the water to binder ratio and total binder content. According to Fig. 1, an increase in the w/b from 0.3 to 0.4 significantly increased the slump flow. However, at fixed HRWRA% the slump flow range got limited with the increase of binder content. For example, when the HRWRA% was fixed at 0.75 % and the binder content was increased to 550 kg/m^3, the maximum predicted slump flow was limited to 700 mm. This was due to the increased demand of HRWRA in order to maintain same slump flow diameter with higher binder content.

The combined effects of w/b and HRWRA have significant influence on the slump flow diameter as shown in Fig. 2. An increase in the HRWRA from 0.3 to 1.2 % (by total content of binder) and w/b from 0.3 to 0.4 significantly increased the slump flow when high binder content (480 kg/m^3) was used.

Great positive effect of the coupled parameters (w/b and HRWRA) in increasing the slump flow was observed with the ESH–LWSCC mixtures. For example, when both parameters (w/b and HRWRA) were maximized at 1.2 % and 0.40, the maximum predicted slump flow for ESH mixtures was 850 mm. This can be attributed to the aggregate shape/gradation and packing density because a lower amount of fluidity is needed to achieve high workability for high-packing density mixture, as in the case of ESH aggregates. According to Assaad and Khayat (2006), the w/b is closely related to flowability of concrete and an increase in w/b improves the flowability of the concrete. Sonebi et al. (2007) state that the SCC fresh properties are significantly influenced by the dosage of water and HRWRA. It is expected that LWSCC mixtures will exhibit similar behaviour compared with normal weight SCC mixtures under the influence of HRWRA.

4.1.2 Influence on the V-funnel Flow Time
An increase of the w/b from 0.3 to 0.4 significantly reduced the V-funnel flow time whereas an increase of HRWRA from 0.3 to 1.2 % only slightly reduced the

Design-Expert® Software

Slump Flow (mm)
● Design Points
850

365

X1 = A: **w/b**
X2 = C: B

Actual Factor
B: HRWRA = 0.75

A: w/b

B: HRWRA

C: Binder content (B)

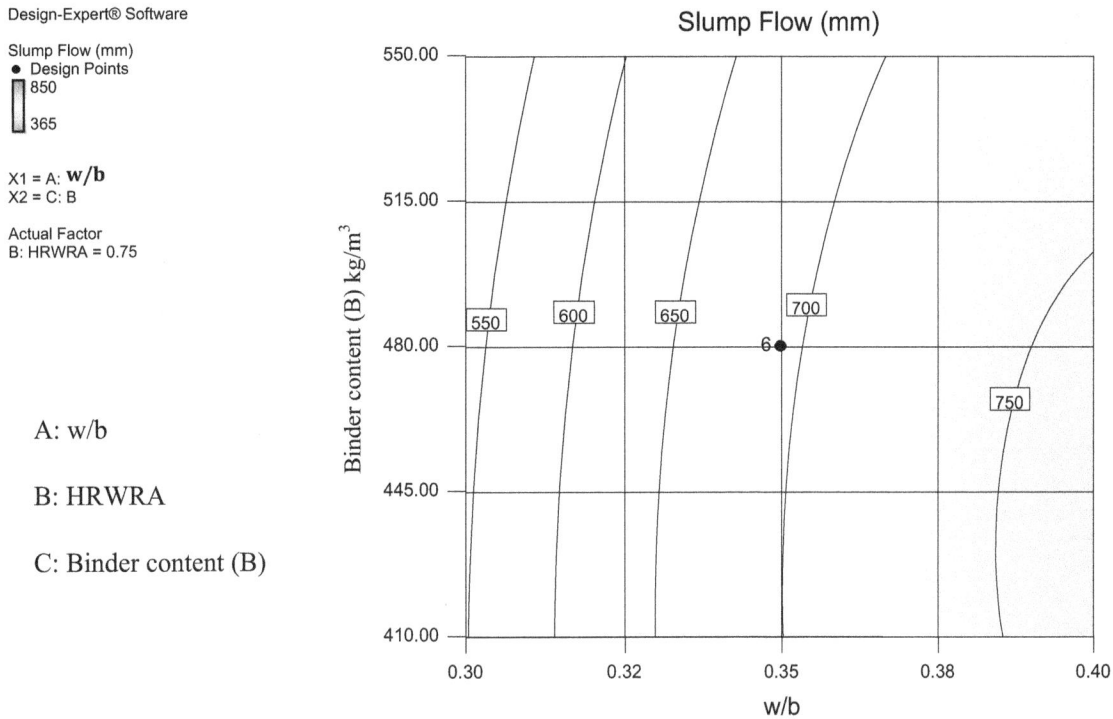

Fig. 1 Contours of slump flow changes of ESH–LWSCCs with w/b, total binder content and HRWRA at 0.75 %.

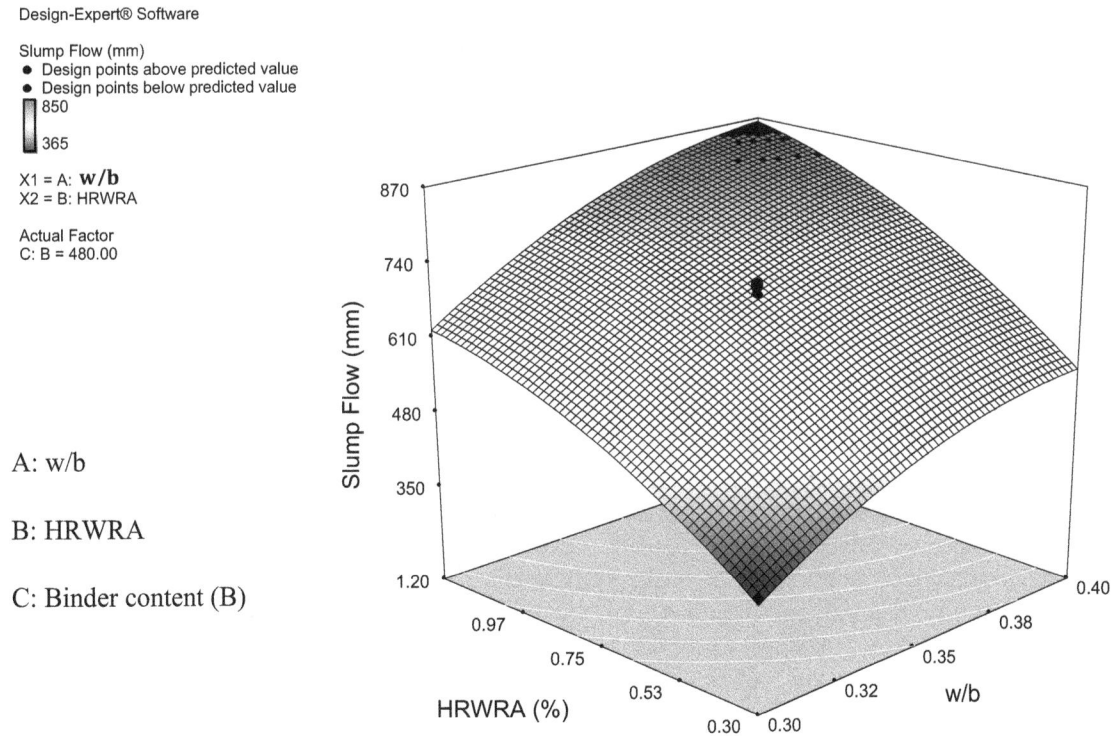

Design-Expert® Software

Slump Flow (mm)
● Design points above predicted value
● Design points below predicted value
850

365

X1 = A: **w/b**
X2 = B: HRWRA

Actual Factor
C: B = 480.00

A: w/b

B: HRWRA

C: Binder content (B)

Fig. 2 Effect of w/b, HRWRA and total binder content at 480 kg/m³ on the slump flow of ESH–LWSCCs.

V-funnel flow time. However, combined maximum increase of both w/b and HRWRA parameters resulted in a substantial reduction of the V-funnel flow time (below 2 s) at given binder content. This observation is in agreement with the conclusion of previous SCC statistical workability study (Sonebi et al. 2007). The V-funnel flow time is indicative of the viscosity of the LWSCC mixture—the higher the flow times the more viscous and less workable is the mix. Changes of V-funnel flow time with w/b and HRWRA are depicted in Fig. 3. The effect of w/b and total binder content on the V-funnel flow time of ESH–LWSCC mixtures is plotted in Fig. 4. It can be concluded that an increase of w/b from 0.3 to 0.4 significantly decreased the V-funnel flow time. However, only a slight increase in flow time was

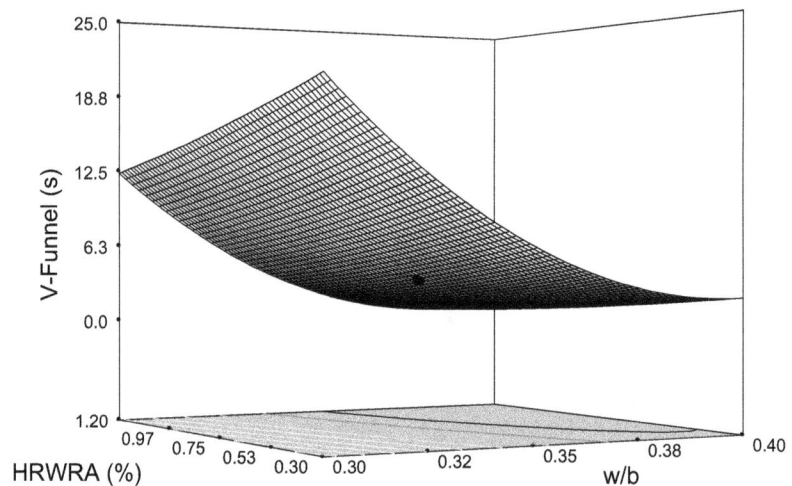

Fig. 3 Effect of w/b, HRWRA and total binder content at 480 kg/m^3 on the V-funnel time of ESH–LWSCCs.

observed with the increase of binder content at a given HRWRA%. This can be attributed to low internal friction (higher excess paste volume) in the ESH mixes.

4.1.3 Influence on the L-Box Ratio

The L-box ratio showed a similar trend of variation as slump flow. An increase of w/b from 0.3 to 0.4 and HRWRA from 0.3 to 1.2 % significantly increased the L-box ratio when a high binder content of 480 kg/m^3 was used. Figure 5 presents the slump flow changes of ESH–LWSCC mixtures depending on the w/b and HRWRA. According to Hwang et al. (2006), a combination of the slump flow and the L-box ratio can be used to assess filling capacity of SCC for quality control and design of SCC for placement in restricted sections or congested elements.

Figure 6 presents contour diagrams of the L-box ratio of ESH–LWSCC mixtures depending on the w/b and total binder, respectively. It can be suggested that as the total binder content is increased, the L-box ratio is reduced for a given HRWRA%. Previous research demonstrated the relationship between w/b, HRWRA, volume of coarse aggregate and L-box ratio for normal weight SCC mixtures where all three parameters are found to significantly influence the L-box ratio (Sonebi et al. 2007).

4.1.4 Influence on the Segregation Resistance

Figure 7 shows that the increase of the binder content appeared to be very effective in increasing the segregation resistance. The increase in binder content enhanced the packing density of mixtures and resulted in a reduction in segregation. This is also attributed to the increased cohesiveness and viscosity of the concrete mixture at high binder content. Similar conclusions were drawn in previous normal weight SCC statistical studies (Patel et al. 2004; Khayat et al.

2000). Figure 8 illustrates the trade-off between variation of the w/b and HRWRA on the segregation resistance of ESH–LWSCC mixtures at a given binder content (480 kg/m^3). These contours show that increasing one or both parameters w/b and HRWRA (from 0.3 to 0.4 and from 0.3 to 1.2 %, respectively), would significantly reduce the segregation resistance of ESH–LWSCC mixtures.

4.1.5 Influence on Other Properties

For all mixes, the filling capacity and J-ring flow/J-ring height difference were positively influenced by w/b and HRWRA. An increase of either or both parameters led to an increase in the measured responses/properties. However, an increase in the binder content alone affects the results negatively—showing a decrease in the measured responses.

The aggregate density played a major role in affecting the fresh unit weight of the mixes. As for the influence of the examined parameters on the response, the fresh unit weight was influenced mainly by the binder content—as the binder content increased the fresh unit weight increased and vice versa. Only the total binder content affected the results of the 28-day air and oven dry unit weights of ESH mixtures. An increase in the total binder content increased both unit weights. This behavior might be attributed to the high absorption rate of aggregates (above 13 %) that slowed the evaporation rate of water from the mixture. The HRWRA% did not have an effect on the results.

For all developed mixes, 7-day compressive strengths were affected by all three parameters (w/b, HRWRA and total binder content). As the binder increased, the 7-day strength increased. In contrast, as the either or both HRWRA (%) and w/b increased the 7-day strength decreased. Nevertheless, it was expected that HRWRA% should not have

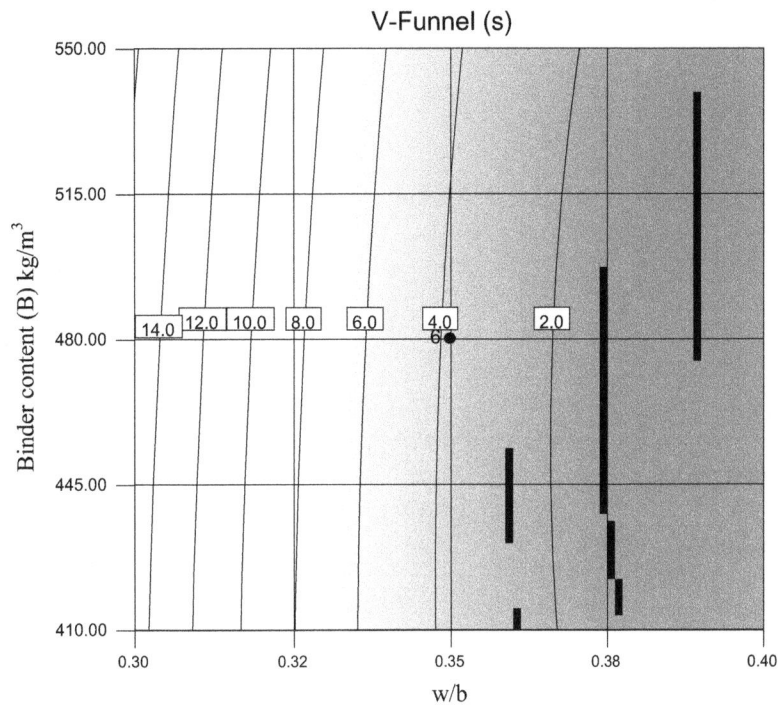

Fig. 4 Contours of V-funnel changes of ESH–LWSCC mixes with w/b, total binder content and HRWRA at 0.75 %.

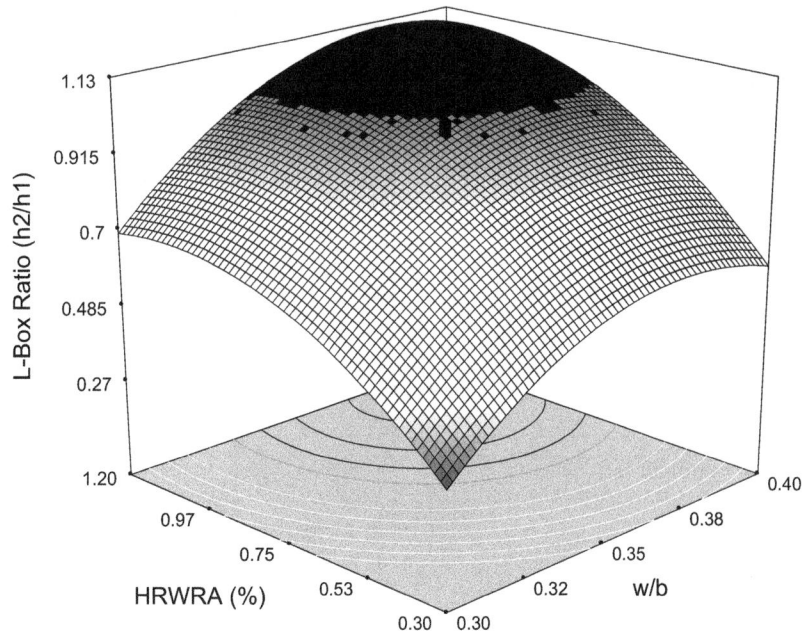

Fig. 5 Effect of w/b, HRWRA and total binder content at 480 kg/m^3 on the L-box of ESH–LWSCCs.

any influence on the 7-day strength. This is because HRWRA% effect is typically weakened away after 24–48 h. On the other hand, the 28-day compressive strengths were mainly affected by w/b and total binder content. An increase in w/b decreased the 28-day strengths, while an increase in total binder content increased the compressive strength

which is agreement with basic knowledge of concrete technology regardless of the concrete type.

4.2 Statistical Evaluation of Test Results

A model analysis of the response was carried out to determine the effectiveness of test parameters in controlling

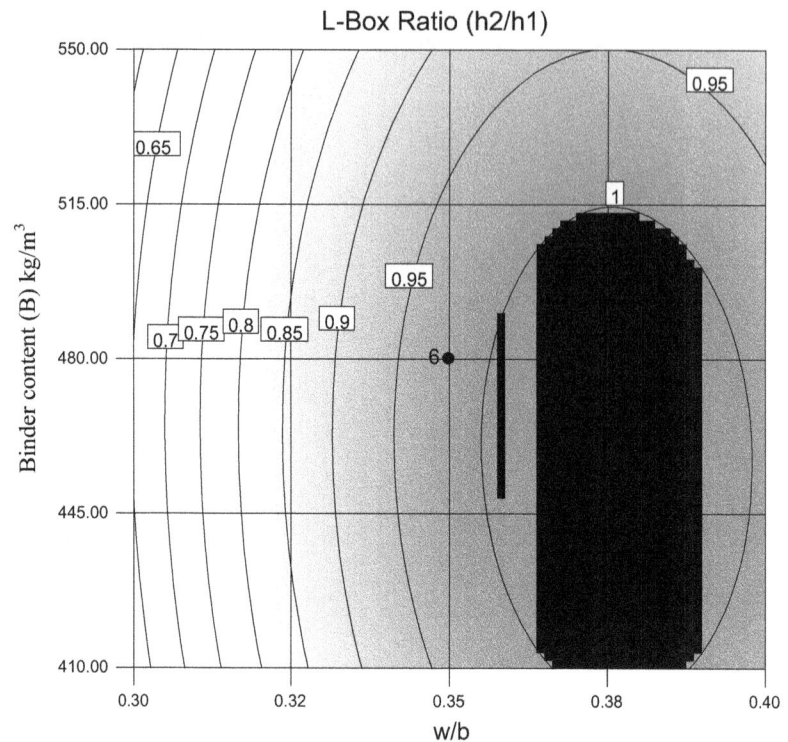

Fig. 6 Contours of L-box ratio changes of ESH–LWSCCs with w/b, total binder content and HRWRA at 0.75 %.

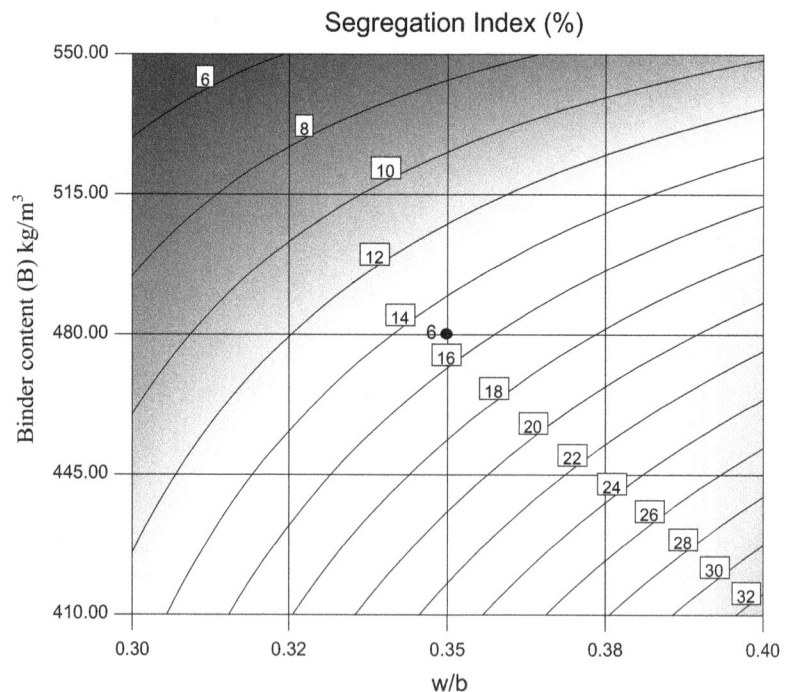

Fig. 7 Contours of segregation resistance changes of ESH–LWSCC mixes with w/b, total binder content and HRWRA at 0.75 %.

the ESH–LWSCC properties. Using GLM-ANOVA, the measured fresh and hardened properties of ESH–LWSCCs such as slump flow, V-funnel flow time, etc., were given as the dependent variables while the experimental test parameters ("w/b", "HRWRA%", and "B") were selected as the independent factors/variables. The general linear model analysis of variance was performed and the effective test

parameters and their percent contributions on the above mentioned properties of ESH–LWSCCs were determined. Table 6 summarizes all the relevant data from statistical evaluation.

The p value in Table 6 shows the significance of the given test parameters on the test results. If a system has a p value (Probabilities) of ≤ 0.05 it is accepted as a significant factor

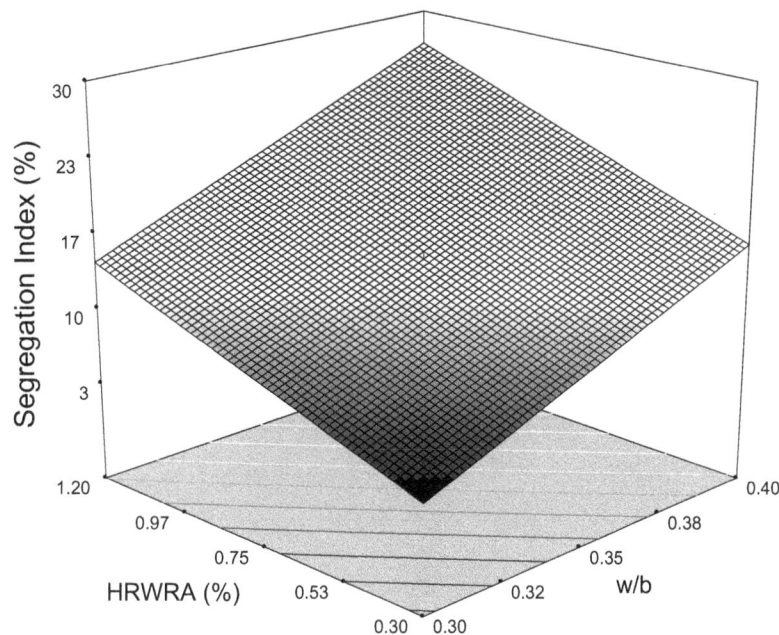

Design-Expert® Software

Segregation Index (%)
● Design points below predicted value

38

4.41073

X1 = A: **w/b**
X2 = B: HRWRA

Actual Factor
C: B = 480.00

A: w/b

B: HRWRA

C: Binder content (B)

Fig. 8 Effect of w/b, HRWRA and total binder content at 480 kg/m³ on the SSR of ESH–LWSCC mixes.

on the test result, as evidence indicates that the parameter is not zero; that is, the contribution of the proposed parameter has a highly significant influence on the measured response (Patel et al. 2004; Sonebi 2004a, b). The contributions of the each parameters on the measured test results are presented in Table 6, where the effectiveness of the independent parameters on the measured response is calculated. The higher the contribution, the higher the effectiveness of the parameter on the response, equally, the lower the contributions the lower effect on the response.

Analysis of the statistical parameters of the derived model, along with the relative significance, and the contribution % of each parameter on the results are given in Table 6. The R^2 values of the ESH–LWSCC response models for the slump flow, V-funnel flow, J-Ring flow, J-Ring height difference, L-box, filling capacity, sieve segregation resistance, 7-day compressive strength, 28-day compressive strength, fresh unit weight, 28-day air dry unit weight, and 28-day oven dry unit weight were found to be 0.96, 0.97, 0.96, 0.94, 0.94, 0.95, 0.90, 0.88, 0.93, 0.73, 0.56, and 0.75, respectively.

Statistically significant models for ESH–LWSCCs with a high correlation coefficient $R^2 > 0.90$ were established for the slump flow, V-funnel, J-ring, J-ring height difference, L-box, filling capacity, sieve segregation resistance and 28-day compressive strength. A relatively lower R^2 values of 0.88, 0.73 and 0.75 were obtained for the 7-day compressive strength fresh and 28-day oven dry unit weights, respectively. Low R^2 of 0.56 was obtained for 28-day air dry unit weight (Table 6).

As for the significance of the parameters on the responses, for example for the slump flow; the order of influence of the test variables is: the dosage of HRWRA, w/b, and the binder

content. The dosage of HRWRA had the greatest effect on the slump flow. The effect of binder content was insignificant to the response. This can be attributed to the fact that flowability is driven by HRWRA dose and w/b rather than the binder content. In fact, to secure the same slump flow with more binder content, an increase of both HRWRA and w/b is necessary.

As for the V-funnel time, the order of influence of the test variables on the response is: w/b, the dosage of HRWRA and then binder content. Whereas the dosage of HRWRA, w/b, and the binder content in this order of influence, are contributing to the responses of J-ring flow, J-ring height different, L-box and filling capacity. The sieve segregation resistance response is greatly influenced by the total binder content, followed by w/b and then the dosage of HRWRA. The contribution % of each parameter on the rest of the results is given in Table 6.

The high correlation coefficient of responses demonstrates excellent correlation, where it can be considered that at least 95 % of the measured values can be accounted for with the proposed models (Patel et al. 2004; Sonebi 2004a, b).

4.3 Mathematical Formulation of ESH–LWSCC Properties

The mathematical relationship between the independent variables and the responses can be estimated using the model. Linear or quadratic relationships are simplified by using a backward stepwise technique. Evaluating the contribution of each parameter and its significant influence on the response is a key tool used in accepting certain contribution (Whitcomb and Anderson 2004; Pradeep 2008).

Table 6 Analysis of GLM-ANOVA model.

Dependent variable	Source of variation	Statistical parameters					Significant	Contribution (%)
		DOF	Sum of Square	Mean square	F	p value		
Slump flow	w/b	1	1.409E+05	1.409E+05	69.88	0.0,001	Y	36.0
	HRWRA	1	2.475E+05	2.475E+05	122.73	0.0001	Y	63.2
	B	1	3101.39	3101.39	1.54	0.2432	N	0.8
V-funnel	w/b	1	571.98	571.98	192.66	0.0001	Y	92.3
	HRWRA	1	46.72	46.72	15.74	0.0027	Y	7.5
	B	1	1.09	1.09	0.37	0.5574	N	0.2
J-ring flow	w/b	1	1.302E+05	1.302E+05	82.89	0.0001	Y	36.4
	HRWRA	1	2.277E+05	2.277E+05	144.89	0.0001	Y	63.6
	B	1	283.95	283.95	0.18	0.6798	N	0.1
J-ring height	w/b	1	70.55	70.55	45.19	0.0001	Y	38.6
	HRWRA	1	103.91	103.91	66.55	0.0001	Y	56.9
	B	1	8.28	8.28	5.30	0.0441	Y	4.5
L-box	w/b	1	0.34	0.34	35.61	0.0001	Y	37.5
	HRWRA	1	0.56	0.56	58.53	0.0001	Y	61.6
	B	1	7.722E−03	7.722E−03	0.81	0.3896	N	0.9
Filling capacity	w/b	1	3663.64	3663.64	53.31	0.0001	Y	37.7
	HRWRA	1	5964.65	5964.65	86.79	0.0001	Y	61.4
	B	1	87.97	87.97	1.28	0.2843	N	0.9
Sieve segregation resistance	w/b	1	464.58	464.58	31.88	0.0001	Y	29.9
	HRWRA	1	357.16	357.16	24.51	0.0003	Y	22.9
	B	1	734.64	7.34.64	50.41	0.0001	Y	47.2
7-Day compressive strength	w/b	1	357.62	357.62	82.24	0.0001	Y	69.3
	HRWRA	1	28.85	28.85	6.63	0.0203	Y	5.6
	B	1	129.55	129.55	29.79	0.0001	Y	25.1
28-Day compressive strength	w/b	1	637.51	637.51	84.72	0.0001	Y	69.0
	HRWRA	1	41.52	41.52	5.52	0.0407	Y	4.5
	B	1	244.55	244.55	32.50	0.0002	Y	26.5
Fresh unit weight	w/b	1	622.78	622.78	0.70	0.4212	N	7.4
	HRWRA	1	683.54	683.54	0.77	0.4001	N	8.1
	B	1	7092.69	7092.69	8.01	0.0178	Y	84.5
28-Day air dry unit weight	w/b	1	75.35	75.35	0.061	0.8093	N	0.9
	HRWRA	1	733.33	733.33	0.59	0.4561	N	8.3
	B	1	8057.00	8057.00	6.48	0.0244	Y	90.9
28-Day oven dry unit weight	w/b	1	546.73	546.73	0.53	0.4852	N	6.6
	HRWRA	1	784.87	784.87	0.75	0.4055	N	9.4
	B	1	6989.34	6989.34	6.72	0.0269	Y	84.0

DOF degree of freedom, *F* statistic test, *p value* probabilities.
Significant: $p < 0.050$ (*Y* yes), $p > 0.050$ (*N* no).

Table 7 Mathematical formulation of ESH–LWSCC properties.

Parameters	Slump flow	V-funnel	J-ring flow	J-ring height	L-box	Filling capacity	SSR
Constant	−2631.74	282.93946	−2803.42	130.55	−10.61	−1020.89	−183.568
w/b	14376.546	−1391.542	13859.24	−479.9	48.58	4597.46	632.705
HRWRA	356.7146	−32.51110	387.503	−51.01	0.77	136.18	38.001
B	0.95820	5.826E−03	1.788	0.051	8.8E−03	0.800	0.3084
w/b × HRWRA	416.66667	59.47019	333.333	55.55	1.16	7.820	8.2980
w/b × B	1.25000	−0.13444	1.428	0.142	−1.8E−3	−0.203	−1.070
HRWRA × B	0.29762	9.043E−03	0.39683	0.023	7.3E−04	0.0455	−0.060
$(w/b)^2$	−18735.07	1818.468	−18149.1	458.64	−64.59	−5369.08	–
$(HRWRA)^2$	−217.9190	1.98610	−259.424	9.077	−0.7061	−74.20	–
$(B)^2$	−1.925E−3	4.035E−05	−2.6E−03	−2.8E−5	−9.5E−6	−8.35E−4	–
R^2	0.96	0.97	0.96	0.94	0.94	0.95	0.90

Parameters	Comp strength		Fresh unit weight	28-day air dry weight	28-day oven dry weight
	7-day	28-day			
Constant	48.71	−79.42	1697.63	−11.93	1995.35
w/b	−109.54	319.19	973.54	3681.54	2279.59
HRWRA	−3.439	3.38	527.89	724.93	563.21
B	0.046	0.328	−1.430	2.920	−4.42
w/b × HRWRA	–	−1.038	−915.71	−1165.45	−1082.20
w/b × B	–	0.229	−5.35	−5.953	−4.81
HRWRA × B	–	−1.9E−3	−0.579	−0.624	−0.52
$(w/b)^2$	–	−821.05	3053.463	–	1012.11
$(HRWRA)^2$	–	−4.14	58.42	–	55.48
$(B)^2$	–	−3.6E−4	4.25E−03	–	7.12E−03
R^2	0.88	0.93	0.73	0.56	0.75

When determining the model for each response, a regression analysis is performed on the basis of a partial model containing only the terms which are statistically significant at a 0.05 level of significance. Then, t-statistics are calculated and the terms that are statistically insignificant are eliminated. This process is repeated until the partial model contains only the significant terms. The experimental data are fed to a mathematical model through multiple linear regression analysis which consisted of the terms which are statistically significant at a 0.05 level. R^2 statistic, which gives a correlation between the experimental data and the predicted response, should be high enough for a particular model to be significant (Muthukumar and Mohan 2004).

The derived equations of the modelled responses are summarized in Table 7 for ESH–LWSCC mixtures. In this Table, mixture variables expressed in actual factored values present a comparison of various parameters as well as the interactions of the modelled responses. The model constants are determined by multi-regression analysis and are assumed to be normally distributed. A negative estimate signifies that an increase of the given parameter results in a reduction of the measured response. For any given response, the presence of parameters with coupled terms, such as $(w/b)^2$ and $(w/b)^3$ indicates that the influence of this parameter (w/b) is quadratic and cubic, respectively.

4.4 Repeatability of the Test Parameters

The repeatability of test parameters at central points is given in Table 8. ESH–LWSCC mixtures 15–20 (center point mixes) are found to satisfy LWSCC performance criteria. This table shows the mean results, standard deviation and coefficient of variance (COV), as well as the standard errors and the relative errors, with 95 % confidence limit of measured response of the six repeated mixes. The relative errors at the 95 % confidence limit for slump flow, V-funnel flow time, J-ring flow, L-box, filling capacity, sieve segregation resistance test, fresh unit weight, 28-day air dry unit weight, 28-day oven dry unit weight, and 7- and 28-day compressive strength in ESH–LWSCC model are found to be limited to 0.6–9.7 %. On the other hand, the relative error

Table 8 Repeatability of test parameters for ESH–LWSCC mixtures.

Test method	Mean ($n = 6$)	SD	COV (%)	Estimated error (95 % CI)	Relative error (%)
Slump flow (mm)	692.50	12.55	1.8	12.26	1.8
V-funnel (s)	3.77	0.23	6.2	0.23	6.1
J-ring flow (mm)	691.67	13.29	1.9	12.99	1.9
J-ring height (mm)	1.50	0.55	36.5	0.54	35.7
L-box (ratio)	0.99	0.01	1.3	0.01	1.2
Filling capacity (%)	98.33	1.21	1.2	1.18	1.2
Sieve segregation resistance (%)	11.83	1.17	9.9	1.14	9.7
7-Day comp strength (MPa)	32.00	1.26	4.0	1.24	3.9
28-Day comp strength (MPa)	45.00	1.79	4.0	1.75	3.9
Fresh unit weight (kg/m^3)	1790.67	10.78	0.6	10.54	0.6
28-Day air dry unit (kg/m^3)	1674.17	11.44	0.7	11.18	0.7
28-Day oven dry unit (kg/m^3)	1614.33	13.85	0.9	13.53	0.8

for J-ring height difference is found 35.7 %. The relative error was defined as the value of the error with 95 % confidence limit divided by the mean value.

5. Phase III: Optimization-Validation of the Statistical Models and Development of Industrial ESH–LWSCC

This phase included the validation of the statistical model and mix proportion optimization process. The optimization was performed to develop mixtures that satisfy EFNARC industrial classifications for SCC (EFNARC 2005). Moreover, this phase also presents the results of additional experimental study to validate whether the theoretically proposed optimum mix design parameters such as w/b, HRWRA%, and total binder (B) can yield the desired fresh and hardened properties for ESH–LWSCCs.

5.1 Verification of Statistical Models

The accuracy of the proposed model was determined by comparing predicted-to-measured values obtained with mixes prepared at the centre of the experimental domain and five other random mixes. Mixes 1–5 were randomly selected to cover a wide range of mixture proportioning within the modelled region, while mixes 6–10 were the centre points of the models. Mixture proportioning and measured responses of these ESH–LWSCC mixtures are presented in Tables 9 and 10, respectively.

Comparisons between predicted and measured values for various ESH–LWSCC responses are illustrated in Figs. 9 and 10 where the dashed lines present the upper and lower estimated error at 95 % confidence limit. Points found above the 1:1 diagonal line indicates that the statistical model overestimates the measured response.

On average, the predicated-to-measured ratios of slump flow, J-Ring flow, L-box ratio, V-funnel flow time, J-Ring height difference, filling capacity %, SSR index %, fresh unit weight, 28-day air-dry unit weight, 28-day oven dry unit weight, and 7- and 28-day compressive strengths were 1.02, 1.01, 1.0, 0.99, 0.98, 1.02, 1.02, 1.0, 1.0, 1.0, 1.02 and 1.02, respectively, indicating an accurate prediction of measured responses within the modelled region. The majority of the data for the measured responses lie close to the 1:1 diagonal line, resulting in the mean value of ratio between predicated-to-measured responses to be 1.00 ± 0.02. This indicates a high accuracy of the derived model to predicate the response.

On the other hand, the majority of the predicated slump flow, J-Ring flow, L- box ratio, V-funnel flow time, J-ring height difference, filling capacity, SSR index, fresh unit weight, 28-day air-dry unit weight, 28-day oven dry unit weight, and 7- and 28-day compressive strengths values (Figs. 9, 10) are within the acceptable limit of ± 12.26, ± 12.99 mm, ± 0.01, ± 0.23 s, ± 0.54 mm, ± 1.18, ± 1.14 %, ± 10.54, ± 11.18, ± 13.53 kg/m^3, ± 1.24 and ± 1.75 MPa, respectively. These limits constitute experimental errors for responses determined from the repeatability tests.

Table 9 Mixture proportions for ESH–LWSCC.

Mix no.	X1 (w/b)	X2 (HRWRA)	X3 (B)	Cement (kg/m³)	FA (kg/m³)	SF (kg/m³)	HRWRA (l/m³)	Water (l/m³)	ESH-aggregate (kg/m³)	
									Coarse	Fine
ESH1	0.4	0.60	520	416	65	39	2.9	208	400	640
ESH2	0.36	0.88	430	344	54	32	3.6	155	455	733
ESH3	0.32	0.94	550	440	69	41	4.9	176	415	665
ESH4	0.37	0.30	420	336	53	32	1.2	155	462	738
ESH5	0.33	1.00	450	360	56	34	4.2	148	455	730
ESH6	0.35	0.75	480	384	60	36	3.6	168	438	698
ESH7	0.35	0.75	480	384	60	36	3.6	168	438	698
ESH8	0.35	0.75	480	384	60	36	3.6	168	438	698
ESH9	0.35	0.75	480	384	60	36	3.6	168	438	698
ESH10	0.35	0.75	480	384	60	36	3.6	168	438	698

Table 10 Test results of ESH–LWSCC mixes used to validate statistical models.

Mix no.	Slump flow (mm)	V-funnel (s)	J-ring Flow (mm)	J-ring height diff (mm)	L-box ratio	Filling capacity (%)	SSR (%)
ESH1	688	1.6	698	0.5	0.86	88	13
ESH2	715	2.6	698	1.5	1.00	100	22
ESH3	636	9.4	655	1.0	0.82	82	7
ESH4	562	3.7	542	5.5	0.69	68	19
ESH5	708	5.8	695	1.5	0.95	96	19
ESH6	705	3.7	710	2.0	0.98	100	11
ESH7	685	4.0	680	1.0	1.00	99	12
ESH8	700	3.7	700	1.0	0.97	97	13
ESH9	685	3.5	680	1.0	1.00	97	10
ESH10	705	4.1	700	2.0	0.99	99	12

Mix no.	Comp strength		Unit weight (kg/m³)		
	7-Day	28-Day	Fresh	28-Day air dry	28-Day oven dry
ESH1	27	38	1,806	1,702	1,630
ESH2	27	39	1,781	1,675	1,611
ESH3	36	50	1,853	1,733	1,688
ESH4	27	37	1,782	1,662	1,616
ESH5	30	44	1,797	1,692	1,622
ESH6	34	48	1,789	1,676	1,604
ESH7	32	44	1,779	1,667	1,611
ESH8	31	45	1,782	1,670	1,614
ESH9	33	46	1,787	1,662	1,597
ESH10	31	43	1,800	1,675	1,625

As can be seen from the validation investigation, the derived model offers adequate predication of workability, unit weight and compressive strength response within the experimental domain of the modelled mixture parameters. It is important to note that the absolute values of the predicated values are expected to change with the changes in raw

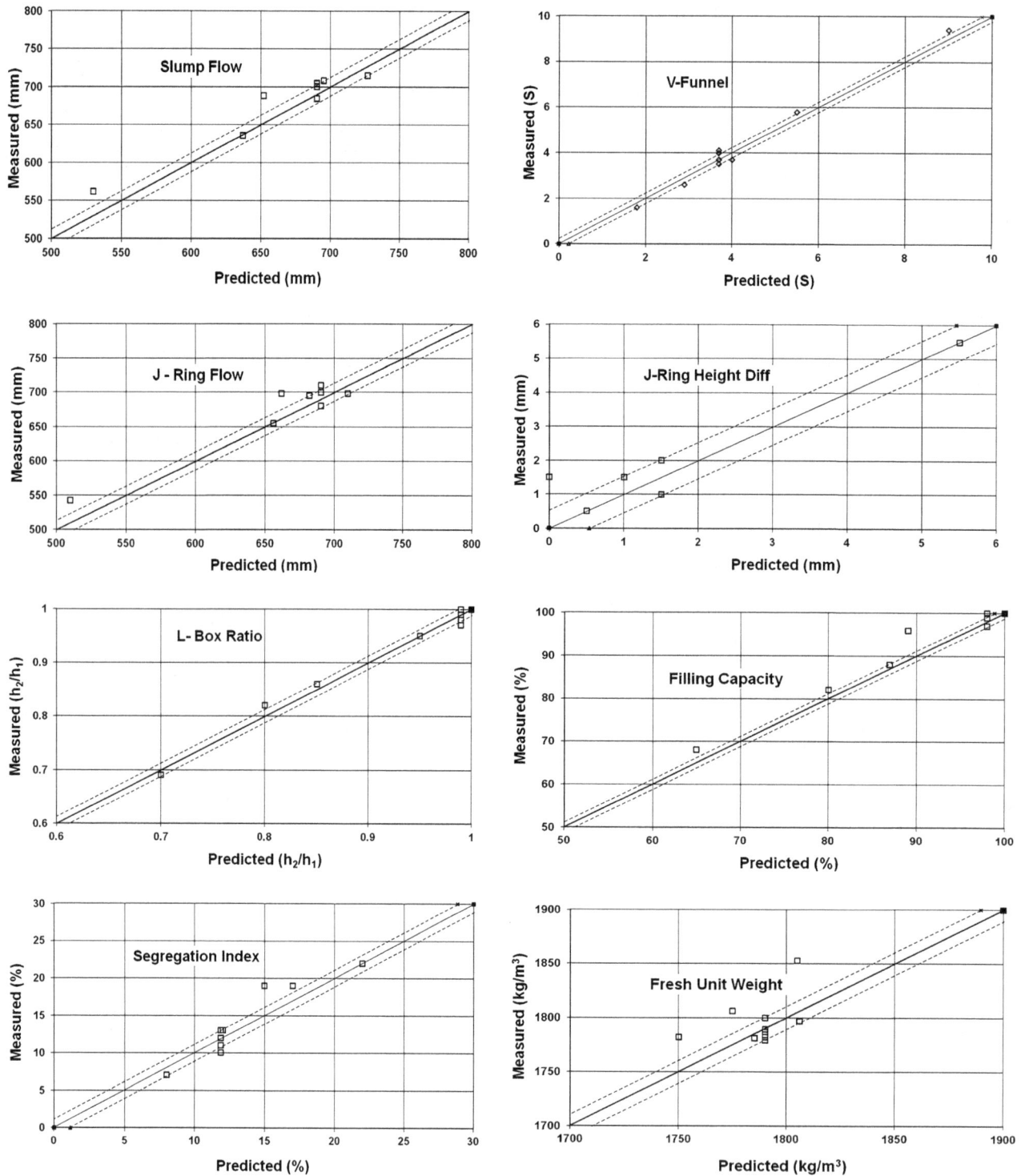

Fig. 9 Predicted versus measured fresh state properties of ESH–LWSCC.

material characteristics. However, the relative contributions of the various parameters are expected to be the same, thus facilitating the mix design protocol.

5.2 ESH–LWSCC Mixture Optimization

Based on the developed statistical model and the outlined relationships between mix design variables and the responses as shown in Table 9, all independent variables are varied simultaneously and independently in order to optimize the response. The objective of the optimization process is to obtain the "best fit" for particular response, considering

alternating multiple responses concurrently. In this study, optimization was performed to develop mixtures that satisfy EFNARC industrial classifications for SCC (EFNARC 2005). The fresh properties of SCC as per EFNARC are presented in Table 11.

The mix proportions (independent variables) were optimized to yield three ESH–LWSCC mixtures with the following fresh properties/classes:

(1) SF1 + VF1 + PA2 + SR2 (Casting by a pump injection system e.g. tunnel linings): ESH–LWSCC1

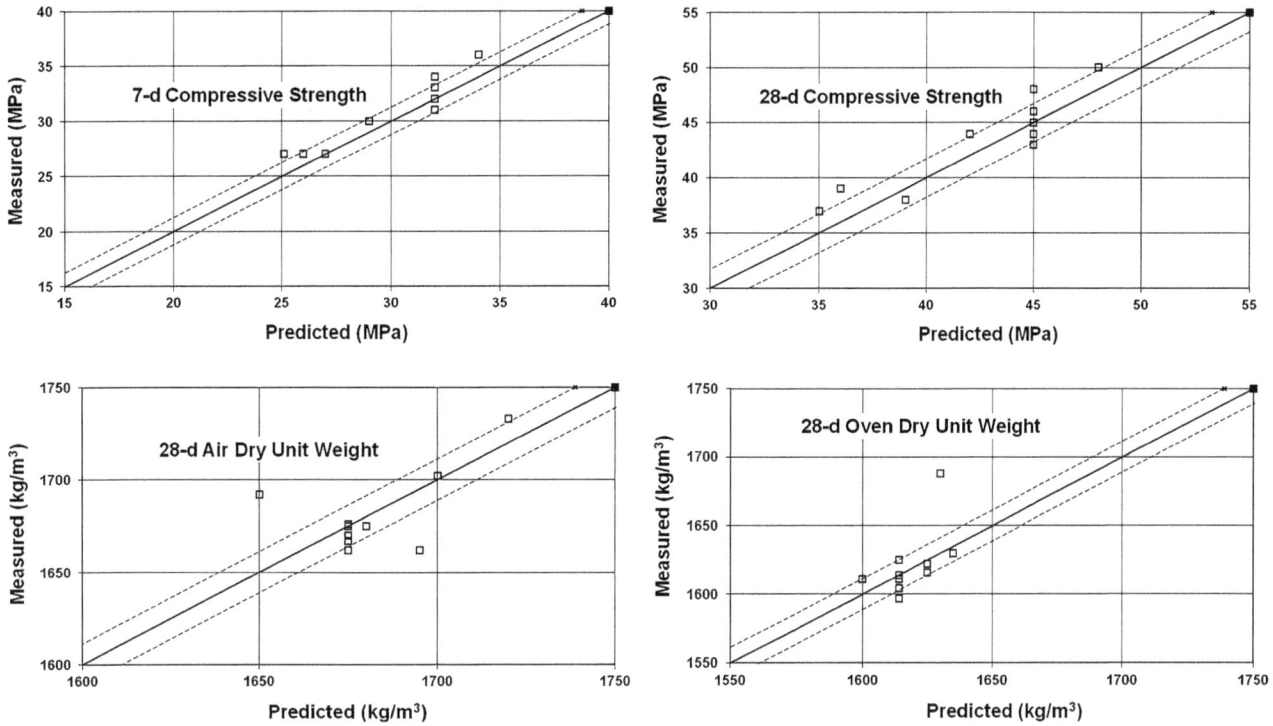

Fig. 10 Predicted versus measured hardened properties of ESH–LWSCC.

Table 11 EFNARC SCC classification.

Slump flow	Slump flow (mm)	
SF1	550–650	
SF2	660–750	
SF3	760–850	
Viscosity	T500 (s)	V-funnel (s)
VS1/VF1	≤2	≤8
VS2/V2	>2	9–25
Passing ability (L-box)	Passing ability ratio (h_2/h_1)	
PA1	≥0.80 with two rebars	
PA2	≥0.80 with three rebars	
Sieve segregation resistance	Segregation resistance (%)	
SR1	≤20	
SR1	≤15	

(2) SF2 + VF1 + PA2 + SR2 (Suitable for many normal applications e.g. walls, columns): ESH–LWSCC2

(3) SF3 + VF1 + PA2 + SR1 (Suitable for vertical applications in very congested structures, structures with complex shapes, or for filling under formwork): ESH–LWSCC 3

VF1 limits were constrained tighter as 4–8 s for ESH–LWSCC 1 and 2 to ensure density stability during application and placement. A numerical optimization technique, using desirability functions (d_j) defined for each target response, was utilized to optimize the responses (Whitcomb and Anderson 2004; Pradeep 2008; Ozbay et al. 2011). Desirability is an objective function that ranges from 0 to 1, where 0 indicates it is outside the range and 1 indicates the goal is fully achieved. The numerical optimization finds a point that maximizes the desirability function. The characteristics of a goal may be altered by adjusting the weight or importance (Ozbay et al. 2011). In this research, target responses were assigned equal weight and importance. All target responses were combined into a desirability function and the numerical optimization software was used to maximize this function (Ozbay et al. 2011; Nehdi and Summer 2002). The goals seeking begin at a random starting point and proceeds up the steepest slope to a maximum. To perform the optimization process, goals, upper and lower limits for the factors and responses were defined as in Table 12.

In order to have an equal importance, five predefined responses (slump flow, J-ring flow, V- funnel, L-box and SSR index) in addition to the goal to minimize both J- ring height difference and fresh unit weight response were considered and optimized simultaneously. Furthermore, filling capacity, 28-day air dry unit weight, 28-day oven dry unit weight, and 7- and 28-day compressive strengths were defined as in the experimental study range.

After runing the numerical optimization process for ESH–LWSCC-1 mixture, 29 solutions were obtained, satisfying the set limits and constrains. The desirability of the proposed solutions ranged from 0.732 to 0.810. As for ESH–LWSCC-2 and 3 mixtures, 25 and 30 solutions were obtained, with desirability ranging from 0.798 to 0.864 and 0.800 to 0.908, respectively. The highest desirability functions value 0.810, 0.864 and 0.908 for achieving the set, goals and limits are given in Table 12. The desirability function changed based

Table 12 Classification of responses goal and limits.

Name of responses	Goal	Lower limit	Upper limit	Lower limit	Upper limit	Lower limit	Upper limit
		ESH–LWSCC-1		ESH–LWSCC-2		ESH–LWSCC-3	
Slump flow (mm)	In range	550	650	660	750	760	850
V-funnel (S)	In range	4	8	4	8	0.0	8
J-ring flow (mm)	In range	550	650	660	750	760	850
J-ring height (mm)	Minimize	0.0	14.0	0.0	14.0	0.0	14.0
L-box ratio (h_2/h_1)	In range	0.8	1.0	0.8	1.0	0.8	1.0
Filling capacity (%)	In range	80	100	80	100	80	100
Sieve segregation (%)	In range	0.0	15	0.0	15	0.0	20
7-Day comp strength (MPa)	In range	20	40	20	40	20	40
28-Day comp strength (MPa)	In range	28	53	28	53	28	53
Fresh unit weight (MPa)	Minimize	1,742	1,892	1,742	1,892	1,742	1,892
28-Day air dry unit (kg/m^3)	In range	1,611	1,765	1,611	1,765	1,611	1,765
28-Day oven dry unit (kg/m^3)	In range	1,566	1,729	1,566	1,729	1,566	1,729

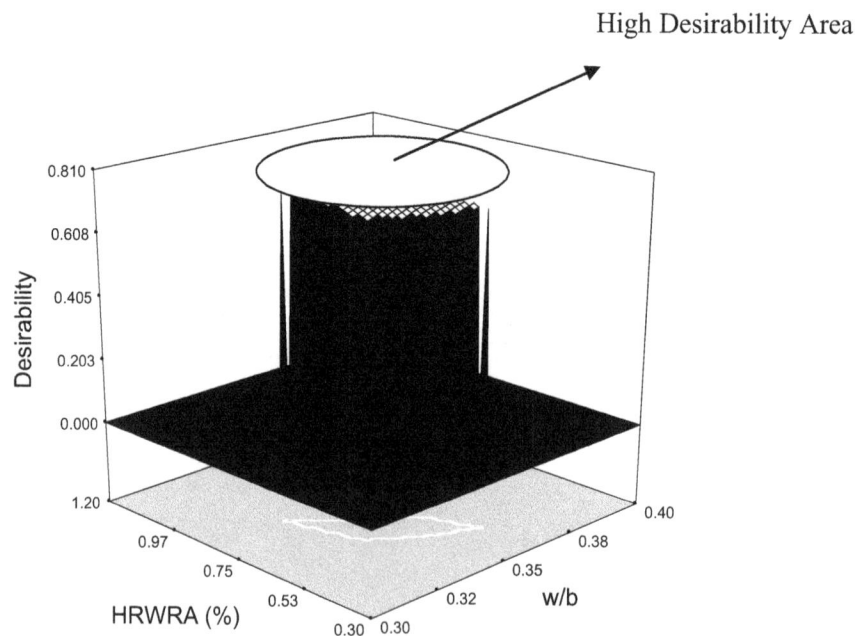

Fig. 11 Effect of w/b, HRWRA and total binder content at 476 kg/m^3 on the desirability function of ESH–LWSCC-1 mixture (EFNARC SCC class 1).

on the optimization process and is graphically presented in Figs. 11 and 12. For ESH–LWSCC mixes of classes 1 and 2 (when keeping the binder content constant at 476, 486 kg/ m^3, respectively), it was found that the desirability function increased only for very limited area (highlighted in the figures), and when the w/b and HRWRA% are between certain

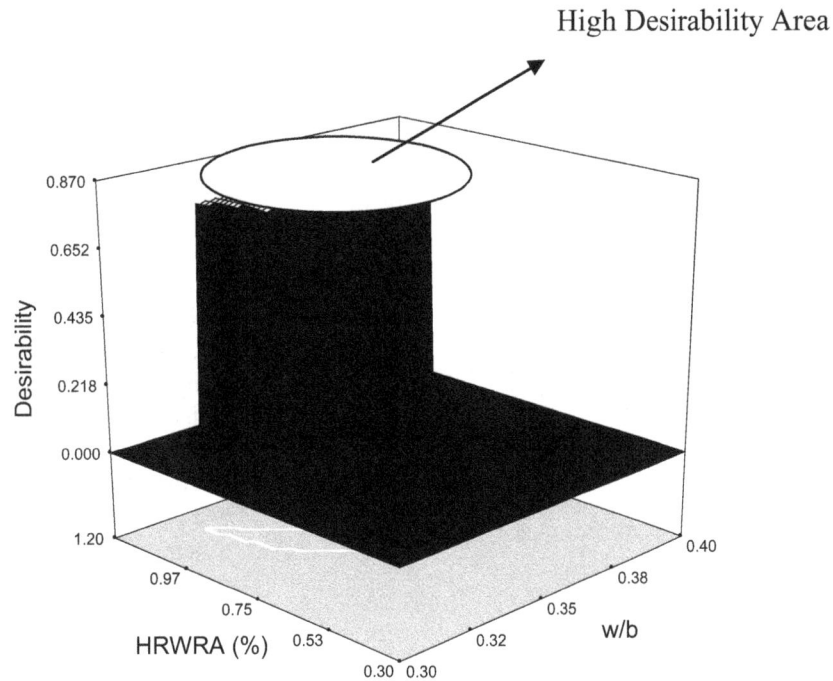

Fig. 12 Effect of w/b, HRWRA and total binder content at 486 kg/m^3 on the desirability function of ESH–LWSCC-2 mixture (EFNARC SCC class 2).

Table 13 Theoretically optimum mix proportions and experimental results.

Mix no.	ESH–LWSCC-1		ESH–LWSCC-2		ESH–LWSCC-3	
Mix parameters and responses	Opt values and expected response	Experimental results for opt mix proportions	Opt values and expected response	Experimental results for opt mix proportions	Opt values and expected response	Experimental results for opt mix proportions
w/b	0.35	0.35	0.35	0.35	0.40	0.40
HRWRA	0.61	0.61	0.83	0.83	0.78	0.78
B	476	476	486	486	504	504
Slump flow (mm)	650	645	708	725	760	770
V-funnel (s)	4	4.9	4	3.8	1.32	2.1
J-ring flow (mm)	650	635	709	715	765	760
J-ring height (mm)	2.2	2	0.83	0	0	0
L-box (%)	0.91	0.87	1	0.98	0.99	0.99
Filling capacity (%)	90	88	99.4	98	99.99	98
Sieve segregation (%)	13.1	12.1	14	13	17.75	18.5
7-Day comp strength (MPa)	30.2	32.2	30.6	33	25.8	24.5
28-Day comp strength (MPa)	44.6	45.75	45.4	47.75	36.95	35.1
Fresh unit weight (kg/m^3)	1,784	1,810	1,791	1,763	1,790	1,780
28-Day air dry unit (kg/m^3)	1,688	1,653	1,695	1,708	1,689	1,650
28-Day oven dry unit (kg/m^3)	1,606	1,585	1,614	1,602	1,610	1,590
Desirability	0.81	–	0.86	–	0.91	–

values. However, desirability value decreased drastically to zero outside this limited area indicating that very specific parameter range is needed to achieve high desirability above 0.8 for ESH–LWSCC mixtures. High desirability only can be achieved for ESH–LWSCC mixes of class 3 when the w/b is kept at 0.4 and for binder content above 500 kg/m^3.

5.3 Verification Experiment for an Optimum Mix Design

Utilizing the established high statistical confidence of the developed models, an experimental study was used to validate whether the theoretically proposed optimum mix design parameters, w/b, HRWRA%, and total binder could yield the desired responses. The test was carried out with the same materials and under the same testing conditions. The results are presented in Table 13. As it can be seen from the optimization/validation process, the model satisfactorily derived the three desired EFNARC-SCC industrial class mixtures. The optimized mixes satisfy the ranges for slump flow, V-funnel time, L-box ratio and segregation resistance percentage.

The derived statistical models can therefore be used as useful and reliable tools in understanding the effect of various mixture constituents and their interactions on the fresh properties of LWSCC. The analysis of the derived models enables the identification of major trends and predicts the most promising direction for future mixture optimization. This can reduce the cost, time, and effort associated with the selection of trial batches.

6. Conclusions

The properties of lightweight self-consolidating concrete (LWSCC), developed with expanded shale (ESH) lightweight aggregates (ESH–LWSCC) were investigated. This research included comprehensive laboratory investigations leading to the development of statistical design model for ESH–LWSCC mixtures accompanied by fresh and hardened performance evaluation of the developed ESH–LWSCC mixtures having varying water to binder ratio (w/b), high range water reducing admixture (HRWRA%) and total binder content (B). This research involved statistical modelling, mix design development, performance evaluation of ESH–LWSCCs, development/validation of statistical models and development of industrial class ESH–LWSCCs. The following conclusions were derived from the results of the comprehensive series of investigations:

1. The w/b has significant influence on the overall performance of ESH–LWSCCs, including fresh and hardened properties. In terms of fresh properties, the w/b has high influence on workability and HRWRA demand. The passing ability and filling capacity increase with the increases of w/b. The segregation resistance decreases with increase in w/b. ESH–LWSCCs with low w/b (0.35) required high dosage of HRWRA for flowability. It is noted that ESH–

LWSCC mixtures proportioned with w/b of less than 0.33 (regardless of HRWRA% or the total binder content), produced unsatisfactory fresh properties, and disqualified to be a LWSCC. On the other hand a balanced LWSCC mixture with w/b of around 0.35 made with ESH lightweight aggregates exhibited satisfactory workability, passing ability, filling capacity and segregation resistance.

2. Similar to normal weight SCC, the w/b has significant influence on the compressive strength of ESH–LWSCC mixtures—mixes with w/b of 0.35 developed higher compressive strength than those with w/b of 0.40.

3. In terms of fresh properties, the total binder content had influence on workability and static stability (segregation resistance) of ESH–LWSCCs. For a given w/b, the HRWRA demand decreased with the increase of total binder content. On the other hand, segregation resistance increased with the increase of total binder content. In contrast, at fixed HRWRA% and w/b, the workability/passing ability/filling capacity decreased and segregation resistance increased with the increase of total binder content.

4. The HRWRA% had significant influence on the workability and static stability of ESH–LWSCC mixtures. For a given w/b and total binder content, the workability/passing ability/filling capacity increased significantly and segregation resistance decreased with the increase of HRWRA%.

5. The established relation between the slump flow and the segregation index confirmed the commonly held notion that ESH–LWSCCs with less than 500 mm slump flow should not exhibit segregation. The chances of ESH–LWSCC segregation are very high beyond a slump flow of 750 mm as the segregation index tends to be more than 20 %. It is always desirable to keep the slump flow between 550 and 750 mm for a stable and homogenous ESH–LWSCC mixture.

6. Generally, use of fine and coarse ESH lightweight aggregates in mix proportioning yielded concretes with a 28-day air dry unit weight of less than 1,840 kg/m^3, classifying them as LWSCC.

7. From ANOVA statistical analysis, it was found that both w/b and (%) of HRWRA had significant impact on the fresh properties of LWSCC mixtures. The total binder content had insignificant impact on the workability, passing ability and filling capacity of ESH–LWSCC mixtures with high aggregate packing density. The effect of the total binder content on the segregation resistance and compressive strength of all ESH–LWSCC mixtures was classified as statistically significant.

8. The established model using the fractional factorial design approach are valid for ESH–LWSCC mixtures with w/b ranging between 0.30 and 0.40, total binder content between 410 and 550 kg/m^3 and HRWRA dosages between 0.3 and 1.2 % by mass of total binder content.

9. It was possible to produce robust ESH–LWSCC mixtures that satisfy the EFNARC criteria for SCC. Three industrial classes of ESH–LWSCC mixtures with wide range of workability performance were successfully developed. These mixtures can cover various ranges of applications, such as tunnel linings, walls, columns, vertical applications in very congested structures, and structures with complex shapes.

10. The statistical analysis and validation results of the derived statistical models indicate that this model can be used to design ESH–LWSCCs and to facilitate the protocol for optimization of ESH–LWSCCs. The theoretical optimum mix proportions can be used to derive desirable fresh properties and compressive strength of ESH–LWSCCs. The developed models and guidelines will ensure a speedy mix design process and reduce the number of trials needed to achieve LWSCC mix specifications.

Overall, this research established a technology which will guide engineers, researchers and manufacturers to develop high performance ESH–LWSCC mixtures. However, additional research is needed to validate the applicability of the model with varying gradation and shapes of aggregates.

References

ACI Committee 213R. (2003). Guide for structural lightweight-aggregate concrete (p. 38). Farmington Hills, MI: American Concrete Institute.

Andiç-Çakır, Ö., & Hızal, S. (2012). Influence of elevated temperatures on the mechanical properties and microstructure of self-consolidating lightweight aggregate concrete. *Construction and Building Materials, 34,* 575–583.

Assaad, J. J., & Khayat, K. H. (2006). Effect of viscosity-enhancing admixtures on formwork pressure and thixotropy of self-consolidating concrete. *ACI Materials Journal, 103*(4), 280–287.

ASTM C138/C138M. (2010). Standard test method for density (unit weight), yield, and air content (gravimetric) of concrete. West Conshohocken, PA: American Society for Testing and Materials.

ASTM C39. (2011). Standard test method for compressive strength of cylindrical concrete specimens. West Conshohocken, PA: American Society for Testing and Materials.

ASTM C567. (2011). Standard test method for determining density of structural lightweight concrete. West Conshohocken, PA: American Society for Testing and Materials.

Bogas, J. A., Gomes, A., & Pereira, M. F. C. (2012). Self-compacting lightweight concrete produced with expanded clay aggregate. *Construction and Building Materials, 35,* 1013–1022.

Bouzoubaa, N., & Lachemi, M. (2001). Self-compacting concrete incorporating high volumes of class F fly ash preliminary results. *Cement and Concrete Research, 31*(2), 413–420.

EFNARC. (2005). *The European guidelines for self compacting concrete: Specification, production and use.* Cambridge, UK: The Self-Compacting Concrete European Project Group.

ESCSI. (2004). Expanded clay, shale and slate, a world of application (p. 5). Worldwide, Salt Lake City, UT, Publication No. 9349.

Fragoulis, D., Stamatakis, M. G., Chaniotakis, E., & Columbus, G. (2003). The Utilization of clayey diatomite in the production of lightweight aggregates and concrete. *Tile and Brick International, 19*(6), 392–397.

Fragoulis, D., Stamatakis, M. G., Chaniotakis, E., & Columbus, G. (2004). Characterization of lightweight aggregates produced with clayey diatomite rocks originating from Greece. *Materials Characterization, 53*(2–4), 307–316.

Ghezal, A., & Khayat, K. H. (2002). Optimizing self-consolidating concrete with limestone filler by using statistical factorial design methods. *ACI Materials Journal, 99*(3), 264–272.

Holm, T. A. (1994). Lightweight concrete and aggregates. STP 169C: Concrete and concrete: Making materials (pp. 522–532). Philadelphia, PA: American Society for Testing and Materials.

Hossain, K. M. A. (2004). Properties of volcanic pumice based cement and lightweight concrete. *Cement and Concrete Research, 34*(2), 283–291.

Hwang, C. L., & Hung, M. F. (2005). Durability design and performance of self-consolidating lightweight concrete. *Construction and Building Materials, 19*(8), 619–626.

Hwang, C.-L., Bui, L. A.-T., Lin, K.-L., & Lo, C.-T. (2012). Manufacture and performance of lightweight aggregate from municipal solid waste incinerator fly ash and reservoir sediment for self-consolidating lightweight concrete. *Cement & Concrete Composites, 34*(10), 1159–1166.

Hwang, S., Khayat, K., & Bonneau, O. (2006). Performance-based specifications of self-consolidating concrete used in structural applications. *ACI Materials Journal, 103*(2), 121–129.

Karahan, O., Hossain, K. M. A., Ozbay, E., Lachemi, M., & Sancak, E. (2012). Effect of metakaolin content on the properties self-consolidating lightweight concrete. *Construction and Building Materials, 31*(6), 320–325.

Khayat, K. H., Ghezal, A., & Hadriche, M. S. (1998). Development of factorial design models for proportioning self-consolidating concrete. In V. M. Malhotra (Ed.) Nagataki Symposium on Vision of Concrete: 21st Century (pp. 173–197).

Khayat, K. H., Ghezal, A., & Hadriche, M. S. (2000). Utility of statistical models in proportioning self-consolidating concrete. In *Proceedings of the First International RILEM Symposium on Self-Compacting Concrete* (pp. 345–359), Stockholm

Khayat, K. H., Lovric, D., Obla, K., & Hill, R. (2002). Stability optimization and performance of self-consolidating concrete made with fly ash. In First North American Conference on the Design and Use of Self-consolidating Concrete (pp. 215-223). Chicago, IL: ACI. November 12–13

Kim, Y. J., Choi, Y. W., & Lachemi, M. (2010). Characteristics of self-consolidating concrete using two types of lightweight coarse aggregates. *Construction and Building Materials, 24*(1), 11–16.

Lachemi, M., Bae, S., Hossain, K. M. A., & Sahmaran, M. (2009). Steel–concrete bond strength of lightweight self-consolidating concrete. *Materials and Structures, 42*(7), 1015–1023.

Müller, H. S., & Haist, M. (2002). Self-compacting lightweight concrete—Technology and use. *Concrete Plant Precast Technology, 71*(2), 29–37.

Muthukumar, M., & Mohan, D. (2004). Optimization of mechanical properties of polymer concrete and mix design recommendation based on design of experiments. *Journal of Applied Polymer Science, 94*(3), 1107–1116.

Nagataki, S., & Fujiwara, H. (1995). Self-compacting property of highly flowable concrete. In V. M. Malhotra (Ed.) ACI SP (SP-154) (pp. 301–314). Farmington Hills, MI: American Concrete Institute.

Nehdi, M. L., & Summer, J. (2002). Optimization of ternary cementitious mortar blends using factorial experimental plans. *Materials Structure Journal, 35*(8), 495–503.

Ozbay, E., Gesoglu, M., & Guneyisi, E. (2011). Transport properties based multi-objective mix proportioning optimization of high performance concretes. *Journal of Materials and Structures, 44*(1), 139–154.

Patel, R., Hossain, K. M. A., Shehata, M., Bouzoubaâ, N., & Lachemi, M. (2004). Development of statistical models for mixture design of high-volume fly ash self-consolidating concrete. *ACI Materials Journal, 101*(4), 294–302.

Pradeep, G. (2008). *Response surface method.* Saarbrücken: VDM Verlag Publishing. 76 p.

Schmidt, S. R., & Launsby, R. G. (1994). In M. J. Kiemele (Ed.), *Understanding industrial designed experiments* (4th ed., pp. 1–48). Colorado Springs, CO: Air Academic Press.

Sonebi, M. (2004a). Medium strength self-compacting concrete containing fly ash: Modelling using factorial experimental plans. *Cement and Concrete Research, 34*(7), 1199–1208.

Sonebi, M. (2004b). Applications of statistical models in proportioning medium-strength self-consolidating concrete. *ACI Materials Journal, 101*(5), 339–346.

Sonebi, M., Bartos, P. J. M., Zhu, W., Gibbs, J., & Tamimi, A. (2000). Final Report Task 4, Hardened Properties of SCC. Brite-EuRam, Contract No. BRPRTC96-0366, Hardened Properties of SCC (p. 75). Advanced Concrete Masonry Center, University of Paisley.

Sonebi, M., Grünewald, S., & Walraven, J. (2007). Filling ability and passing ability of self-consolidating concrete. *ACI Materials Journal, 104*(2), 162–170.

Stamatakis, M. G., Bedelean, M., Gorea, H., Alfieris, D., Tziritis, E., & Kavouri, S. (2011). Clay-rich rocks and mining wastes for the production of lightweight aggregates with thermal insulation properties. *Refractories Worldforum, 3*(1), 85–92.

Topçu, I. B., & Uygunoğlu, T. (2010). Effect of aggregate type on properties of hardened self-consolidating lightweight concrete (SCLC). *Construction and Building Materials, 24*(7), 1286–1295.

Whitcomb, P. J., & Anderson, M. J. (2004). *RSM simplified: Optimizing processes using response surface methods for design of experiments* (p. 292). New York, NY: Productivity Press.

Wu, Z., Zhang, Y., Zheng, J., & Ding, Y. (2009). An experimental study on the workability of self-compacting lightweight concrete. *Construction and Building Materials, 23*(5), 2087–2092.

8

Chloride Penetration in Circular Concrete Columns

M. Morga[1], and G. C. Marano[2],*

Abstract: Most of the diffusion models of chloride ions in reinforced concrete (RC) elements proposed in literature are related to an isotropic homogeneous semi-infinite medium. This assumption reduces the mathematical complexity, but it is correct only for plane RC elements. This work proposes a comparison between the diffusion model of chloride ions in RC circular columns and in RC slab elements. The durability of RC cylindric elements estimated with the circular model instead of the plane model is shown to be shorter. Finally, a guideline is formulated to properly use the standard and more simple plane model instead of the circular one to estimate the time to corrosion initiation of cylindrical RC elements.

Keywords: chloride ions diffusion, circular column, time to corrosion initiation, pitting corrosion of rebars.

List of Symbols

$erf(\bullet)$	Error function
t	Time from the beginning of the exposure to chloride solution
t_{cr}	Time to corrosion initiation
x	Perpendicular distance from external surface in the slab model
w/c	Water-cement ratio
$C(x, t)$ or $C(\rho_1, t)$	Chloride concentration in RC structural member at the distance x or ρ_1 from the external surface at time t from the first exposure to chloride solution
C_0	Chloride concentration on the external surface of the RC structural member
C_{cr}	Chloride threshold concentration for the beginning of the depassivation
D (cm^2/year)	Diffusion coefficient
R	Radius of the circular cross-section
ρ	Radial distance from the centre in a circular model
$\rho_1 = R - \rho$	Radial distance from the external surface in a circular model
$erf\left(\frac{x}{2\sqrt{Dt}}\right)$	Diffusion element of the slab model
$\frac{2}{R}\sum_{m=1}^{\infty}\frac{1}{\alpha_m}\frac{J_0(\rho\alpha_m)}{J_1(R\alpha_m)}e^{(-D\alpha_m^2 t)}$	Diffusion element of the circular model
$J_0(\rho\alpha_m)$	Bessel's first function with order zero
$J_1(R\alpha_m)$	Bessel's first function with order one

[1]Mobility Department – Transportation Infrastructures Technologies, AIT Austrian Institute of Technology, Vienna, Austria.

[2]Department of Civil Engineering and Architecture, Technical University of Bari, Bari, Italy.

*Corresponding Author; E-mail: g.c.marano@gmail.com

1. Introduction

The reinforced concrete (RC) is one of the most used construction materials. During the hydration process the pH becomes highly alkaline inside the concrete matrix due to the formation of the products of the cement hydration. This alkaline pH produces a passive film of iron oxide or hydroxide on the surface of the steel bars that protect the bars themselves from the corrosion. Therefore, in RC element the reinforcements are not corroded until this layer stays. During its service life a RC structure may be exposed to aggressive species that migrate inside the concrete from their exposed surface and reduce the alkaline protective pH. When the concentration of these aggressive species reaches a threshold value and consequently the pH in the matrix concrete is sufficiently low, the oxidation of the steel bars begins. This process is referred to as "depassivation" (Val and Trapper 2008; Martin-Perez et al. 2001; Saetta et al. 1995; Izquierdo et al. 2004) of the reinforcement bars. The reduction of the pH in the concrete matrix is produced by two main mechanisms: the CO_2 diffusion from the atmosphere and the diffusion of active ions, like chloride ions (Martin-Perez et al. 2001; Saetta et al. 1995). The diffusion of chloride ions is the principal cause of the reinforcement bar corrosion in marine environment or cold climates.

Chloride ions are sometimes present inside the hydration water used to mix the concrete. In that case these ions slow down or prevent the formation of the protective passive layer on the steel surface. In the other cases the chloride ions migrate into the contrete matrix from the external enviroment. The time at which the critical chloride concentration is reached on the passive layer is called "time to initiation", according to the Tuutti model (Tuutti 1982) (see Fig. 1). This time devides the initiation phase from the propagation phase. During the initiation phase the chloride ions diffuse in the concrete matrix, but the steel corrosion is absent. In the propagation phase the corrosion started and proceed to a critical value of steel loss that causes failure of the structural element or all the structure. The propagation phase is considerably shorter than the initiation phase. Therefore, the prediction of the time to corrosion initiation is an important aspect in reliability analysis of RC structures because it indicates the beginning of the strenght reduction of the structural elements (Nogueira and Leonel 2013; Marano et al. 2008). For this reason the estimation of the time to corrosion initiation has been investigated by several researchers.

The chloride penetration in the concrete matrix occurs through either permeation and absorbition of a chloride solution or diffusion of free chloride ions into the saturated concrete (Val and Trapper 2008; Martin-Perez et al. 2001). These mechanisms may also occur sequentially. The absorption runs out in short time and after that the diffusion begins due to the saturation of the concrete pores (Martin-Perez et al. 2001). Each transport process is modelled by a different mathematical law depending on the forces that are at stake. In the convection process the motion of the solution inside the concrete pores (and then the motion of the chloride ions) is caused by the moisture/humidity gradient. The gradient of chloride concentration between the external surface and the inside of saturated concrete matrix causes the chloride ions motion in the diffusion process (Val and Trapper 2008; Nogueira and Leonel 2013; Basheer et al. 2002). This paper proposes the investigation of the chloride ingress in piers partially submerged in seawater, but fully saturated: the part of the pier out of the water is subject to continuous wetting cycles, therefore it is constantly wet and saturated. In these conditions the diffusion phenomenon is prevalent. The diffusion process of chloride ions in the concrete matrix is not irreversible in case of inversion of the concentration gradient and is history independent, so it is a normal diffusion process (Nogueira and Leonel 2013). For that reason the Fick's Second Law is used to model the time-variant chloride ions diffusion. The validity of the Fick's second law to model the diffusion of chloride ions in concrete is determined on an empirical basis (Chatterji 1995).

Different elements influence the chloride penetration into RC elements as combination of air or water pressure, humidity and concetration differences or temperature differences of solutions (Izquierdo et al. 2004; Nogueira and Leonel 2013; Basheer et al. 2002). Some studies investigated the influence of the different factors on the chloride threshold value that lead to the depassivation of the rebars (Izquierdo et al. 2004; Azad 1998). Most of those studies are based on the simple plane monodirectional diffusion model, called "slab model". In this model the concrete matrix is assumed to be an isotropic homogeneous semi-infinite medium with higher chloride concentration on its external plane surface than the inside concentration. In this model the concentration gradient causes the diffusion of chloride ions inside the concrete matrix along the direction perpendicular to the external surface. The Fick's Second Law (Chatterji 1995) describing the diffusion process in this slab model depend only on one spatial variable x and the time t (Collepardi et al. 1972):

$$\left(\frac{\partial C}{\partial t}\right) = D\frac{\partial^2 C}{\partial x^2} \tag{1}$$

The spatial variable x is the distance of the inside point, where the concentration is estimated, from the external surface, while the constant D is the diffusion coefficient. During the diffusion process the chloride concentration C changes inside the concrete matrix and the concentration gradient decreases.

The parameters affecting the chloride diffusion inside the concrete matrix are initial chloride concentration in the concrete mix, type of cement used in the mix, water-cement ratio (w/c ratio), curing conditions, external temperature and external concentration of the chloride ions and variation of the external concentration in time. The effect of the type of cement, the w/c ratio and the curing conditions on the chloride diffusion are all included in the value of the diffusion coefficient D (in m^2/s) (Azad 1998). In the most simple studies the chloride diffusion coefficient is kept constant

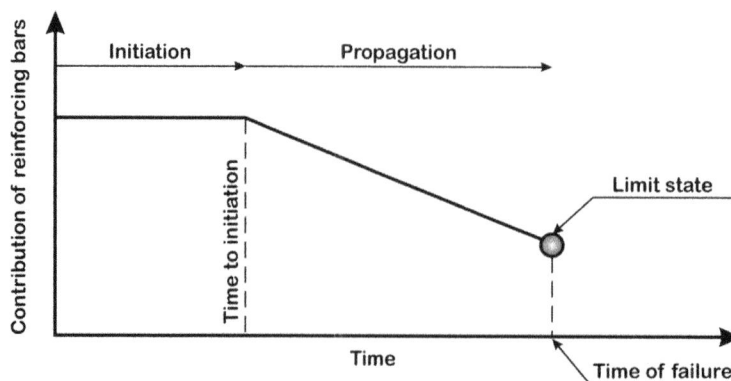

Fig. 1 Deterioration in reinforcing bar due to corrosion process.

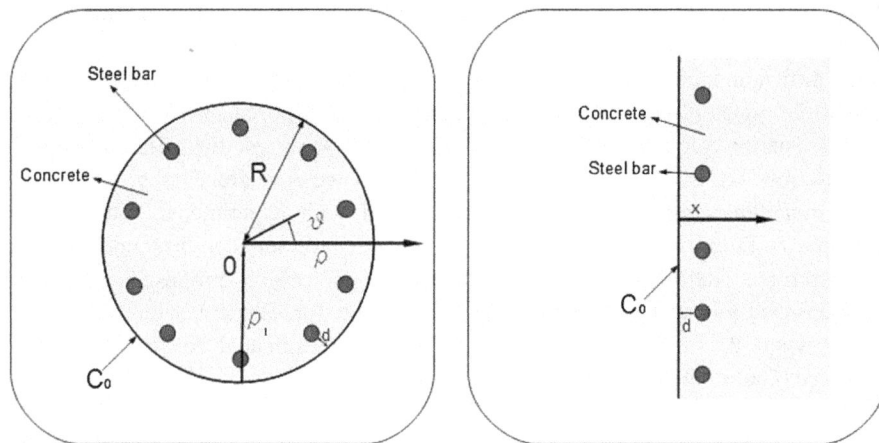

Fig. 2 *Left* cross-section of a circular RC column; *right* cross-section of an RC semi-infinite plane member.

during the diffusion process (Collepardi et al. 1972). Some scholars presented works in which the diffusion coefficient is estimated as a function of different parameters, like the w/c ratio (Lin 1990). Different laws to estimate the diffusion coefficient of ordinary portland concrete (OPC) Type I were compared in (Vu and Stewart 2000). The diffusion coefficients estimated according to these laws resulted to be similar for w/c ratio in the range 0.3–0.5. Furthermore the diffusion coefficient is highly dependent from the temperature, but the w/c ratio influences the value of this coefficient in the same way at different temperatures, as showed in (Kirkpatrick et al. 2002). Other studies demonstrated the dependence of diffusion coefficient on the age the temperature and the humidity at the same time (Val and Trapper 2008; Martin-Perez et al. 2001). It is important to notice that in literature the diffusion coefficient is always evaluated by assuming the occurrence of diffusion in a semi-infinite medium with plane external surface. This assumption is questionable in case the estimation of the diffusion coefficient is supported by experimental data: both laboratory samples and several real RC structural member from which samples are taken are not semi-infinite plane elements.

Although the deterministic approach to investigate the chloride diffusion in RC structural members is the most adopted, some studies proposed a probabilistic assessment of the time to corrosion initiation in RC elements. The time-dependent probability of steel corrosion initiation in partially saturated RC members after exposure to chloride ions is calculated in (Val and Trapper 2008) by taking into account both the diffusion and convection phenomena. In (Nogueira and Leonel 2013) and (Saassouh and Lounis 2012) a probabilistic model of chlorides diffusion is coupled to reliability algorithms to determine the probability of failure of an RC structure. Both numerical approach, like Monte Carlo simulation, (Nogueira and Leonel 2013; Saassouh and Lounis 2012) and analytical approaches, like FORM (Nogueira and Leonel 2013; Saassouh and Lounis 2012) and SORM (Saassouh and Lounis 2012), were tested by scholars to estimate the limit state of corrosion in RC structures. The uncertainties affecting the diffusion coefficient D have a large

scatter in the first life years of an RC structure, while after that time the overall uncertainty of the diffusion coefficient results reduced (Saassouh and Lounis 2012). The time and space invariant diffusion coefficient is an questionable assumption, but it is often supported by measures on the real concrete elements (Saassouh and Lounis 2012). It is clear that some uncertainty could be reduced by increasing the knowledge of the corrosion mechanism or testing the structure: in this case the uncertainty should be modelled as epistemic instead of aleatory (Saassouh and Lounis 2012; Do et al. 2005). The different kind of uncertainty could be handled by random, fuzzy or fuzzy random variables (Sobhani and Ramezanianpour 2011). In (Do et al. 2005) the uncertainty of the parameters on which the time to corrosion initiation depends is treated with the fuzzy logic. The defuzzification of the time to corrosion initiation could be considered a deterministic value to use in the maintenance planning of the structure. The fuzzy logic is an effective tool also in case the model defined to assess the reliability of RC structures subject to pitting reinforcement corrosion depends on both probabilistic and nonprobabilistic parameters (Marano et al. 2008). In complex models for the life cycle prediction and service life design of RC structures the uncertainties of the model parameters and the degradation mechanisms are not negligible. The interdependence among both parameters and mechanisms is also critical for the life cycle prediction of RC structure, therefore also in the assessment of the time to corrosion initiation. In (Sobhani and Ramezanianpour 2011) the high complexity of this problem depending on several parameters modelled as fuzzy random variables and interdependent mechanisms was managed by proposing an algorithm with a soft computing core.

In previous deterministic studies (Val and Trapper 2008; Tuutti 1982; Basheer et al. 2002) the time to corrosion initiation is evaluated with the Fick's second law formulated for the slab model (Eq. (1)). The solution of this equation is (Crank 1975)

$$C(x,t) = C_0 \left[1 - erf\left(\frac{x}{2\sqrt{Dt}} \right) \right] \qquad (2)$$

where $erf(\bullet)$ is the error function, C_0 is the chloride concentration on the external surface (expressed in kg/m^3 or in %), D is the apparent chloride diffusion coefficient and t is the duration of the exposure to chlorides (Tang and Joost 2007). Assuming D as time-invariant, the complement of the error function $(1 - erf(\bullet))$ describes the evolution of the chloride diffusion front in a monodimensional model in dependence to the exposure time t (Tang and Joost 2007; Tuutti 1982) and the distance from the external surface x.

Generally, the solution of the Fick's second law depends on the geometry of the element where the diffusion occurs, as well on the shape of the source (Crank 1975). Therefore, the time to corrosion initiation due to chloride ions diffusion depends on the geometry of the RC element. Some scholars proposed solutions of the Fick's second law for aggressive species in the two-dimensional (2-D) models (Val and Trapper 2008; Martin-Perez et al. 2001; Frier and Sørensen 2007) and one-dimensional not plane (1-D) model of RC elements (Arora et al. 1997). First Martin-Perez et al. (2001) and after Val et al. (Val and Trapper 2008) used a 2-D ingress of chloride ions into partially saturated RC members with rectangular cross-section. In these studies the chloride penetration is due to both the diffusion and the convection, while in (Frier and Sørensen 2007) only the chloride diffusion is estimated. In these 2-D chloride diffusion models it is assumed that the chloride ions penetrate at the same into the concrete section time along two directions perpendicular to the exposed external surfaces. This kind of model is applicable to RC elements with rectangular cross-section. Val and Trapper (2008) proved that the total chloride content increases faster near the reinforcement in a corner of a 2-D cross-section than in the proximity of the rebar in a 1-D model. For this reason the life-time of a rectangural RC element is shorter than the life-time of a RC element with geometry approximable to a plane semi-infinite element.

The piers cross-section of bridges or quays could be not only rectangular, but also circular. In (Arora et al. 1997)

the chloride diffusion process was investigated for columns with a circular cross-section. In that article the diffusion in a model of one steel bar covered by saturated concrete and exposed to a chloride solution was analysed. Sensitivity analyses for different initial and boundary conditions were made. Finally, the analytical results were compared with experimental data taken on circular columns with radii smaller than ones used in real structures. Although that article proposed the solution of the Fick's second law for element with circular cross section, the results were estimated for RC members with unrealistic dimensions.

This paper proposes the analytical solution of Fick's second law for RC members with circular cross section calculated for RC cylindrical columns with realistic dimensions. These results are compared with the results of the chloride diffusion in the slab model to highlight the difference between the time to corrosion initiation of the RC elements estimated with the model of a cylindrical column and the time estimated with the semi-infinite plane model. This comparison leads to indications about the limit of use the simple "slab model" to estimate the time to corrosion initiation in case of RC members with circular cross-section.

2. Chloride Diffusion in a Cylindrical Column

For the 1-D semi-infinite chloride diffusion model the Fick's Second Law is function of only one spatial variable x expressed in Cartesian coordinates. This variable corresponds to the direction perpendicular to the external surface exposed to the chloride solution (Crank 1975). In 2-D models of the chloride diffusion into RC elements the ions propagate along two different directions perpendicular to the external surfaces exposed to the chloride solution. In this case the Fick's Second Law is function of two spatial variables expressed in Cartesian coordinates (Crank 1975) (x and y):

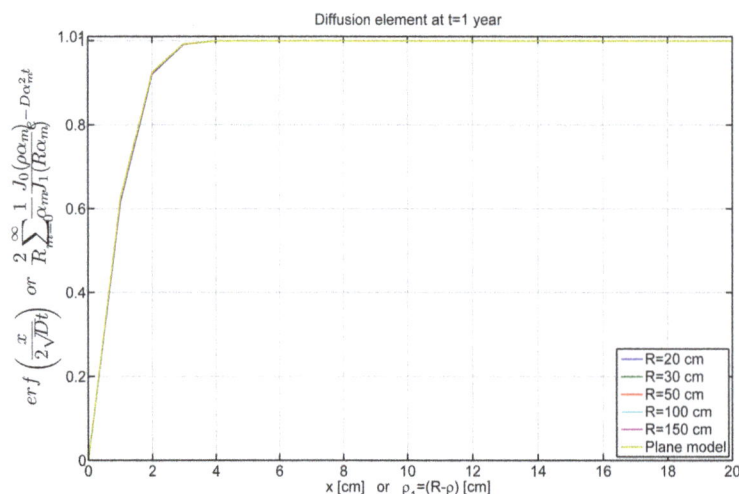

Fig. 3 'Diffusion elements' of the slab model and the circular model. The values of the diffusion element of the circular model is estimated for columns characterized by radii in the range (20–150 cm). The diffusion elements are evaluated at time 1 year from the first exposure of the concrete matrix to chloride ions and for different distance (x and ρ_1) in the concrete matrix from the exposed external surface.

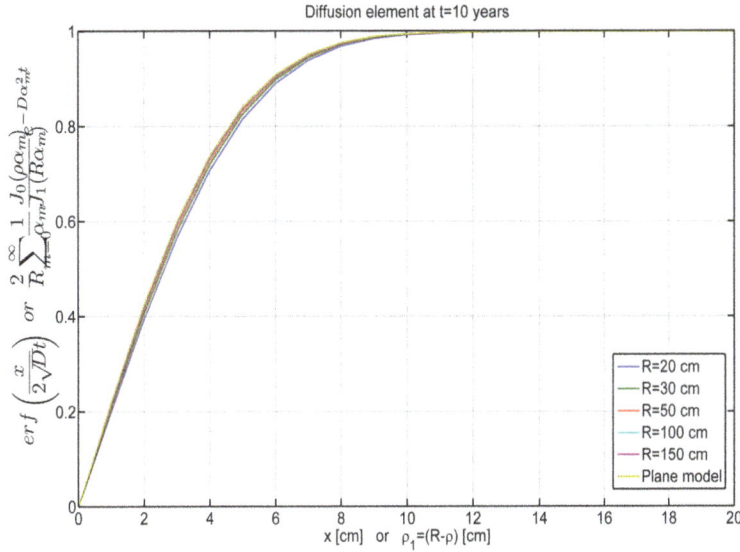

Diffusion element at t=10 years

$erf\left(\frac{x}{2\sqrt{Dt}}\right)$ or $\frac{2}{R}\sum_{m=1}^{\infty}\frac{1}{\alpha_m}\frac{J_0(\rho\alpha_m)}{J_1(R\alpha_m)}e^{-D\alpha_m^2 t}$

x [cm] or ρ_1=(R-ρ) [cm]

- R=20 cm
- R=30 cm
- R=50 cm
- R=100 cm
- R=150 cm
- Plane model

Fig. 4 'Diffusion elements' of the slab model and the circular model. The values of the diffusion element of the circular model is estimated for columns characterized by radii in the range (20–150 cm). The diffusion elements are evaluated at time 10 year from the first exposure of the concrete matrix to chloride ions and for different distance (x and ρ_1) in the concrete matrix from the exposed external surface.

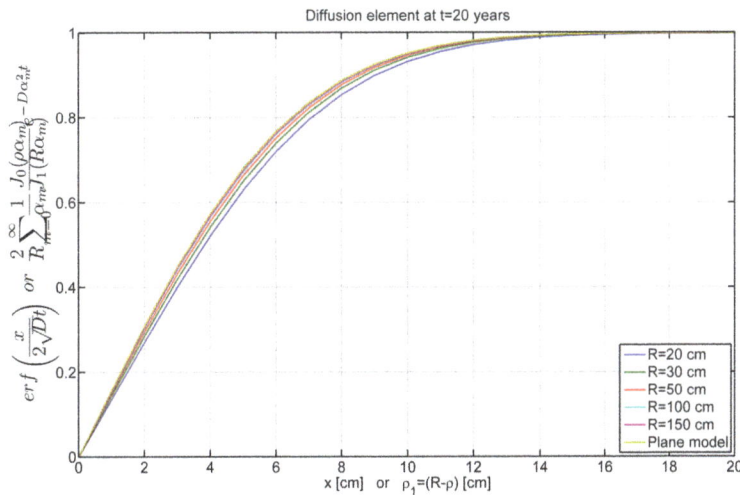

Diffusion element at t=20 years

$erf\left(\frac{x}{2\sqrt{Dt}}\right)$ or $\frac{2}{R}\sum_{m=1}^{\infty}\frac{1}{\alpha_m}\frac{J_0(\rho\alpha_m)}{J_1(R\alpha_m)}e^{-D\alpha_m^2 t}$

x [cm] or ρ_1=(R-ρ) [cm]

- R=20 cm
- R=30 cm
- R=50 cm
- R=100 cm
- R=150 cm
- Plane model

Fig. 5 'Diffusion elements' of the slab model and the circular model. The values of the diffusion element of the circular model is estimated for columns characterized by radii in the range (20–150 cm). The diffusion elements are evaluated at time 20 year from the first exposure of the concrete matrix to chloride ions and for different distance (x and ρ_1) in the concrete matrix from the exposed external surface.

$$\left(\frac{\partial C}{\partial t}\right) = D\left(\frac{\partial^2 C}{\partial x^2} + \frac{\partial^2 C}{\partial y^2}\right) \tag{3}$$

The chloride diffusion in RC elements with circular cross-section is described by the Fick's Second Law expressed in cylindrical coordinates

$$\left(\frac{\partial C}{\partial t}\right) = D\left(\frac{1}{\rho}\frac{\partial}{\partial \rho}\left(\rho\frac{\partial C}{\partial \rho}\right) + \frac{1}{\rho^2}\frac{\partial^2 C}{\partial \theta^2} + \frac{\partial^2 C}{\partial z^2}\right) \tag{4}$$

The RC cylindrical elements are axial-symmetrical, so in them the chloride diffusion occurs along the radial direction from the external surface exposed to the chloride concentration C_0 to the inside of the concrete matrix.

Therefore, the Fick's Second Law expressed with cylindrical coordinates (4) becomes (Crank 1975)

$$\frac{\partial C}{\partial t} = D\left(\frac{1}{\rho}\frac{\partial C}{\partial \rho} + \frac{\partial^2 C}{\partial \rho^2}\right) \tag{5}$$

where the chloride concentration depends only on the time and one spatial variable: the radial coordinate ρ (Fig. 2).

The boundary conditions in the cylindrical model are

$$C(R,t) = C_0 \qquad t \in [0, +\infty] \tag{6}$$

$$\frac{\partial(C(0,t))}{\partial \rho} = 0 \qquad t \in [0, +\infty] \tag{7}$$

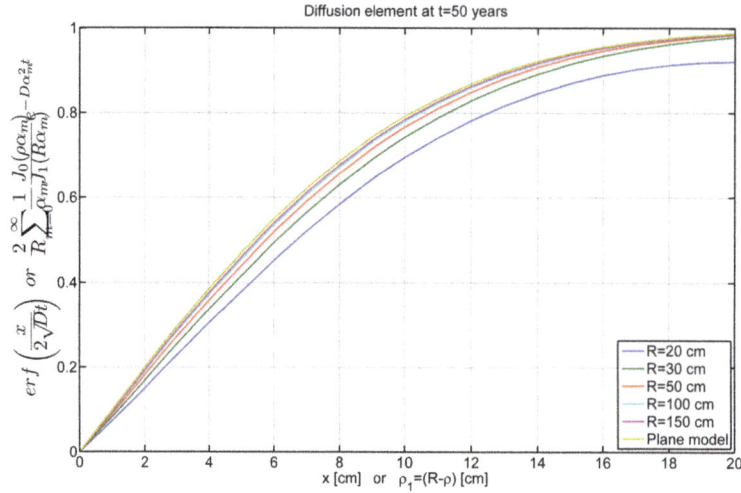

Fig. 6 Diffusion elements of the slab model and the circular model. The values of the diffusion element of the circular model is estimated for columns characterized by radii in the range (20–150 cm). The diffusion elements are evaluated at time 50 year from the first exposure of the concrete matrix to chloride ions and for different distance (x and ρ_1) in the concrete matrix from the exposed external surface.

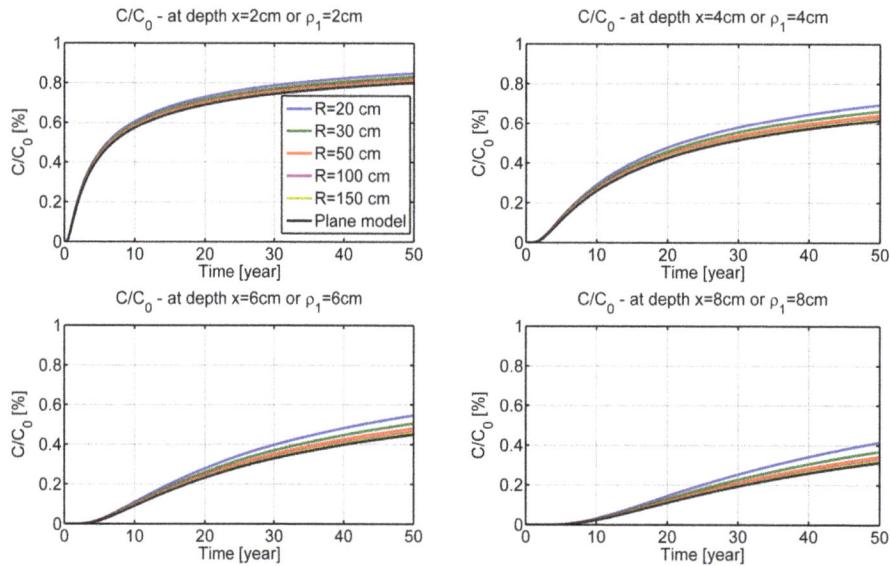

Fig. 7 Ratio of the chloride concentration at depth x or ρ_1 from the external exposed surface C and the chloride concentration on the external surface C_0 as function of time of exposure to chloride t. Value extimated for slab model and circula model characterized by radii in the range (20–150 cm).

Table 1 Ratio of the chloride concentration in a circular RC cross-section and the chloride concentration of the plane RC element C/C_{slab} estimated at depth $\rho_1 = 2$ cm from the external surface.

C/C_{slab}	$t = 1$ year	$t = 5$ years	$t = 10$ years	$t = 20$ years	$t = 50$ years
$R = 20$ cm	1,054	1,055	1,056	1,057	1,060
$R = 30$ cm	1,035	1,036	1,036	1,036	1,037
$R = 50$ cm	1,021	1,021	1,021	1,021	1,021
$R = 100$ cm	1,010	1,010	1,010	1,010	1,010
$R = 150$ cm	1,007	1,007	1,007	1,007	1,007

Table 2 Ratio of the chloride concentration in a circular RC cross-section and the chloride concentration of the plane RC element C/C_{slab} estimated at depth $\rho_1 = 4$ cm from the external surface.

C/C_{slab}	$t = 1$ year	$t = 5$ years	$t = 10$ years	$t = 20$ years	$t = 50$ years
$R = 20$ cm	1,119	1,120	1,121	1,124	1,130
$R = 30$ cm	1,074	1,075	1,075	1,076	1,078
$R = 50$ cm	1,043	1,043	1,043	1,043	1,044
$R = 100$ cm	1,021	1,021	1,021	1,021	1,021
$R = 150$ cm	1,014	1,014	1,014	1,014	1,014

Table 3 Ratio of the chloride concentration in a circular RC cross-section and the chloride concentration of the plane RC element C/C_{slab} estimated at depth $\rho_1 = 6$ cm from the external surface.

C/C_{slab}	$t = 1$ year	$t = 5$ years	$t = 10$ years	$t = 20$ years	$t = 50$ years
$R = 20$ cm	1,196	1,198	1,200	1,204	1,216
$R = 30$ cm	1,118	1,119	1,120	1,121	1,124
$R = 50$ cm	1,066	1,066	1,067	1,067	1,068
$R = 100$ cm	1,031	1,031	1,032	1,032	1,032
$R = 150$ cm	1,021	1,021	1,021	1,021	1,021

Table 4 Ratio of the chloride concentration in a circular RC cross-section and the chloride concentration of the plane RC element C/C_{slab} estimated at depth $\rho_1 = 8$ cm from the external surface.

C/C_{slab}	$t = 1$ year	$t = 5$ years	$t = 10$ years	$t = 20$ years	$t = 50$ years
$R = 20$ cm	–	1,295	1,298	1,304	1,323
$R = 30$ cm	–	1,169	1,170	1,172	1,176
$R = 50$ cm	–	1,091	1,092	1,092	1,093
$R = 100$ cm	–	1,043	1,043	1,043	1,043
$R = 150$ cm	–	1,028	1,028	1,028	1,028

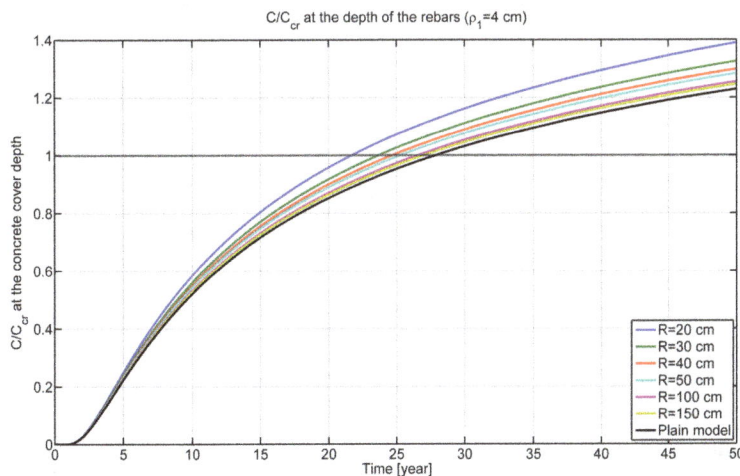

Fig. 8 Ratio of the chloride concentration estimated at the depth of rebars in the concrete matrix (x or $\rho_1 = 4$ cm) C and the chloride threshold concentration for the beginning of the depassivation C_{Cr} as function of the exposure time t. The *horizontal black line* indicate the reaching of the threshold chloride, so the beginning of the depassivation of the rebars. (Color figure online).

Table 5 Time to corrosion initiation of circular cross-section or plane element.

Time [year]	$R = 20$ cm	$R = 30$ cm	$R = 40$ cm	$R = 50$ cm	$R = 100$ cm	$R = 150$ cm	Slab
$w/c = 0.50$	21.8	23.7	24.7	25.4	26.6	27.1	27.9
$w/c = 0.45$	43.5	47.4	49.4	50.7	53.2	54.1	55.8
$w/c = 0.40$	86.9	94.8	98.8	101.3	106.4	108.1	111.5

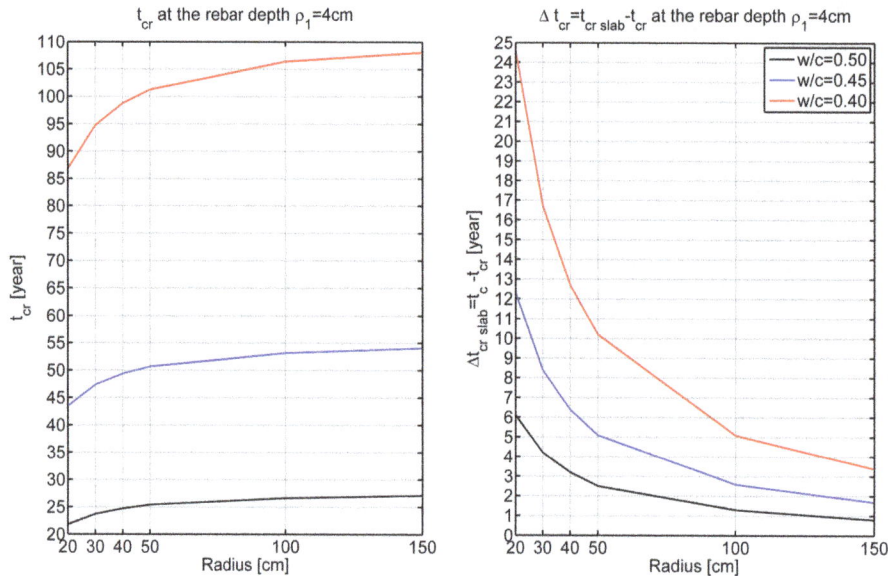

Fig. 9 *Left* time to corrosion initiation t_{Cr} of the rebars rebar estimated for circular model as function of the radius of cross-section R. Calculation proposed for different water-concrete ratios w/c. *Right* difference between the time to corrosion initiation of a cylindrical RC column and the time to corrosion initiation of the plane RC element $\Delta t_{Cr-slab}$ as function of the radius of the circular cross-section R.

Table 6 Difference of the critical time to corrosion initiation $\Delta t_{Cr-slab}$ of the RC plane element and of the RC element with circular cross-section $t_{Cr-circular}$.

Δt (year)	$R = 20$ cm	$R = 30$ cm	$R = 40$ cm	$R = 50$ cm	$R = 100$ cm	$R = 150$ cm
$w/c = 0.50$	6.1	4.2	3.2	2.5	1.3	0.8
$w/c = 0.45$	12.3	8.4	6.4	5.1	2.6	1.7
$w/c = 0.40$	24.6	16.7	12.7	10.2	5.1	3.4

while the initial condition is

$$C(\rho, 0) = f(\rho) \qquad \rho \in \,]0, R[\tag{8}$$

The first boundary condition (Eq. (6)) defines a constant chloride concentration C_0 on the external surface (where $\rho = R$), while the second one (Eq. (7)) is related to the axial-symmetry of the diffusion. The initial condition (Eq. (8)) defines the chloride concentration inside the RC column at the time of the first concrete exposure to the chloride solution.

Equation (5) with the boundary conditions (Eqs. (6) and (7)) and the initial condition (Eq. (8)) can be solved as indicated in the appendix. The solution is

$$C = C_0 \left(1 - \frac{2}{R} \sum_{m=1}^{\infty} \frac{1}{\alpha_m} \frac{J_0(\rho \alpha_m)}{J_1(R \alpha_m)} e^{(-D\alpha_m^2 t)} \right) \tag{9}$$

where J_n is Bessel's first function and α_m depends on it.

3. Comparison Between Cylindrical Model and Slab Model

As afore-said, the depassivation process of the rebars occurs when the chloride concentration in the concrete matrix at the depth of the rebar location exceeds the threshold concentration C_{cr}:

- For the cylindrical columns $C(t_{cr}, \rho_s) = C_{cr}$
- For the element with plane surface $C(t_{cr}, x_s) = C_{cr}$

Clearly, the circular model is equivalent to the slab model for $R \to \infty$.

In the slab model the chloride concentration at depth x and time t depends on the error function:

$$erf \left(\frac{x}{2\sqrt{Dt}} \right) \tag{10}$$

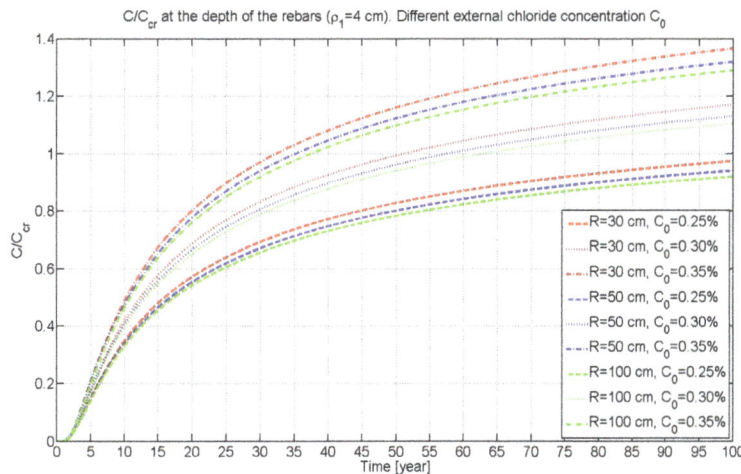

Fig. 10 The ratio of chloride concentration at the depth of steel bars in the concrete matrix C and the threshold chloride concentration C_{cr} as function of the exposure time t. Calculation made for chloride concentrations on the external surface $C_0 = [0.25, 0.30, 0.35\ \%]$ and radii of the circular cross section $R = [30, 50, 100\ cm]$.

In the circular model the chloride concentration at depth ρ at the time t is a function of the sum of products of an exponential term and a ratio of Bessel's first functions:

$$\frac{2}{R}\sum_{m=1}^{\infty}\frac{1}{\alpha_m}\frac{J_0(\rho\alpha_m)}{J_1(R\alpha_m)}e^{(-D\alpha_m^2 t)} \tag{11}$$

To be compared to Eq. (2), Eq. (9) should be expressed in term of $\rho_1 = R - \rho$, i.e. depth from the external surface, instead of ρ, i.e. distance from the centre of the cross section. Hereafter both Eqs. (10) and (11) are called "diffusion element". In the first part of the study the diffusion coefficient used is $D = 0.63\ cm^2/year$: it is a characteristic value for the young OPC Type I with cement-ratio $w/c = 0.5$ and compressive strength of 30–40 MPa.

Figures 3, 4, 5 and 6 show the values of the diffusion element estimated for both the slab model and the circular model with different radii as function of depth in the concrete element from the external surface. In these figures the value of the diffusion element is drawn for different time of exposure to chloride solution: 1, 10, 20 and 50 years. Comparing Figs. 3, 4, 5 and 6, it is clear that the value of the diffusion element of the circular model is smaller than the value of the diffusion element of the slab model for exposure time of 20 or more years. This difference increases with the decrease of the radius of the circular cross-section and the increase of the depth at which the ions diffusion is assessed.

Figure 7 shows the ratio of the chloride concentration inside the element and the chloride concentration on the exposed external surface as function of the depth at with it is estimated. From this figure it is clear that after around 10 years of exposure to the chloride solution the chloride concentration at a given depth in the concrete matrix from the external surface is higher for circular columns than for plane elements. This concentration difference that is clearly visible already at 10 years of exposure grows with the radius and the increase of depth at which the ratio is evaluated. The

observation deduced from Fig. 7 is confirmed by the data collected in Tables 1, 2, 3 and 4.

As consequence of the results presented in the previous figure, Fig. 8 shows that the depassivation process begins earlier in the cylindrical columns than in the plane elements. Therefore, it is clear that for RC columns with small round cross-sections the life-time is shorter than the life-time of plane RC members and this difference cannot be overlooked.

As said in the introduction section, the time to corrosion initiation t_{cr} the instant at which the chloride concentration C at the depth of the rebars in the concrete matrix exceeds the chloride threshold concentration for the depassivation C_{cr}. In this paper the threshold chloride concentration used for the simulations is $C_{cr} = 0.2$, while the chloride concentration on the external surface is $C_0 = 0.4$. The final aim of this study is to give indications about the correct and judicious use of the standard slab model instead of the circular one for the chloride diffusion model in real RC elements. For this purpose the time to corrosion initiation of both circular columns with different radii and flat RC members is estimated and the values are collected in Table 5. Furthermore, the values of the critical time for the circular model are plotted on the left side of Fig. 9. The difference between the time to corrosion initiation estimated with the slab model and that time estimated with the circular model is shown on the right side of Fig. 9. In this figure the time to corrosion initiation is also estimated for three values of the diffusion coefficient of young concrete D (0.63, 0.32, 0.16 $cm^3/year$) corresponding to three different w/c ratios (0.50, 0.45, 0.40). From Fig. 9 and the data collected in Tables 5 and 6 it is clear the importance of the model used for the estimation of the chloride concentration for columns with radius smaller than 50 cm and a w/c ratio lower than 0.5 in case of young OPC Type I. In fact, the model for the chloride diffusion in circular domain determines shorter critical time than the model for the diffusion in a semi-infinite plane. This difference cannot be overlooked both for the evaluation of the life time of single RC elements with circular cross-section and for the evaluation of the life time of whole RC structures with circular columns.

The chloride concentration on the external surface influences the estimation of the critical time because the velocity of the chloride ions diffusion increases with higher chloride concentration gradient. For low external concentrations the difference of the critical time estimated for circular columns with the different radii is lower, so in this case it is possible to use the diffusion model in a semi-infinite plane instead of the diffusion model in a circular domain. This result is shown in Fig. 10. By the observation of Fig. 10, an important conclusion can be drawn: the critical time is strongly influenced by the column radius for chloride concentration on the external surface higher than 0.25. Therefore, in that case the model used to define the chloride diffusion has a significant influence.

Finally, all the data collected in all the tables and plotted in the figures suggest the following indication about the use of the slab model to estimate the time to corrosion initiation for cylindrical columns. In case of chloride concentration higher than 0.25 on the external surface of cylindrical RC column characterized by radius smaller than 50 cm the model to use necessarily to estimate the critical time to corrosion is the circular one. The error produced by the use of the slab model is large and should not be overlooked in the reliability analysis of existing RC columns. This practical and clear indication was missing in the literature.

4. Conclusions

The estimation of the time to corrosion initiation in aggressive environment is still a matter of study in life time assessment of RC structures. Many different mathematical models of the chloride diffusion in the concrete matrix were proposed in literature. These models are characterized by different complexity and number of parameters to be estimated, but most of them are related to the diffusion in a semi-infinite plane medium. This assumption is correct for RC plane members with only one surface exposed to a solution of chloride ions. It is inadequate for RC elements with different geometrical shapes, such as cylindrical elements. In RC elements with a circular cross-section the chloride penetration is faster than in plane elements, so the slab model used in literature overestimates the real time to corrosion initiation in case of cylindrical RC columns. In this paper an exact analytical model for the chloride penetration in RC elements with circular cross-section is used to evaluate the real time to corrosion initiation of these RC elements. In this study real values of the parameters characterizing the chloride diffusion in saturated condition and real dimensions of RC structural members are used to calculate the time to corrosion initiation of the rebars. The times to corrosion initiation calculated for plane and cylindrical RC members show the importance of the geometry for the correct assessment of the life time of a structure. In particular the results show that the circular model for the diffusion has to be used for columns with radius smaller than

50 cm and for the external chloride concentration higher than 0.25. This practical indication is directed to modify the standard approach conventionally used to estimate the critical time in RC circular columns. For this reason this study is an advance in the technical knowledge.

Appendix

Equation (5) can be solved with the method of the separation of variables:

$$C(\rho, t) = \varphi(\rho)\psi(t) \tag{12}$$

so the PDE (Eq. (12)) becomes

$$\frac{\partial(\varphi(\rho)\psi(t))}{\partial t} = D\left(\frac{1}{\rho}\frac{\partial(\varphi(\rho)\psi(t))}{\partial \rho} + \frac{\partial^2(\varphi(\rho)\psi(t))}{\partial \rho^2}\right) \tag{13}$$

The PDE (Eq. (13)) can be written as

$$\frac{\varphi(\rho)\psi'(t)}{D} = \psi(t)\left(\frac{1}{\rho}\varphi'(\rho) + \varphi''(\rho)\right) \tag{14}$$

Separating the functions $\psi(t)$ and $\varphi(\rho)$ and re-arranging the equation two differential equations equal to the constant $-\lambda^2$ are obtained:

$$\frac{\psi'(t)}{\psi(t)}\frac{1}{D} = \frac{1}{\rho}\frac{\varphi'(\rho)}{\varphi(\rho)} + \frac{\varphi''(\rho)}{\varphi(\rho)} = -\lambda^2 \tag{15}$$

The equation can be written as a system of equations:

$$\begin{cases} \frac{\psi'(t)}{\psi(t)}\frac{1}{D} = -\lambda^2 \\ \frac{1}{\rho}\frac{\varphi'(\rho)}{\varphi(\rho)} + \frac{\varphi''(\rho)}{\varphi(\rho)} = -\lambda^2 \end{cases} \tag{16}$$

The first equation of Eq. (16) is a linear first order differential equation with the time as independent variable, while the second one is an ordinary second order differential equation with radial coordinate as independent variable and it is called Bessel's equation.

Inserting the boundary and initial conditions, Eqs. (6–8), the solution of Eq. (5) is (Saassouh and Lounis 2012)

$$C = C_0\left(1 - \frac{2}{R}\sum_{m=1}^{\infty}\frac{1}{\alpha_m}\frac{J_0(\rho\alpha_m)}{J_1(R\alpha_m)}e^{(-D\alpha_m^2 t)}\right)$$
$$+ \frac{2}{R^2}\sum_{m=1}^{\infty}\frac{1}{\alpha_m}\frac{J_0(\rho\alpha_m)}{J_1^2(R\alpha_m)}e^{(-D\alpha_m^2 t)}\int \rho f(\rho)J_0(\rho\alpha_m)d\rho \tag{17}$$

where α_m are computed from the zeros of Bessel's first function with zero order

$$J_0(\alpha_m R) = 0 \tag{18}$$

and J_n is the Bessel's first function with order n.

When the initial chloride concentration in the circular cross-section is constant $f(\rho) = C_{in}$, Eq. (17) becomes

$$C = (C_0 - C_{in})\left(1 - \frac{2}{R}\sum_{m=1}^{\infty}\frac{1}{\alpha_m}\frac{J_0(\rho\alpha_m)}{J_1(R\alpha_m)}e^{\left(-D\alpha_m^2 t\right)}\right) + C_{in} \tag{19}$$

Finally, in case of concrete matrix with no chloride content at the time of exposure beginning $C_{in} = 0$ the solution of Eq. (19) of the diffusion equation is

$$C = C_0\left(1 - \frac{2}{R}\sum_{m=1}^{\infty}\frac{1}{\alpha_m}\frac{J_0(\rho\alpha_m)}{J_1(R\alpha_m)}e^{\left(-D\alpha_m^2 t\right)}\right) \tag{20}$$

References

Arora, P., Popov, B. N., Haran, B., Ramasubramanian, M., Popova, S., & White, R. E. (1997). Corrosion initiation time of steel reinforcement in a chloride environment—A one dimensional solution. *Corrosion Science, 39*(4), 739–759.

Azad, A. K. (1998). Chloride diffusion in concrete and its impact on corrosion of reinforcement. In *Proceedings of symposium on performance of concrete structures in the Arabian Gulf environment* (pp. 262–273).

Basheer, L., Kropp, J., & Cleland, D. J. (2002). Assessment of the durability of concrete from its permeation properties: A review. *Construction and Building Materials, 15*(2–3), 93–103.

Chatterji, S. (1995). On the applicability of the Fick's second law to chloride ion migration through Portland cement concrete. *Cement and Concrete Research, 25*(2), 299–303.

Collepardi, C. M., Marcialis, A., & Turriziani, R. (1972). Penetration of chloride ions into cement pastes and concrete. *Journal of the American Ceramic Society, 55*(10), 534–535.

Crank, J. (1975). *The mathematics of diffusion* (2d ed.). London, UK: Oxford Press.

Do, J., Song, H., So, S., & Soh, Y. (2005). Comparison of deterministic calculation and fuzzy arithmetic for two prediction model equation of corrosion initiation. *Journal of Asian Architecture and Building Engineering., 4*(2), 447–454.

Frier, C, & Sørensen, J. D. (2007). Stochastic analysis of the multi-dimensional effect of chloride ingress into reinforced concrete. In *10th international conference of applications of statistics and probability in civil engineering* (pp. 135–136), Tokyo, Japan, 31 July–3 August 2007. London, UK: Marcel Dekker.

Izquierdo, D., Alonso, C., Andrade, C., & Castellote, M. (2004). Potenziostatic determination of chloride threshold values for rebar depassivation. *Experimental and Statistical Study, Electrochimica Acta, 49*, 2731–2739.

Kirkpatrick, T. J., Weyers, R. E., Sprinkel, M. M., & Anderson-Cook, C. M. (2002). Impact of specification changes on chloride-induced corrosion service life of bridge decks. *Cement and Concrete Research, 32*(8), 1189–1197.

Lin, S. H. (1990). Chloride diffusion in a porous concrete slab. *Corrosion, 46*, 964–967.

Marano, G. C., Quaranta, G., & Mezzina, M. (2008). Fuzzy Time-dependent reliability analysis of RC beams subject to pitting corrosion. *Journal of Materials in Civil Engineering, 20*(9), 578–587.

Martin-Perez, B., Pantazopoulou, S. J., & Thomas, M. D. A. (2001). Numerical solution of mass transport equations in concrete structures. *Computers & Structures, 79*(13), 1251–1264.

Nogueira, C. G., & Leonel, E. D. (2013). Probabilistic models applied to safety assessment of reinforced concrete structures subjected to chloride ingress. *Engineering Failure Analysis, 31*, 76–89.

Saassouh, B., & Lounis, Z. (2012). Probabilistic modelling of chloride-induced corrosion in concrete structures using first- and second-order reliability methods. *Cement & Concrete Composites, 34*, 1082–1093.

Saetta, A. V., Schrefler, B. A., & Vitaliani, R. V. (1995). 2-D model for carbonation and moisture/heat flow in porous materials. *Cement and Concrete Research, 25*(8), 1703–1712.

Sobhani, J., & Ramezanianpour, A. A. (2011). Service life of the reinforced concrete bridge deck in corrosive environments: A soft computing system. *Applied Soft Computing, 11*, 3333–3346.

Tang, L., & Joost, G. (2007). On the mathematics of time-dependent apparent chloride diffusion coefficient in concrete. *Cement and Concrete Research, 37*, 589–595.

Tuutti, K. (1982). Corrosion of steel in concrete. Swedish Cement and Concrete Research Institute, Stockholm, Sweden. Report No. CBI Research FO 4:82.

Vu, K. A. T., & Stewart, M. G. (2000). Structural reliability of concrete bridges including improved chloride-induced corrosion models. *Structural Safety, 22*(4), 313–333.

Val, D. V., & Trapper, P. A. (2008). Probabilistic evaluation of initiation time of chloride-induced corrosion. *Reliability Engineering and System Safety, 93*(3), 364–372.

9

Effect of Adding Scoria as Cement Replacement on Durability-Related Properties

Aref Mohamad al-Swaidani[1],*, and Samira Dib Aliyan[2]

Abstract: A lot of reinforced concrete (RC) structures in Syria went out of service after a few years of construction. This was mainly due to reinforcement corrosion or chemical attack on concrete. The use of blended cements is growing rapidly in the construction industry due to economical, ecological and technical benefits. Syria is relatively rich in scoria. In the study, mortar/concrete specimens were produced with seven types of cement: one plain Portland cement (control) and six blended cements with replacement levels ranging from 10 to 35 %. Rapid chloride penetration test was carried in accordance with ASTM C 1202 after two curing times of 28 and 90 days. The effect on the resistance of concrete against damage caused by corrosion of the embedded steel has been investigated using an accelerated corrosion test by impressing a constant anodic potential. The variation of current with time and time to failure of RC specimens were determined at 28 and 90 days curing. In addition, effects of aggressive acidic environments on mortars were investigated through 100 days of exposure to 5 % H_2SO_4, 10 % HCl, 5 % HNO_3 and 10 % CH_3COOH solutions. Evaluation of sulfate resistance of mortars was also performed by immersing in 5 % Na_2SO_4 solution for 52 weeks. Test results reveal that the resistance to chloride penetration of concrete improves substantially with the increase of replacement level, and the concretes containing scoria based-blended cements, especially CEM II/B-P, exhibited corrosion initiation periods several times longer than the control mix. Further, an increase in scoria addition improves the acid resistance of mortar, especially in the early days of exposure, whereas after a long period of continuous exposure all specimens show the same behavior against the acid attack. According to results of sulfate resistance, CEM II/B-P can be used instead of SRPC in sulfate-bearing environments.

Keywords: durability, corrosion resistance, blended cement, natural pozzolans, acid attack, scoria, sulfate attack.

1. Introduction

Use of natural pozzolan in production of blended Portland cements makes important effects on physical, chemical, mechanical and durability properties of mortar and concrete depending on its substitution ratio and its fineness (Al-Chaar et al. 2013; Senhadji et al. 2012; Hossain 2009; Ghrici et al. 2006; Cavdar and Yetgin 2007; Turanli et al. 2005; Colak 2003; Rodriguez-Camacho and Uribe-Afif 2002). In addition, since these materials enter the cement production after kiln process, they also provide important economical and ecological benefits (Mehta and Monteiro 2006). According to Mehta and Monteiro (2006), the manufacture of one tonne of Portland cement (PC) clinker consumes energy of about

4GJ, and releases nearly one tonne of CO_2 in the atmosphere. For this reason, a particular attention was recently given to the exploitation of natural pozzolan which is broadly abundant in Syria. More than 30,000 km^2 of the country is covered by Tertiary and Quaternary-age volcanic rocks (GEGMR 2011), among which scoria occupies important volume with estimated reserves of more than 600 million tonnes (GEGMR 2007). The cement produced in the country is almost of CEM I, although an addition of natural pozzolan up to 5 % was frequently used in most local cement plants. Hence, less than 300,000 tonnes of these pozzolans are only exploited annually (the annual production of PC in Syria is about 6 million tonnes) (GOCBM 2011).

The premature deterioration of reinforced concrete (RC) structures due to corrosion is a major problem in Syria, especially, in the marine environment. A lot of structures went out of service after a few years of construction. Although chloride ions in concrete do not directly cause severe damage to the concrete, they contribute to the corrosion of steel bars embedded in concrete. Therefore, study of chloride penetrability of concrete is an important for evaluating reinforcing steel corrosion in RC structures. In addition, chemical attacks on concrete structures subjected to sulfate- and acid-bearing environments have caused serious damage to these concretes. This has prompted the search for

---

[1]Faculty of Architectural Engineering, Arab International University (AIU), Damascus, Syria.

*Corresponding Author; E-mail: aydlswaidani@yahoo.fr; a-swaidani@aiu.edu.sy

[2]Syrian Arab Organization for Standardization and Metrology (SASMO), Damascus, Syria.

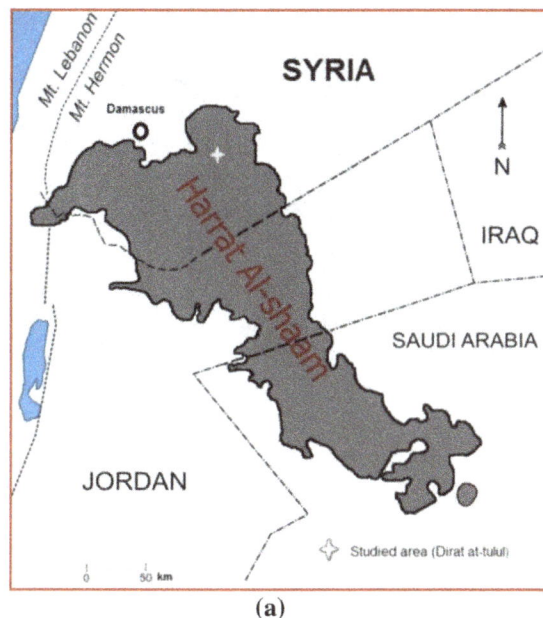

Fig. 1 Map of Harrat Al-Shaam, photo of the studied quarry & the used scoria aggregate. **a** Map of Harrat Al-Shaam and the studied area. **b** The studied scoria quarry, some volcanic scoria cones are shown behind. **c** The studied scoria aggregate.

economical methods of extending the service life of structure. One of these methods was the use of blended cements, which is growing rapidly in the construction industry. Although there are numerous studies on using natural pozzolan as cement replacement, no detailed research was conducted in the past to investigate the potential use of scoria in production of blended cements in Syria.

This study is part of the first detailed research in Syria to investigate the potential utilization of scoria as cement replacement in producing Portland-pozzolan cements, and its effects on the performance of mortar and concrete. The study is of particular importance not only for the country but also for other areas of similar geology, e.g. Harrat Al-Shaam, a volcanic field covering a total area of some 40,000 km^2, third of which is located in the country. The rest is covering parts from Jordan and Saudi Arabia.

The objective of this paper is to report a part of this ongoing research on the effect of different amount of scoria when adding as cement replacement on some durability-related properties. Penetrability of chloride ions, corrosion of reinforcing steel, acid and sulfate attacks have been particularly investigated. Some chemical, physical and mechanical properties have also been reported.

2. Experimental Procedure

2.1 Scoria

The scoria used in the experiments was collected from a Tal Dakwa' quarry, at 70 km southeast of Damascus as shown in Fig. 1. The petrographical examination showed the scoria is consisted of amorphous glassy ground mass, vesicles, plagioclase and olivine with the following percentages (based on an optical estimate): 20, 35, 20 and 25 %, respectively. The scoria is dark black to blackish-grey in color with some red-brown spots, mostly due to its iron oxides content. Figure 2 shows thin sections of the used scoria. The chemical analysis of scoria used in the study is summarized in Table 1. This analysis was carried out by means of wet chemical analysis specified in EN 196-2 (1989).

2.2 Cement Samples

Three types of binder were prepared, one plain PC CEM I (control), and two blended cements: CEM II/A-P and CEM II/B-P (EN 197-1), each of them with 3 replacement levels of scoria: (10, 15 and 20 %) and (25, 30, 35 %), respectively. 5 % of gypsum was added to all these cements. The clinker used for producing the binders was obtained from

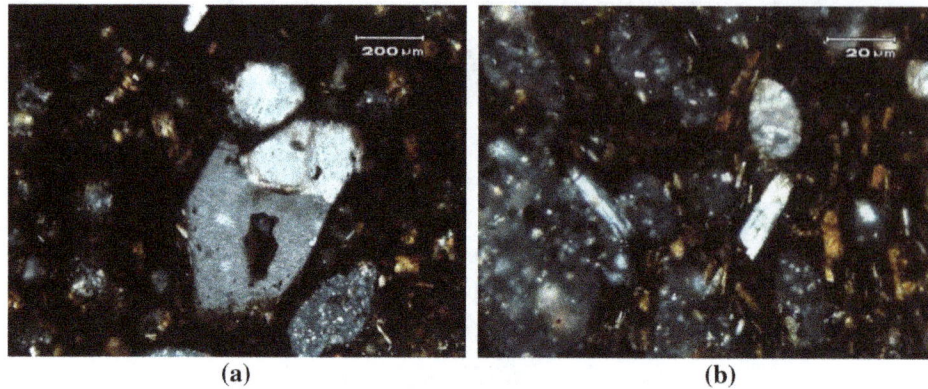

Fig. 2 Thin sections of the scoria. **a** Microphenocryst of olivine in volcanic glass matrix with vesicles, some of which are *filled with white minerals*. **b** Microphenocrysts of elongated plagioclase in volcanic glass matrix with vesicles, some of which are *filled with white minerals*.

Table 1 Chemical composition of the used materials.

Chemical composition (by mass, %)	Materials				
	Scoria[a]	Clinker	Gypsum	Dolomite aggregate	Natural sand
SiO_2	46.52	21.30	0.90	0.42	93.39
Al_2O_3	13.00	4.84	0.07	0.38	0.57
Fe_2O_3	11.40	3.99	0.10	0.10	0.24
CaO	10.10	65.05	32.23	31.40	1.70
CaO_f	–	2.1	–	–	–
MgO	9.11	1.81	0.20	20.46	0.20
SO_3	0.27	0.25	45.29	0.18	1.15
Loss on ignition	2.58	–	21.15	46.48	2.52
Na_2O	2.14	0.60	–	0.06	0.06
K_2O	0.77	0.28	–	0.30	0.05
Cl^-	<0.1	0.05	–	0.021	0.017
Pozzolan activity index [ASTM C 618]	79 (at 7 days) 85 (at 28 days)				

[a] $SiO_{2(reactive)}$ content in the studied scoria = 42.22 % (determined in accordance with EN 196-2).

Adra Cement Plant, Damascus, Syria. Chemical analysis of clinker and gypsum is shown in Table 1. All binders were interground by a laboratory grinding mill to a Blaine fineness 3200 ± 50 cm^2/g. All replacements were made by mass of cement. Table 2 shows the chemical, physical and mechanical properties of the binders produced. CEM I (the control sample) was designated as C1, whereas blended cements were designated according to the replacement level. For instance, C2/10 % and C7/35 % refer to the blended cements containing 10 and 35 % of scoria, respectively. In acid and sulfate attack tests, sulfate-resisting PC was employed for comparison. It was designated as C8.

2.3 Mortar Mixtures

Eight mortar mixtures were prepared using these binders and sand meeting the requirements of ASTM C 778. In all mixtures, binder: sand ratio was kept constant as 1:2.75 by weight. Mixtures containing CEM I and SRPC were prepared with a w/b ratio of 0.485. Mixtures containing scoria-based blended cements were prepared by changing the w/b ratio in order to obtain a flow within ±5 of that of the CEM I mortar. For acid attack tests, a set of three cubes from each mixture was cast for each of acid solution and curing age. For sulfate attack test, prismatic mortar bars and cubes were cast from each mixture in accordance with ASTM C1012 (2004).

2.4 Concrete Mixes

Seven concrete mixes were prepared using a grading of aggregate mixtures kept constant for all concretes. Aggregates used in the study were crushed dolomite with river bed natural sand added. Both aggregates were obtained from

Table 2 Chemical, physical and mechanical properties of plain and blended cements.

Chemical composition (%)	C1/CEMI	C2/10 %	C3/15 %	C4/20 %	C5/25 %	C6/30 %	C7/35 %	C8/SRPC
Chemical properties of plain and blended cements								
SiO2	20.69	21.59	22.35	23.25	24.00	24.33	24.61	20.72
Al2O3	5.09	5.20	5.68	5.73	6.55	6.80	7.39	4.33
Fe2O3	4.23	4.75	4.79	5.15	5.43	5.47	6.31	5.79
CaO	60.62	58.21	55.18	53.05	50.30	48.00	44.84	61.69
MgO	2.46	2.66	3.23	3.39	3.87	4.11	4.63	1.21
SO3	2.26	2.31	2.20	2.20	2.30	2.26	2.55	2.13
Loss on ignition	1.41	1.40	1.43	1.37	1.47	1.48	1.60	2.90
Na2O	0.60	0.71	0.83	0.94	1.07	1.16	1.31	0.21
K2O	0.35	0.39	0.43	0.46	0.50	0.53	0.57	0.19
Cl⁻	0.023	0.021	0.022	0.019	0.018	0.019	0.019	0.022
Insoluble residue	1.03	1.58	2.09	2.51	3.48	4.08	5.33	0.36
Main compounds of clinker used in cement specimens (Based on Bogue composition)								
C3S	53.36	50.55	47.74	44.94	42.13	39.32	36.51	50.16
C2S	17.76	16.82	15.89	14.95	14.02	13.08	12.15	21.58
C3A	5.78	5.47	5.17	4.86	4.56	4.26	3.95	1.69
C4AF	11.53	10.93	10.32	9.71	9.11	8.50	7.89	17.60
C3S/C2S	3.0	3.0	3.0	3.0	3.0	3.0	3.0	2.32
Physical properties of plain and blended cements								
Specific gravity	3.13	3.09	3.05	3.02	2.99	2.98	2.96	
Initial setting (min)	151	153	153	153	152	153	158	
Final setting (min)	178	179	180	180	179	181	188	
Water demand (%)	25.1	25.2	25.2	25.4	25.4	25.4	25.5	
Soundness (mm)	0.6	0.7	0.8	0.8	0.9	1.1	0.9	
Residue on 45 µm sieve (%)	13.6	14.3	14.8	15.2	16.1	17.0	17.9	
Residue on 90 µm sieve (%)	6.4	6.2	6.4	6.5	6.7	6.9	6.8	
Mechanical properties of plain and blended cement mortars								
Strength of mortars at 28 days curing (MPa)	45.6	44.2	42.3	40.6	37.1	33.7	30.6	

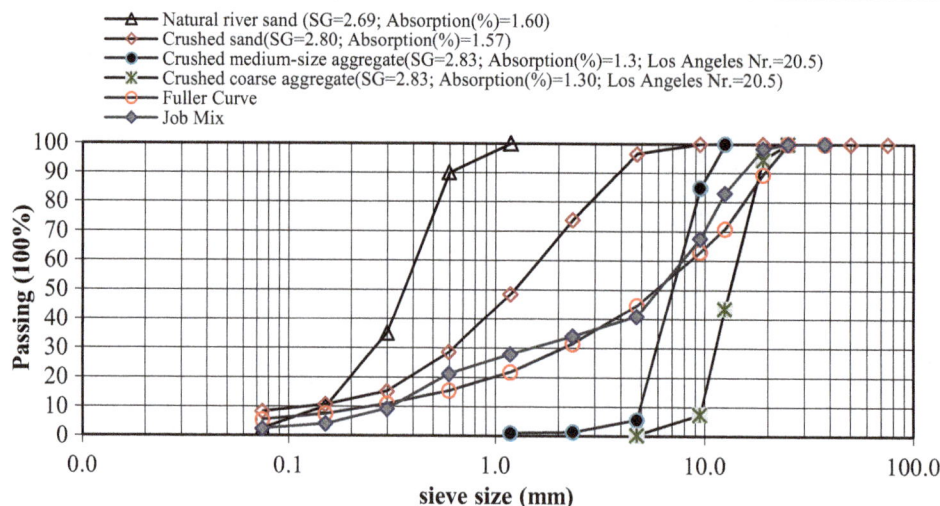

Fig. 3 Grading curves of aggregates used in the concrete mixes with some physical properties.

local sources. Chemical composition and grading with some physical properties of the aggregates are illustrated in Table 1 and Fig. 3, respectively. Their quantities in 1 m³ concrete mix based on the oven-dry condition are as follows: 565.5 kg of coarse aggregate, 565.5 kg of medium-size aggregate, 447.5 kg of coarse sand and 286.5 kg of natural sand. All concrete mixes were designed to have a water-binder ratio of 0.6 and a slump of 150 ± 20 mm. This w/b ratio was kept constant in order to directly observe the effects of adding scoria as cement partial replacement on properties of concrete.

Six concrete cubes (150 mm) were cast for each of binders and curing ages for the determination of compressive strength. Three plain concrete cylinders (100 mm × 200 mm) for each of binder type and curing age were cast, for evaluating the chloride penetrability. The RC specimen for the accelerated corrosion tests was 100 mm × 200 mm concrete cylinder in which 12 mm diameter steel bar was centrally embedded. The steel bar was embedded into the concrete cylinder such that its end was at least 45 mm from the bottom of the cylinder, and it was coated with epoxy at the exit from the concrete cylinder in order to eliminate crevice corrosion. Three specimens for each binder type and curing age were cast.

2.5 Compressive Strength Test of Concrete

The compressive strength development was determined on 150 mm cubic concrete specimens, in accordance with ISO 4012 (1978), after 2, 7, 28, 56 and 90 days curing.

2.6 Rapid Chloride Penetrability Test

The test was conducted in accordance with ASTM C 1202 (2001). Three slices of 100 mm in diameter and 50 mm in thickness were cut from the middle portion of each concrete cylinder specimens (100 mm × 200 mm). The set-up of RCPT is illustrated in Fig. 4. The total charge passed through the sample, in coulombs, is determined by calculating the area under the current–time plot during the 6-h test period. It is generally agreed that for low-permeability concretes, the value of the charge, in coulombs, passed through the specimens should not exceed 2000. Three cylinder specimens of each concrete mix were tested after 28 and 90 days curing.

2.7 Accelerated Corrosion Test

A rapid corrosion test was used to compare the corrosion performance of concretes containing binders produced. Similar techniques with little differences were reported by other researchers (Horsakulthai et al. 2011; Parande et al. 2008; Ha et al. 2007; Saraswathy and Song 2007; Guneyisi et al. 2005; Rossignolo and Agesini 2004; Shaker et al. 1997; Khedr and Idriss 1995; Al-Tayyib and Al-Zahrani 1990). In the study, RC specimens were immersed in a 15 % NaCl solution leveling the half of the concrete cylinder and the steel bar (working electrode) was connected to the positive terminal of a DC power source while the negative terminal was connected to a steel plate (counter electrode) placed near the concrete specimen in the solution. The corrosion process was initiated by impressing a relatively high anodic potential of 12 V to accelerate the corrosion process.

Figure 5 shows a schematic representation of the experimental setup for the accelerated corrosion test. The specimens were monitored periodically to see how long it takes for corrosion cracks to appear on the specimen surface. The current readings with time were recorded at 4 h-intervals. Three specimens from each concrete mix were tested after 28 and 90 days curing.

2.8 Acid Attack Test

The relative acid resistance was determined in accordance with ASTM C267 (2001). The aggressive acid environmental conditions were simulated using the following acids: 5 % sulfuric acid (H_2SO_4), 10 % hydrochloric acid (HCl), 5 % nitric acid (HNO_3) and 10 % acetic acid (CH_3COOH) of pH equal to 0.5, 0.25, 0.45 and 2.2, respectively. The 28 and 90 days cured mortar cubes were immersed in these

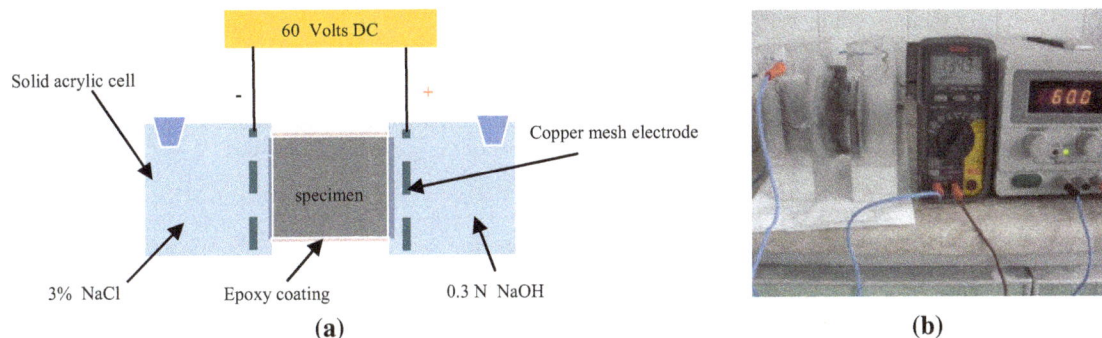

Fig. 4 Experimental setup of rapid chloride penetration test. **a** Schematic representation of experimental setup of rapid chloride penetration test. **b** View of experimental setup (one of current readings for C2/10 % specimen).

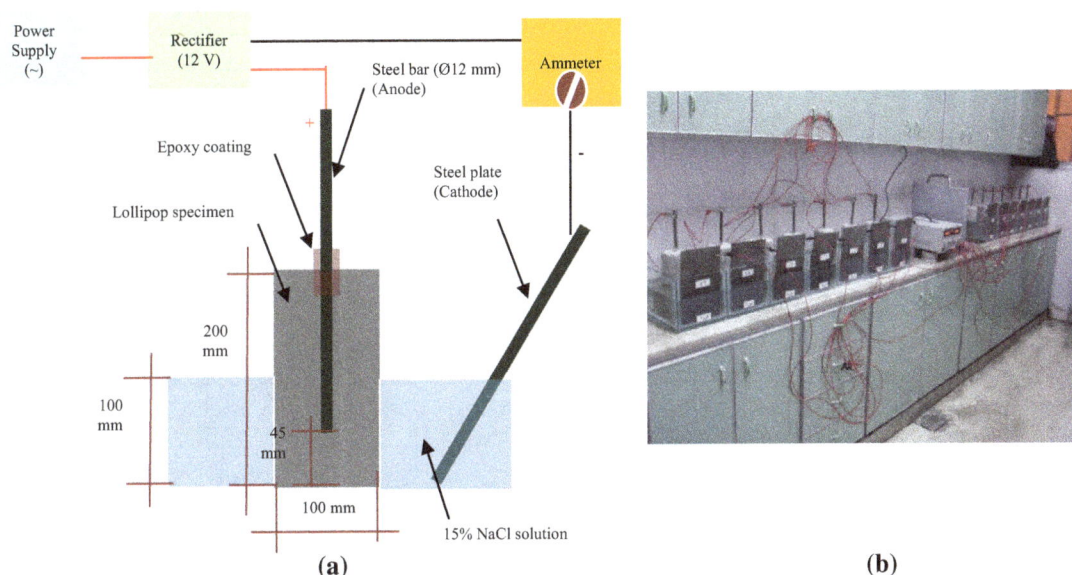

Fig. 5 Schematic representation and view of experimental setup for the accelerated corrosion test. **a** Setup for the accelerated corrosion test. **b**. View of the experimental setup.

aggressive acid environments for 100 days. The Plexiglas containers with immersed mortar specimens were kept covered throughout the testing period to minimize the evaporation. At 2, 7, 14, 28, 56, 84 and 100 days of exposure, the mortar specimens were cleaned with distilled water, then the acid resistance was evaluated through measurement of the weight loss of the specimens determined as follows: Weight loss $(\%) = [(W_1 - W_t)/W_1] \times 100$, where W_1 is the weight (grams) of the specimens before immersion and W_t is the weight (grams) of cleaned specimens after t day immersion. The average weight losses for each mortar cubes have been reported.

2.9 Sulfate Attack Test

The evaluation of sulfate attack resistance of mortars was performed in accordance with ASTM C1012 (2004). Length measurements of the prepared prismatic specimens were performed at 1, 2, 3, 4, 8, 13, 15, 17, 26, 38, 52 weeks after immersing the specimens into the sulfate (5 % Na_2SO_4) solution. The solution was renewed four times throughout the test; at 8, 17, 26 and 38 weeks.

3. Discussion of Results

3.1 Properties of Scoria and Blended Cements

As seen from Table 1, scoria is considered as suitable material for use as pozzolan. It satisfied the standards requirements for such a material by having a combined SiO_2, Al_2O_3 and Fe_2O_3 of more than 70 %, a SO_3 content of less than 4 % and a loss on ignition of less than 10 % (ASTM 2001). $SiO_{2reactive}$ content is more than 25 %, as well (EN 2004). In addition, it has a strength activity index with PC higher than the values specified in ASTM C618 (2001). The chemical and physical properties of scoria-based blended cements are also in conformity with the standards requirements (ASTM 2001). Their contents of MgO and SO_3 are less than 6 and 4 %, respectively. The loss on ignition is also less than 5 % as specified in ASTM C595 (2001). Setting time and soundness values meet the limits specified in ASTM C595 (2001), for all binder types. The compressive strengths of blended cement mortars are lower than that of the plain PC at 28 days curing. This reduction is mainly due to the slowness of pozzolanic reaction between pozzolan and

Table 3 Compressive strength of concrete cubes.

Sample	Compressive strength (MPa)-normalized				
	2 days (%)	7 days (%)	28 days (%)	56 days (%)	90 days (%)
C1/CEM I (control)	13.6–100	24.9–100	35.0–100	40.6–100	42.3–100
C2/10 %	12.8–94	24.4–98	33.6–96	39.8–98	41.6–98
C3/15 %	11.8–87	21.0–84	30.1–86	37.0–91	39.3–93
C4/20 %	11.0–81	20.1–81	29.5–84	35.1–86	38.3–91
C5/25 %	9.9–73	17.6–71	26.5–76	31.8–78	36.5–86
C6/30 %	8.9–65	16.5–66	24.6–70	29.6–73	33.1–78
C7/35 %	8.5–63	16.0–64	23.4–67	28.0–69	30.9–73

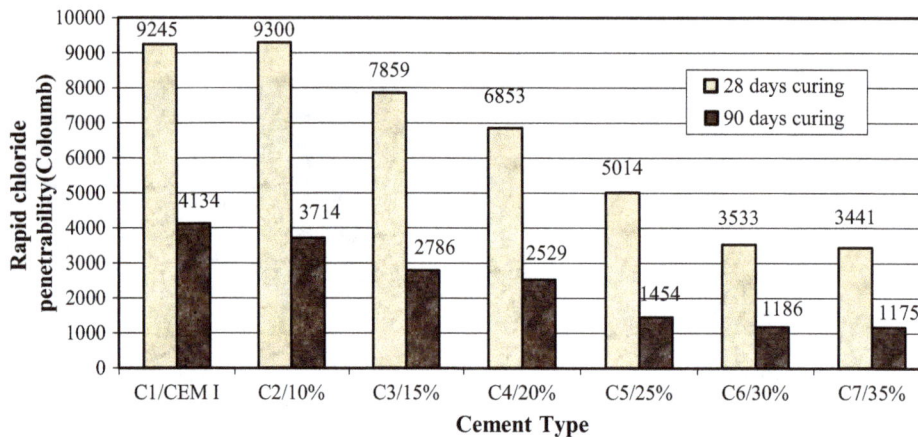

Fig. 6 Influence of scoria content on chloride penetrability of concrete.

CH released during cement hydration (Mindess et al. 2003), and could be due the coarser blended cement particles, as shown in Table 2.

3.2 Compressive Strength of Concretes

Results of compressive strength test are summarized in Table 3. As expected, the compressive strength of the concrete increases with curing time with a high rate of strength gains at early ages which gradually decrease at longer ages. Plain cement concrete specimens have higher compressive strengths at any age when compared with blended cement concretes. The compressive strength at 7 days decreases from 24.9 to 16.0 MPa when CEM I and CEM II/B-P with 35 % of scoria were used, respectively. The relative compressive strength after 2 days curing with 35 % of scoria is only 63 % of that of CEM I specimens. This could be explained by the slowness of the pozzolanic reaction between the glassy phase in scoria and the CH released during cement hydration. However, due to the continuation of this reaction and the formation of a secondary C–S–H, a greater degree of hydration is achieved resulting in strengths after 90 days curing which are comparable to those of CEM I specimens.

3.3 Rapid Chloride Penetrability

Results of rapid chloride penetrability test are shown in Fig. 6. The resistance to chloride penetration was greatly

increased with the use of scoria-based blended cement concretes. The total charge passed is substantially reduced with increase in scoria content and curing time. The total charge passed through concrete mix containing C5/25 % binder type after 28 days curing was about half of the control mix, and decreased to about one third after 90 days curing. None of concretes has a total charge passed less than 2000 coulombs after 28 days curing. This expected result may be due to the high w/b ratio. However, the concretes containing CEM II/B-P with scoria contents of 25, 30 and 35 %, showed the best performance among all the specimens. According to ASTM C1202 (2001), these concretes can be considered as low chloride permeable after 90 days curing. The improvement in resistance to chloride penetration may be related to the refined pore structure of these concretes and their reduced electrical conductivity (Talbot et al. 1995). This which was confirmed by many researchers (Ghrici et al. 2006; Rukzon and Chindaprasirt 2009; Chindaprasirt et al. 2007; Gastaldini et al. 2007; Pourkhorshidi et al. 2010; Ramezanianpour et al. 2010; Hossain et al. 2008; Rukzon and Chindaprasirt 2009; Chindaprasirt et al. 2007; Gastaldini et al. 2007), is due to the secondary pozzolanic reaction which contributes to make the microstructure of concrete denser. If a higher volume of scoria is added, much lower penetrability can be achieved.

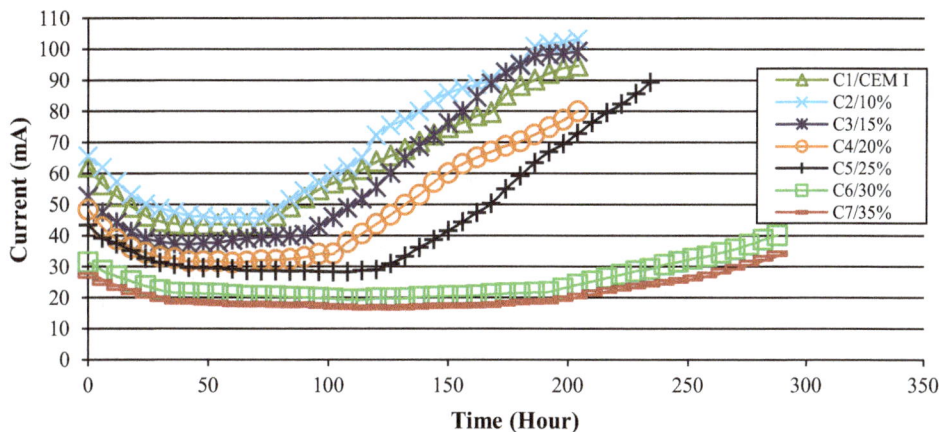

Fig. 7 Typical curve of corrosion current versus time of concrete specimens tested after 28 days curing.

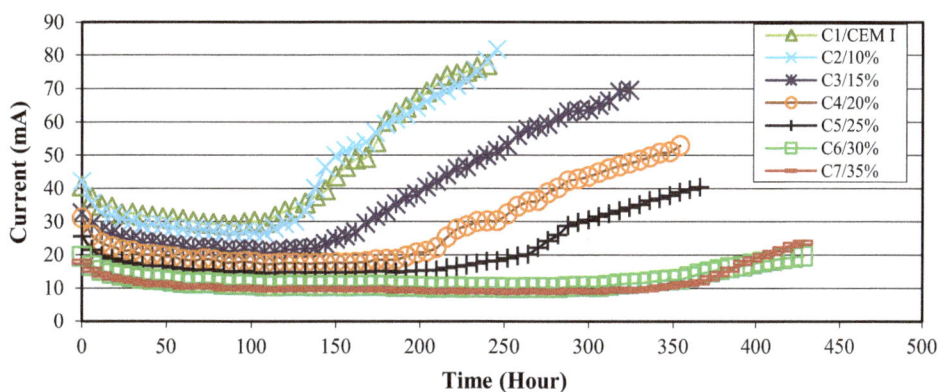

Fig. 8 Typical curve of corrosion current versus time of concrete specimens tested after 90 days curing.

3.4 Corrosion Resistance

Typical curves of corrosion current versus time for the concrete specimens made with CEM I, CEM II/A-P and CEM II/B-P binder types are illustrated in Figs. 7 and 8, respectively. As seen from Figs. 7 and 8, current–time curve initially descended till a time value after which a steady low rate of increase in current was observed, and after a specific time period a rapid increase in current was detected until failure. The decreasing tendency of current at the very early time could be explained by the filling of the pores with salt and other deposits in the salt water (Reddy et al. 2011). Almost a similar variation of the corrosion current with time has also been observed by other researchers (Horsakulthai et al. 2011; Guneyisi et al. 2005). The first visual evidence of corrosion was the appearance of brown stains on the surface of the specimens. Cracking was observed shortly thereafter, and it was associated with a sudden rise in the current. Figure 9 presents the average corrosion times required to crack the specimens made with CEM I, CEM II/A-P and CEM II/B-P. Time to cracking in CEM I concrete specimens was in the range of 70–106 h (3–5 days), whereas that in CEM II/B-P was in the range of 120–370 h (5–16 days), depending on the replacement level and age at testing. The best corrosion resistance was obtained from blended cements with 35 % additive ratio. The test lasted for 194 and 370 h in this specimen after 28 and 90 days curing times,

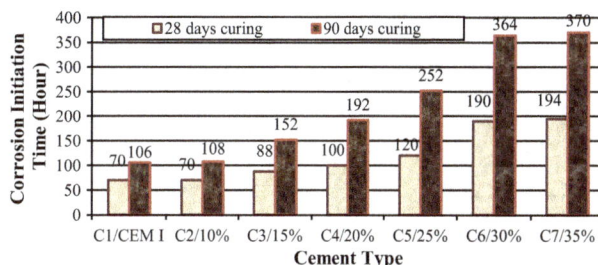

Fig. 9 Average corrosion times of specimens at 28 and 90 days curing.

respectively. Results also demonstrated that the increase in time to cracking with test age (from 28 to 90 days) was greater in almost all of the cases for the scoria-based cement concrete specimens when compared with the plain concrete specimens. This delay in corrosion time when using blended cements may be related to the pozzolanic reaction of scoria which contributes to fill the voids and pores in concrete with an additional C–S–H gel. This leads to decrease of pore size and to a smaller effective diffusivity for chloride (Guneyisi et al. 2005; Hossain 2003). This can improve the long-term corrosion resistance of RC structures, and make concrete denser and less permeable (Guneyisi et al. 2005; Hossain 2003). Also, It was noted from Figs. 7 and 8 the corrosion resistance of most of blended cement concrete specimens

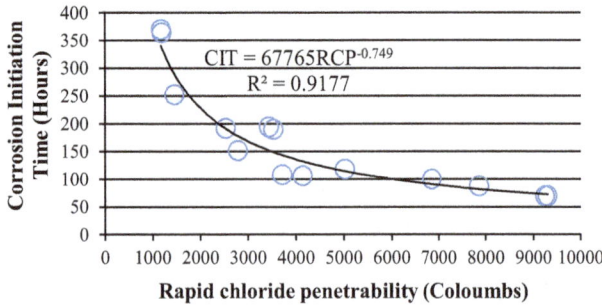

Fig. 10 Variation of corrosion initiation time with rapid chloride penetrability.

increased significantly with age while that of the plain cement concrete had a slightly increase which has also been indicated by other researchers (Guneyisi et al. 2005).

3.5 Corrosion Initiation Time Versus Chloride Penetrability of Concrete

It was observed from Fig. 10 that the corrosion initiation time and chloride penetrability of concrete are closely related to each other. The analysis results showed that there is a good correlation between corrosion initiation time and total charge passed through concrete specimens, with a regression coefficient ($R^2 \approx 0.92$). According to Montgomery and Peck (1982) a regression coefficient, R^2, of more than 0.85 indicates an excellent correlation between the fitted parameters. Increasing the total charge passed through concrete specimens reduces the time to initiate corrosion.

3.6 Acid Attack

The resistance to acid solutions was measured by means of the weight loss of the mortar cubes. Figures 11, 12, 13 and

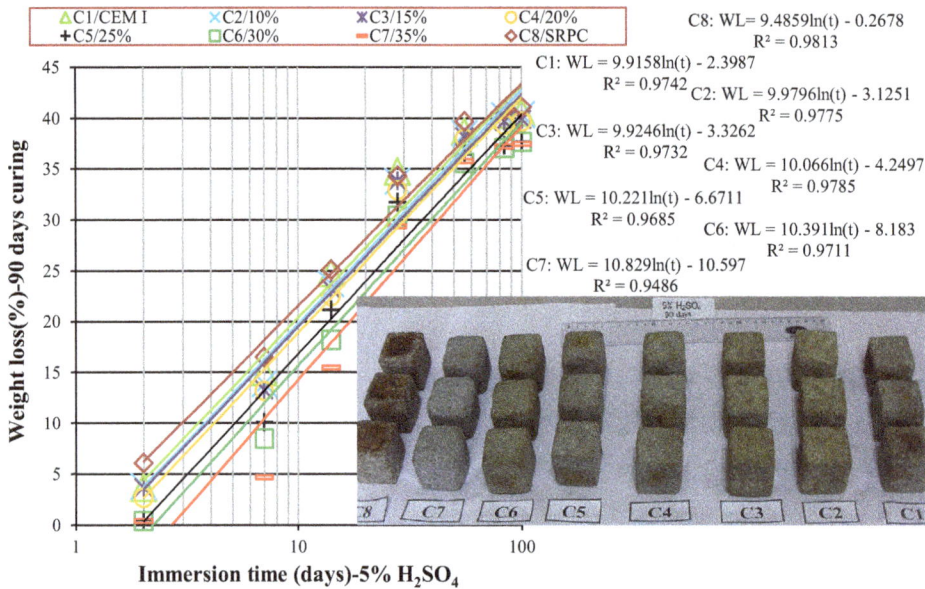

Fig. 11 Weight losses over time of 90 days cured mortars immersed in 5 % H_2SO_4 solution.

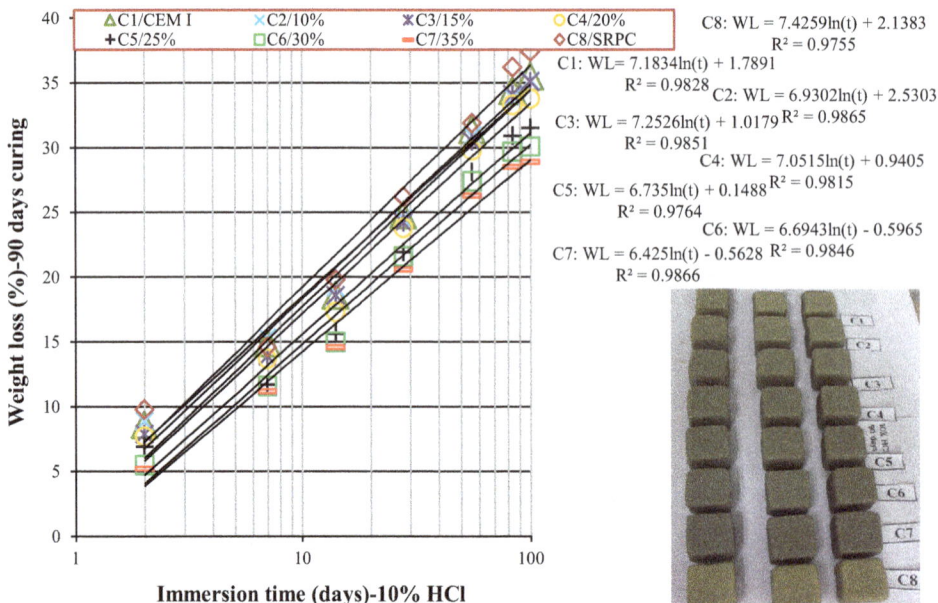

Fig. 12 Weight losses over time of 90 days cured mortars immersed in 10 % HCl solution.

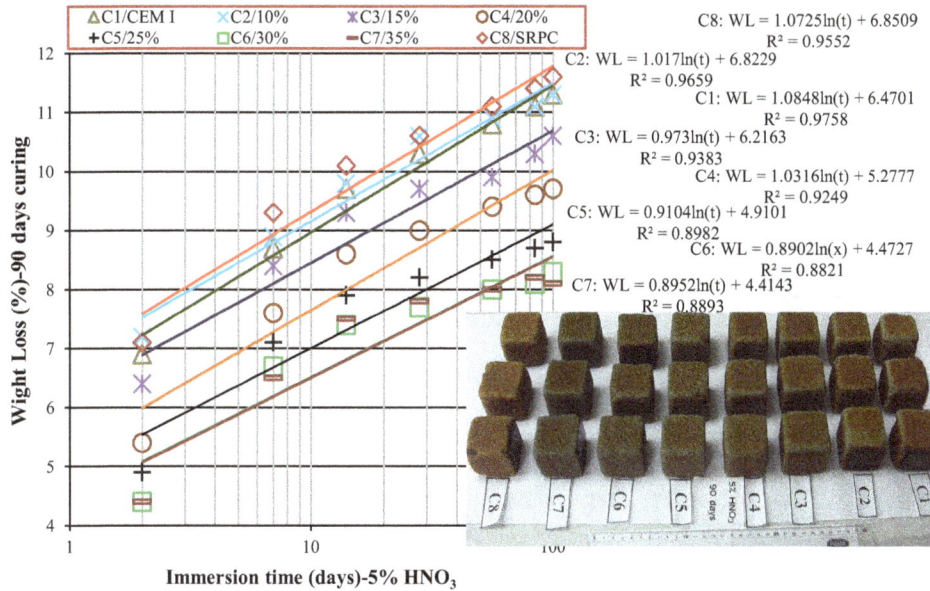

C8: WL = 1.0725ln(t) + 6.8509
R² = 0.9552
C2: WL = 1.017ln(t) + 6.8229
R² = 0.9659
C1: WL = 1.0848ln(t) + 6.4701
R² = 0.9758
C3: WL = 0.973ln(t) + 6.2163
R² = 0.9383
C4: WL = 1.0316ln(t) + 5.2777
R² = 0.9249
C5: WL = 0.9104ln(t) + 4.9101
R² = 0.8982
C6: WL = 0.8902ln(x) + 4.4727
R² = 0.8821
C7: WL = 0.8952ln(t) + 4.4143
R² = 0.8893

Fig. 13 Weight losses over time of 90 days cured mortars immersed in 5 % HNO₃ solution.

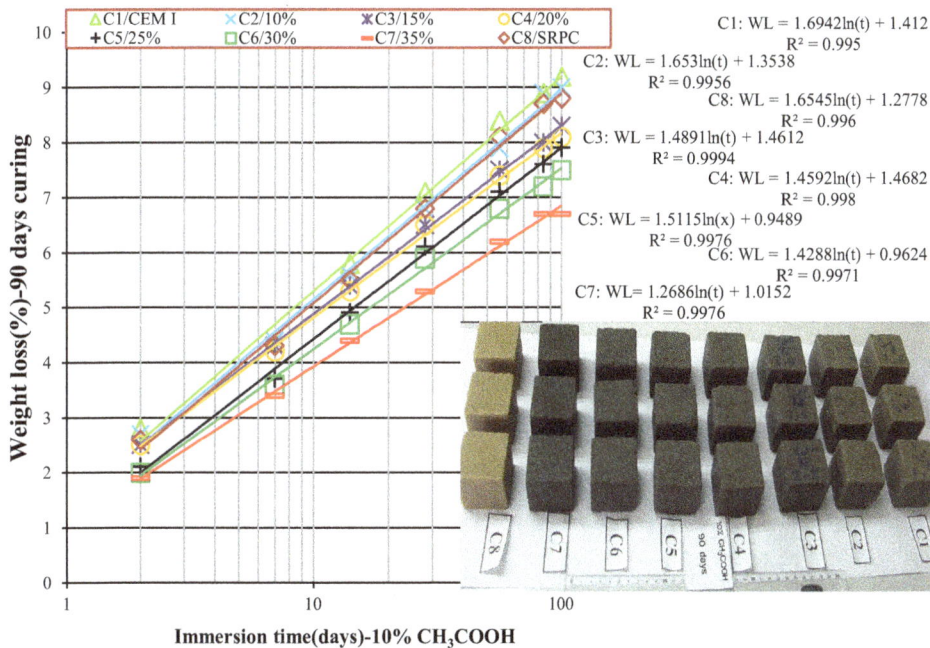

C1: WL = 1.6942ln(t) + 1.412
R² = 0.995
C2: WL = 1.653ln(t) + 1.3538
R² = 0.9956
C8: WL = 1.6545ln(t) + 1.2778
R² = 0.996
C3: WL = 1.4891ln(t) + 1.4612
R² = 0.9994
C4: WL = 1.4592ln(t) + 1.4682
R² = 0.998
C5: WL = 1.5115ln(x) + 0.9489
R² = 0.9976
C6: WL = 1.4288ln(t) + 0.9624
R² = 0.9971
C7: WL = 1.2686ln(t) + 1.0152
R² = 0.9976

Fig. 14 Weight losses over time of 90 days cured mortars immersed in 10 % CH₃COOH solution.

14 show the weight loss of 90 days cured mortar cubes immersed in 5 % H_2SO_4, 10 % HCl, 5 % HNO_3 and 10 % CH_3COOH solutions, respectively.

The weight loss is considered as a function of time. As expected, the acid resistance of the mortars improves with increasing the replacement level of scoria. This improvement of acid resistance is higher at early ages and decreases with increasing the immersion time. Beyond 28 days of exposure, slight improvements in acid resistance have been found. For instance, weight loss in 90 days cured mortars dropped from 34.8 % (CEM I) to 29.4 % (C7/35 %) after 28 days of exposure to 5 % H_2SO_4. From Figs. 11, 12, 13 and 14, SRPC and CEM I suffered the most deterioration in terms of weight loss at earlier ages of exposure to all acid solutions. The

weight loss of the CEM II/B was between half and three-quarters of the weight loss of the CEM I mixture at the first seven days of exposure. The number of days needed to register a 10 % loss in weight is considered in the evaluation. As seen from Table 4, the 10 % weight loss was obtained with the C7/35 % mix at up to 6.2 and 6.70 days of exposure to sulfuric acid; 4.6 and 5.2 days of exposure to hydrochloric acid at 28 and 90 days curing, respectively. The SRPC and CEM I mixtures, however, reached the same weight loss at only 3.0 and 3.5 days when exposed to sulfuric acid; at 3.2 & 2.9 days and 3.7 & 3.2 days when exposed to hydrochloric acid at 28 and 90 days curing, respectively. None of CEM II/B mixtures lost 10 % weight even after 100 days of exposure to nitric and acetic acids.

Table 4 Number of days needed to register a 10 % weight loss of mortar cubes.

Cement type	Number of days to register 10 % weight loss							
	5 % H_2SO_4		10 % HCl		5 % HNO_3		10 % CH_3COOH	
	28 days curing	90 days curing	28 days curing	90 days curing	28 days curing	90 days curing	28 days curing	90 days curing
C1/CEMI	3.5	3.5	3.7	3.2	22.5	25.9	NR	NR
C2/10 %	3.7	3.7	3.7	3.1	25.4	22.8	NR	NR
C3/15 %	3.8	3.8	3.8	3.6	55.5	48.9	NR	NR
C4/20 %	4.0	4.1	3.9	3.7	79.1	97.3	NR	NR
C5/25 %	4.7	5.1	4.1	4.3	NR	NR	NR	NR
C6/30 %	5.2	5.8	4.4	4.9	NR	NR	NR	NR
C7/35 %	6.2	6.7	4.6	5.2	NR	NR	NR	NR
C8/SRPC	3.0	3.0	3.2	2.9	20.1	18.8	NR	NR

NR 10 % weight loss was not reached.

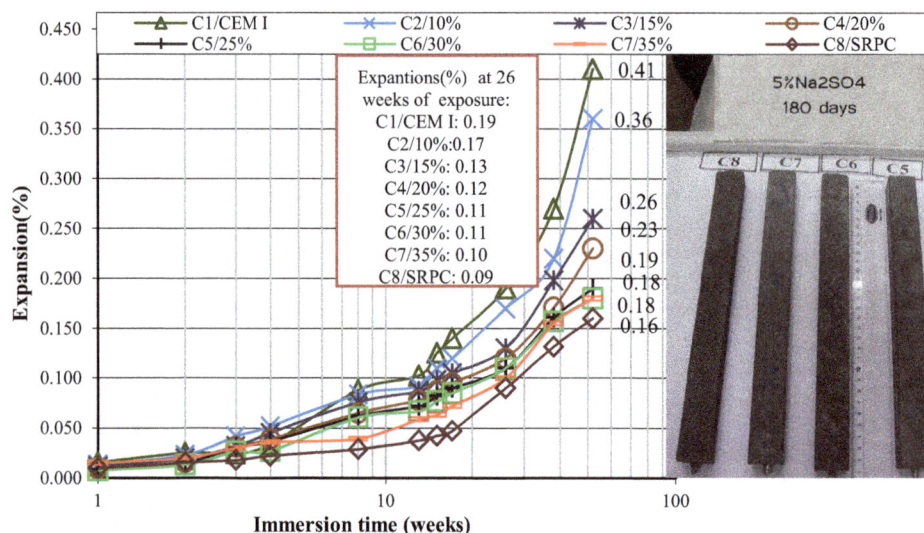

Fig. 15 Length changes over time of prismatic mortars immersed in 5 % Na_2SO_4.

The overall degree of attack tended to be more severe in sulfate and hydrochloric solutions compared with nitric and acetic acids.

The better performance of scoria-based cements can be due to the pozzolanic reaction (Cao et al. 1997; Aydın et al. 2007). This reaction between scoria and calcium hydroxide liberated during the hydration of cement (Aydın et al. 2007), led to a refinement of the pore structure resulting in a highly impermeable matrix (Cao et al. 1997). The pozzolanic reaction also fixes $Ca(OH)_2$, which is usually the most vulnerable product of hydration of cement in so far as acid attack is concerned (Neville 2011). Contrary to expectation, the weight loss of SRPC mortars was very similar to that of CEM I mortars under similar conditions. This is possibly because SRPC and CEM I mortars both contain lime and calcium silicates in large proportions. This was in agreement with results of Fattuhi and Hughes (1988). It should be also noted that after 100 days of exposure to acid solutions, especially sulfuric and nitric acids, SRPC and, to a smaller

degree, CEM I mortar cubes showed a surface layer of brown color. This brown-colored layer, which is probably composed of ferric oxides (Pavlik 1994), can be attributed to the higher content of C_4AF in both cements. Generally, all acid attacks on mortars are associated with erosion and softening due to leaching of Ca and decalcification of C–S–H (Cao et al. 1997). Hence, expansion might be expected to be less with the reduced pH values (Cao et al. 1997). As generally expected, none of all binders used can provide a long-term resistance when exposed to strong acids.

3.7 Sulfate Attack

The results of the expansion test of bars immersed in 5 % Na_2SO_4 solution for up to 52 weeks are shown in Fig. 15. It was clearly seen form Fig. 15, the use of scoria reduced expansion of the mortar bars. This reduction in expansion increased with increase in the scoria replacement level. These results were similar to those of other studies in which pozzolanic materials were used as cement replacements to improve sulfate resistance (Tangchirapat et al. 2009; Irassar et al. 2000). It should be also noted that the amount of expansion for SRPC mortar bars was very similar to that of CEM II/B mortar bars. At 26 and 52 weeks, the amount of expansion of the CEM II/B mortar bars ranged from 0.10 % to 0.11 % and from 0.18 % to 0.19 %, respectively, whereas the expansion of SRPC was 0.09 and 0.16 %, respectively. This was despite the lack of C_3A in SRPC. The slightly elevated expansion noted in SRPC used in the study, which made it unsuitable for severe exposure, is possibly because the SRPC mortar contains lime and C_4AF in large proportions when compared to the CEM II/B mortars. These results confirmed earlier findings that the presence of C_3A is not the only cause of expansion due to sulfate attack (Cao et al. 1997; Tangchirapat et al. 2009).

According to ACI Committee 201 (1991) ettringite formation derived from ferroaluminate phase has also been assumed as a potential sulfate deterioration problem. The improvement of sulfate resistance by adding scoria as cement replacement can be mainly explained by the pore size refinement, the removal of lime by the pozzolanic reaction of scoria with the lime liberated during cement hydration and the C_3A dilution (Irassar et al. 2000; Hooton and Emery 1990; al-Amoudi 2002; Al-Dulaijan et al. 2003). The pore size refinement reduced the permeability of the paste, thus limiting the ingress of sulfate ions (Irassar et al. 2000). In terms of cement composition, C_3A is the main compound involving sulfate resistance, whereas C_4AF, an alumina bearing phase, and CH released from silicates hydration can also affect the sulfate resistance of low C_3A Portland cements (Gonzalez and Irassar 1997). On the other hand, the C_3S content was considered an important parameter on sulfate resistance of PC, too (Irassar et al. 2000). It can be noted that for CEM I of about 6 % C_3A content, blending with 25 % scoria content or more usually resulted in a performance similar to that of SRPC. This result is similar to that reported by Lawrence (1990).

4. Conclusion

From the experimental results, the following conclusions could be drawn:

- The studied scoria is a suitable material for use as a natural pozzolan. It satisfied the ASTM & EN requirements for such a material. The physical properties of binders containing scoria are also in conformity with the standards requirements.
- The compressive strength of concrete containing scoria-based binders was lower than that of plain cement concrete at all ages of concrete in this study. At early ages, the concrete containing CEM II/B-P binder types had compressive strengths much lower than that of plain cement concrete. However, at 90 days curing, the compressive strengths of blended cement concretes are comparable to those of plain cement concrete.
- The chloride penetrability of scoria-based concrete mixes is much lower than that of plain concrete, especially at high replacement levels of scoria.
- According to the results of accelerated corrosion test, concretes produced with scoria-based binders decelerated rebar corrosion. Particularly, CEM II/B-P binder types with 25, 30 and 35 % scoria content were found to delay corrosion significantly. Use of scoria at 30 % cement replacement level delayed significantly initial corrosion times under chloride-bearing environments.
- Based on the results, blending CEM I of 6 % C_3A content, with 25 % scoria content or more resulted in a performance similar to that of SRPC and an enhanced acid resistance, as well.
- Adding scoria as cement replacement reduced the expansion of the mortar bars exposed to sodium sulfate solution. More reduction occurs with increasing the replacement level.
- Blended cement concretes have lower compressive strengths, but greater resistance to chloride penetration, longer corrosion initiation times, greater resistance to acid attack and lower expansion in sodium sulfate solution compared with plain cement concretes after 28 and 90 days curing. So, it would be erroneous to predict durability based on strength.
- Based on the results obtained, it is suggested that scoria can be used up to 30 % as a partial substitute for PC in production of blended cements. This addition ratio can reduce the quantity of CO_2 released by Syrian cement plants, and the consumed energy. So, production of a green concrete could be promoted.

Acknowledgments

The authors gratefully acknowledge the technical and financial support of this research from the management of General Organization for cement and Building Materials/Adra Cement Plant. Thanks are also expressed to Chemist Nazeer Adarnaly, Eng. Amjad Bernieh (Lafarge Co.) and Prof. Tamer al-Hajeh, vice-president of AIU for their appreciated help.

References

ACI Committee 201. (1991). Guide to durable concrete. *ACI Materials Journal*, 88, 551–554.

al-Amoudi, O. S. B. (2002). Attack on plain and blended cements exposed to aggressive sulfate environments. *Cement & Concrete Composites, 24*, 304–316.

Al-Chaar, G. K., Al-Kadi, M., & Asteris, P. G. (2013). Natural pozzolan as a partial substitute for cement in concrete. *The Open Construction and Technology Journal, 7*, 33–42.

Al-Dulaijan, S. U., Maslehuddin, M., Al-Zahrani, M. M., Sharif, A. M., Shameem, M., & Ibrahim, M. (2003). Sulfate resistance of plain and blended cements exposed to varying concentrations of sodium sulfate. *Cement & Concrete Composites, 25*, 429–437.

Al-Tayyib, A. J., & Al-Zahrani, M. M. (1990). Corrosion of steel reinforcement in polypropylene fiber reinforced concrete structures. *ACI Materials Journal, 87*(2), 108–113.

ASTM C1012 (2004). Standard test method for length change of hydraulic-cement mortars exposed to a sulfate solution. West Conshohocken, PA: ASTM International.

ASTM C1202 (2001). Electrical indication of concrete's ability to resist chloride ion penetration. West Conshohocken, PA: ASTM International.

ASTM C267 (2001). Standard test methods for chemical resistance of mortars, grouts, and monolithic surfacings and polymer concretes. West Conshohocken, PA: ASTM International.

ASTM C618 (2001). Standard test methods for coal ash and raw or calcined natural pozzolan for use as a mineral admixture in concrete. West Conshohocken, PA: ASTM International.

ASTM C595 (2001). Standard specification for blended hydraulic cements. West Conshohocken, PA: ASTM International.

Aydın, S., Yazıcı, H., Yigiter, H., & Baradan, B. (2007). Sulfuric acid resistance of high-volume fly ash concrete. *Building and Environment, 24*, 717–721.

Cao, H. T., Bucea, I., Ray, A., & Yozghatlian, S. (1997). The effect of cement composition and pH of environment on sulfate resistance of Portland cements and blended cements. *Cement & Concrete Composites, 19*(2), 161–171.

Cavdar, A., & Yetgin, S. (2007). Availability of tuffs from northeast of Turkey as natural pozzolans on cement, some chemical and mechanical relationships. *Construction and Building Materials, 21*, 2066–2071.

Chindaprasirt, P., Chotithanorm, C., Cao, H. T., & Sirivivatnanon, V. (2007). Influence of fly ash fineness on the chloride penetration of concrete. *Construction and Building Materials, 21*, 356–361.

Colak, A. (2003). Characteristics of pastes from a Portland cement containing different amounts of natural pozzolan. *Cement and Concrete Research, 33*, 585–593.

EN 196-2 (1989). Methods of testing cement, part 2. Chemical analysis of cement. Brussels, Belgium: European Committee for Standardization.

EN 197-1 (2004). Cement: part 1. Composition, specification and conformity criteria for common cements. Brussels, Belgium: European Committee for Standardization.

Fattuhi, N. I., & Hughes, B. P. (1988). SRPC and modified concretes subjected to severe sulphuric acid attack. *Magazine of Concrete Research, 40*, 159–166.

Gastaldini, A. L. G., Isaia, G. C., Gomes, N. S., & Sperb, J. E. K. (2007). Chloride penetration and carbonation in concrete with rice husk ash and chemical activators. *Cement & Concrete Composites, 21*, 356–361.

Ghrici, M., Kenai, S., & Meziane, E. (2006). Mechanical and durability properties of cement mortar with Algerian natural pozzolana. *Journal of Material Science, 41*, 6965–6972.

Gonzalez, M. A., & Irassar, E. F. (1997). Ettringite formation in low C3A Portland cement exposed to sodium sulfate solution. *Cement and Concrete Research, 27*(7), 1061–1072.

Guneyisi, E., Ozturan, T., & Gesoglu, M. (2005). A study on reinforcement corrosion and related properties of plain and blended cement concretes under different curing conditions. *Cement & Concrete Composites, 27*, 449–461.

Ha, T., Muralidharan, S., Bae, J., Ha, Y., Lee, H., Park, K., & Kim, D. (2007). Accelerated short-term techniques to evaluate the corrosion performance of steel in fly ash blended concrete. *Building and Environment, 42*, 78–85.

Hooton, R. D., & Emery, J. J. (1990). Sulphate resistance of a Canadian slag cement. *ACI Materials Journal, 87*(6), 547–555.

Horsakulthai, V., Phiuvanna, S., & Kaenbud, W. (2011). Investigation on the corrosion resistance of bagasse-rice husk-wood ash blended cement concrete by impressed voltage. *Construction and Building Materials, 25*, 54–60.

Hossain, K. M. A. (2003). Blended cement using volcanic ash and pumice. *Cement and Concrete Research, 33*, 1601–1605.

Hossain, K. M. A. (2009). Resistance of scoria-based blended cement concrete against deterioration and corrosion in mixed sulfate environment. *Journal of Materials in Civil Engineering ASCE, 21*(7), 299–308.

Hossain, A. B., Shirazi, S. A., Persum, J., Neithalath, N. (2008). Properties of concrete containing vitreous calcium aluminosilicate pozzolan. In *Proceedings of the 87th transportation research board annual meeting*, January, Washington DC.

Irassar, E. F., Gonzalez, M. A., & Rahhal, V. (2000). Sulfate resistance of type V cements with limestone filler and natural pozzolan. *Cement & Concrete Composites, 22*(5), 361–368.

ISO 4012 (1978). Concrete: Determination of compressive strength of test specimens. London, UK: ISO.

Khedr, S. A., & Idriss, A. F. (1995). Resistance of silica-fume concrete to corrosion-related damage. *ASCE, Journal of Materials in Civil Engineering, 7*(2), 102–107.

Lawrence, C. D. (1990). Sulfate attack on concrete. *Magazine of concrete Research, 42*(153), 249–264.

Mehta, P. K., & Monteiro, P. J. M. (2006). Concrete: Microstructure, properties, and materials (3rd ed.). New York, NY: McGraw-Hill, ISBN 0-07-146289-9.

Mindess, S., Young, J. F., & Darwin, D. (2003). *Concrete* (2nd ed.). Upper Saddle River, NJ: Prentice Hall.

Montgomery, D. C., & Peck, E. A. (1982). *Introduction to linear regression analysis*. New York, NY: Wiley.

Neville, A. M. (2011). *Properties of concrete* (5th ed.). London, UK: Pearson Education.

Parande, A. K., Babu, B. R., Karthic, M. A., Deepak Kumaar, K. K., & Palaniswamy, N. (2008). Study on strength and corrosion performance for steel embedded in metakaolin blended concrete/mortar. *Construction and Building Materials, 22*, 127–134.

Pavlik, V. (1994). Corrosion of hardened cement paste by acetic and nitric acids; Part II: Formation and chemical composition of the corrosion products layer. *Cement and Concrete Research, 24*, 1495–1508.

Pourkhorshidi, A. R., Najimi, M., Parhizkar, T., Jafarpour, F., & Hillemeier, B. (2010). Applicability of the standard specification of ASTM C 618 for evaluation of natural pozzolans. *Cement & Concrete Composites, 32*, 794–800.

Ramezanianpour, A. A., Mirvalad, S. S., Aramun, E., Peidayesh, M. (2010). Effect of four Iranian natural pozzolans on concrete durability against chloride penetration and sulfate attack. In P. Claisse et al. (Ed.), *Proceedings of the 2nd international conference on sustainable construction materials and technology, 28–30 June*, Ancona, Italy.

Reddy, D. V., Edouard, J. B., Sobhan, K., Rajpathak, S.S. (2011). Durability of reinforced fly ash-based geopolymer concrete in the marine environment. In *Proceedings of the 36th Conference on Our World in Concrete & Structures*, August 14–16, Singapore.

Rodriguez-Camacho, R. E., & Uribe-Afif, R. (2002). Importance of using natural pozzolans on concrete durability. *Cement and Concrete Research, 32*, 1851–1858.

Rossignolo, J. A., & Agesini, M. V. C. (2004). Durability of polymer-modified lightweight aggregate concrete. *Cement & Concrete Composites, 26*(4), 357–380.

Rukzon, S., & Chindaprasirt, P. (2009). Effect of grinding on chemical and physical properties of rice husk ash. *Int J Miner Metal Mater, 16*(2), 242–247.

Saraswathy, V., & Song, H.-W. (2007). Corrosion performance of rice husk ash blended concrete. *Construction and Building Materials, 21*, 1779–1784.

Senhadji, Y., Escadeillas, G., Khelafi, H., Mouli, M., & Benosman, A. S. (2012). Evaluation of natural pozzolan for use as supplementary cementitious material. *European Journal of Environmental and Civil Engineering, 16*(1), 77–96.

Shaker, F. A., El-Dieb, A. S., & Reda, M. M. (1997). Durability of styrene-bautadiene latex modified concrete. *Cement and Concrete Research, 27*(5), 7711–7720.

Talbot, C., Pigeon, M., Maarchand, M., & Hornain, J. (1995). Properties of mortar mixtures containing high amounts of various supplementary cementitious materials. In V. M. Malhotra (Ed.), *Proceeding of the fifth international conference on the use of fly ash, silica fume, slag, and natural pozzolana in Concrete, ACI SP 153, Milwaukee* (pp. 125–152). Milwaukee, WI: American Concrete Institute.

Tangchirapat, W., Jaturapitakkul, C., & Chindaprasirt, P. (2009). Use of palm oil fuel ash as a supplementary cementitious material for producing high-strength concrete. *Construction and Building Materials, 23*(7), 2641–2646.

The General Establishemnt of Geology and Mineral Resources in Syria. (GEGMR). (2007). Official document no. (3207/T/9), dated 21.11.2007 (in Arabic).

The General Organisation for Cement & Building Materials. (GOCBM) (2011). www.cemsyria.com. Accessed 2011 (in Arabic).

The General Establishment of Geology and Mineral Resources in Syria (GEGMR). (2011). A Guide for mineral resources in Syria (in Arabic).

Turanli, L., Uzal, B., & Bektas, F. (2005). Effect of large amounts of natural pozzolan addition on properties of blended cements. *Cement Concrete Research, 35*(6), 1106–1111.

10

Test Results and Nonlinear Analysis of RC T-beams Strengthened by Bonded Steel Plates

Wei Ren[1],*, Lesley H. Sneed[2], Yiting Gai[3], and Xin Kang[2]

Abstract: This paper describes the test results and nonlinear analysis of reinforced concrete T-beams strengthened by bonded steel plates under increasing static loading conditions. The first part of this paper discusses the flexural tests on five T-beams, including the test model design (based on similarity principles), test programs, and test procedure. The second part discusses the nonlinear numerical analysis of the strengthened beams, in which a concrete damage plasticity model and a cohesive behavior were adopted. The numerical analysis results are compared with experimental data and show good agreement. The area of bonded steel plate and the anchor bolt spacing were found to have an impact on the cracking load, yield load, and ultimate load. An increase in the area of steel plate and a reduction of the anchor spacing could significantly improve the cracking and ultimate loads and decrease the damage of the beam.

Keywords: bridge, cohesive behavior, concrete damage plasticity, cracking, finite element method, simulation.

1. Introduction

Bonding of a steel plate with epoxy adhesive on the tensile side of a reinforced concrete (RC) beam is one of the most accepted and widely used techniques for flexural strengthening. Over the past several decades, a large number of studies have been conducted by using this technique in different parts of the world (Adhikary and Mutsuyoshi 2002; Aprile et al. 2001; Oehlers et al. 1998). Despite its long history, a unified and rational design method for strengthening RC beams with bonded steel plates has not yet been established.

This paper presents the test results and numerical simulations of RC T-beam strengthened by bonded steel plates under increasing static loading conditions. The numerical simulations were conducted using a concrete damaged plasticity model and cohesive behavior to examine the possible failure mechanisms of the RC T-beam strengthened by bonded steel plates. The established model can be applied in future investigations to study different failure modes and the effects of various parameters on the strength and serviceability of steel-plate-strengthened RC beams. From the results, it is found that the simple model proposed can accurately predict the overall flexural behavior of strengthened beams up to ultimate state.

2. Experiments

2.1 Specimen Preparation

The test specimen design was based on a 16 m long RC T-beam that was designed in accordance with the Chinese Highway Bridge Design Code (1974). The test specimen was designed based on similarity theory (Zhao et al. 2011). Similar constants are the geometric similarity constant $C_l = 1/8$, the similarity constant of the elastic modulus $C_E = 1$, and the similarity constant of Poisson's ratio $C_n = 1$.

According to the reinforcement ratio similarity criterion π_1, $\rho_m = \rho_p$ is obtained, (i.e., the reinforcement ratio of the specimen is the same as that of the original beam). According to the flexural reinforcement ratio similarity criterion π_2, $\rho_{fm} = \rho_{fp}$ is obtained (i.e., the flexural reinforcement ratio obtained from the specimen is the same as that of the original beam). According to the loading ratio similarity criterion π_3, $P_m = P_p/64$ (i.e., the load ratio obtained from the original beam is 64 times more than that of the test specimen). According to the unit weight similarity criterion π_4, $\gamma_m = 8\gamma_p$ (i.e., the load ratio obtained from the original beam is eight times more than that of the specimen).

Based on the above similarity analysis, the similarity parameters are given as: the similarity constant of the dead load $C_w = 8$, and the similarity constant of the load $C_F = 1/64$.

[1]Key Laboratory of Bridge Detection Reinforcement Technology Ministry of Communications, Chang'an University, Xi'an 710075, Shaanxi, China.

*Corresponding Author; E-mail: rw20062@163.com
[2]Department of Civil, Architectural & Environmental Engineering, Missouri University of Science and Technology, Rolla, MO 65409, USA.

[3]CCCC First Highway Consultants CO., LTD, Xi'an 710075, Shaanxi, China.

Table 1 Geometric dimensions of the test specimen.

	l (m)	b (m)	h (m)	h_f (m)	b_f (m)
Bridge beam	16	1.6	1.1	0.11	0.18
Test specimen	2	0.2	0.138	0.034	0.052

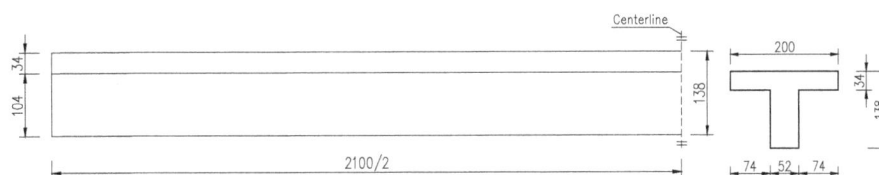

Fig. 1 Geometry of test specimen (dimensions in mm).

Table 2 Measured material properties of steel bar, steel plate, and concrete.

Yield strength (MPa)	Breaking strength (MPa)	Elastic modulus (MPa)
Steel Reinforcing Bar (Φ10)		
456.3	729.1	1.96×10^5

Specimen no.	Compressive strength (MPa)	Elastic modulus (MPa)
Concrete		
I	41.2	3.5×10^4
II	39.1	3.3×10^4

Yield strength (MPa)	Breaking strength (MPa)	Elastic modulus (MPa)	Elongation (%)
Steel Plate			
317.2	463.2	1.24×10^5	5

This test program consisted of testing five 2,100 mm (with a clear span of 2,000 mm) long RC T-beams. The geometric dimensions of the test specimen are shown in Table 1 and Fig. 1, which are determined according to the above-mentioned similarity constants.

2.2 Material Properties

The measured material properties of the steel reinforcing bars, the steel plate, and the concrete are shown in Table 2.

2.3 Experimental Program

A total of five beam specimens were tested. The beam dimensions are shown in Fig. 2. The beams were reinforced with steel plates bonded to the tension surface of the beam. All the beams were designed to fail in flexure, even after strengthening with steel plates, which made it possible to estimate the increase in flexural strength. Beam B-0 was kept as the reference (unstrengthened) beam, and other beams were bonded with steel plates. Each beam was reinforced with (2)–10 mm diameter bars in the flexural tension region, and 6 mm diameter stirrups spaced at 65 mm in the shear span and 100 mm in the region of pure flexure. Concrete was cast in two different batches and was designated as Type I and Type II (Table 2). One 1,800 mm long steel plate with a thickness of 2.0 mm was bonded to the bottom of each of the strengthened beams using epoxy adhesive. Two different

plate widths, 21 and 42 mm, were investigated. The properties of the steel reinforcing bars and the steel plates are summarized in Table 2. Before bonding the steel plates, the bottom surface of the beam was roughened by mechanical grinding and cleaned thoroughly with acetone. The bonding faces of the steel plates were also sand blasted and cleaned thoroughly with acetone. In order to attach the steel plates tightly to the beams, M8 bolts (diameter: 8 mm, length: 40 mm) were used, and within the region of pure flexure, two anchor spacings, 250 mm (Fig. 2a) and 1,000 mm (Fig. 2b), were investigated. Test specimen information is summarized in Table 3. The test results are presented in Sect. 4.

2.4 Testing Procedure and Instrumentation

All beam specimens were tested under four-point loading over a span of 2,000 mm. Strains in the tensile reinforcing bars, the steel plates at different locations, and the concrete were recorded. Strain gauges were attached to the bottom edge and the mid-thickness of the steel plates at the middle of each shear span, as well as at the midspan in the longitudinal direction. Midspan displacement was measured by a linear variable displacement transducer (LVDT). The specimens were loaded under displacement control (Fig. 3). Crack initiation and propagation were monitored by visual inspection during testing, and crack patterns were marked.

(a)

(b)

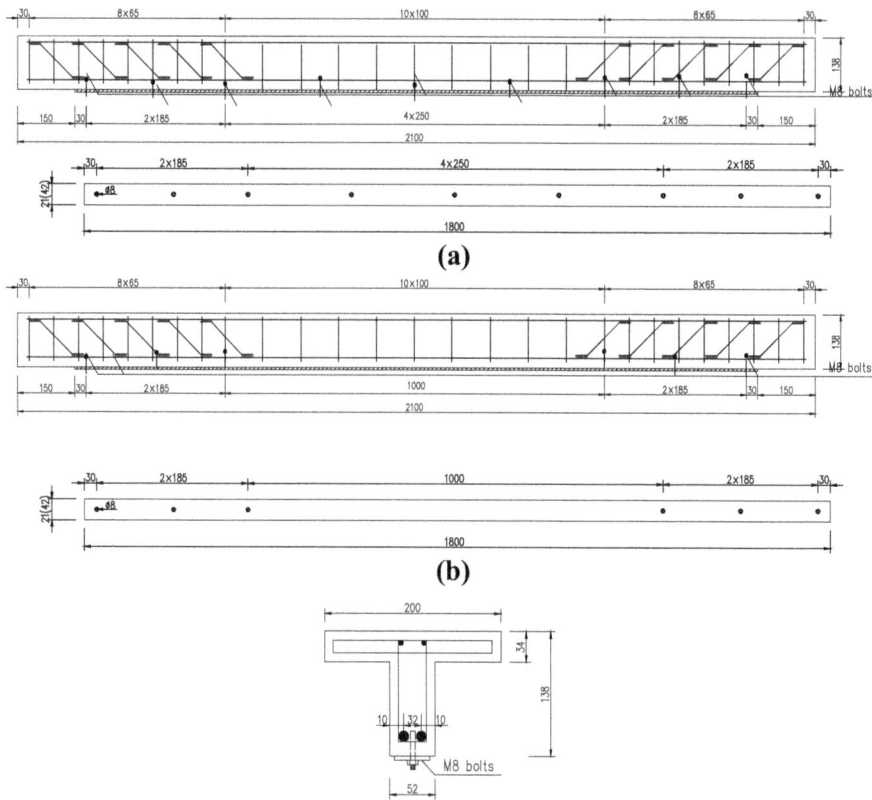

Fig. 2 Reinforcement details of the test specimens (dimensions in mm).

Table 3 Test specimens.

Specimen	Concrete type	Anchorage in pure bending region	Width of steel plate (mm)
B-0		–	–
B-2-2	I	Anchor at 2 points	21
B-5-2		Anchor at 5 points	21
B-2-4	II	Anchor at 2 points	42
B-5-4		Anchor at 5 points	42

3. Numerical Analysis

3.1 Concrete Damage Plasticity Model

In numerical analysis of RC structures, the concrete constitutive relation has a significant impact on the results. The commercial program used in this study, ABAQUS, provides different types of concrete constitutive models including (1) a smeared crack model; (2) a discrete crack model; and (3) a damage plasticity model (ABAQUS 2010). The concrete damaged plasticity model, which can be used for modeling concrete and other quasi-brittle materials, was used in this study. This model combines the concepts of isotropic damaged elasticity with isotropic tensile and compressive plasticity to model the inelastic behavior of concrete. The model assumes scalar (isotropic) damage and can be used for both monotonic and cyclic loading conditions, and has better convergence. Elastic stiffness degradation from plastic straining in tension and compression were accounted in this study (Lubliner et al. 1989, Lee and Fenves 1998). Cicekli

et al. (2007) and Qin et al. (2007) proved that damaged plasticity model could provide an effective method for modeling the concrete behavior in tension and compression.

The main parameters required in the concrete damage plasticity model, including the constitutive relationship of concrete, are defined by the user. The study described in this paper used the constitutive relation of concrete from the Chinese code GB 50010-2002 (China 2002), Guo (2001) and Xue et al. (2010).

The stress–strain relation for concrete under uniaxial tension equation is described using Fig. 4 and Eq. (1). Damage is assumed to occur only after the peak stress is reached.

$$y = \frac{x}{\alpha_t (x-1)^{1.7} + x} \quad x \geq 1 \tag{1a}$$

$$\alpha_t = 0.312 f_t^2 \tag{1b}$$

where α_t is a decline curve parameter of concrete under uniaxial tension. $\alpha_t = 0$ means the concrete constitutive

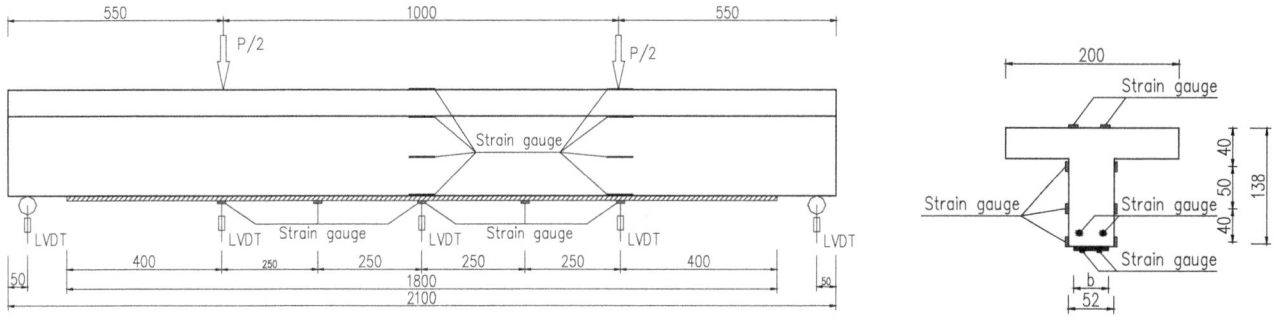

Fig. 3 Loading schematic (dimensions in mm).

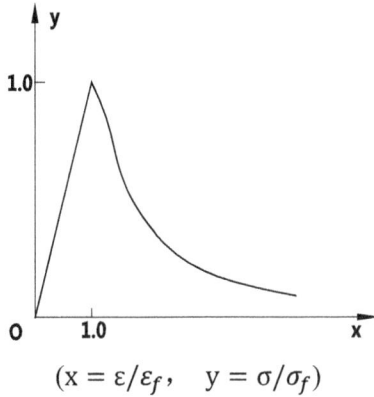

$(x = \varepsilon/\varepsilon_f, \quad y = \sigma/\sigma_f)$

Fig. 4 Concrete uniaxial tensile stress–strain curve.

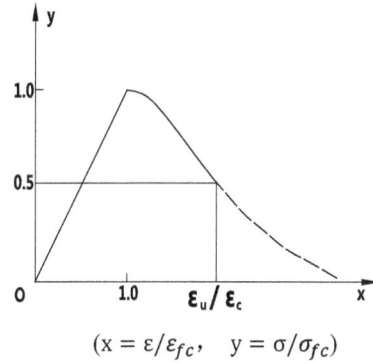

$(x = \varepsilon/\varepsilon_{fc}, \quad y = \sigma/\sigma_{fc})$

Fig. 5 Concrete uniaxial compressive stress–strain curve.

relation becomes a horizontal line (i.e., fully plastic) after peak load, while $\alpha_t = \infty$ means the concrete constitutive relation becomes a vertical line (i.e., fully brittle) after the peak load. f_t is the tensile strength of concrete. Similarly the stress–strain relation for concrete under uniaxial compression can be described using Fig. 5 and Eq. (2).

$$y = \frac{x}{\alpha_d(x-1)^2 + x} \quad x \geq 1 \tag{2a}$$

$$\alpha_d = 0.157 f_c^{0.785} - 0.905 \tag{2b}$$

where α_d is a decline curve parameter of concrete under uniaxial compression and f_c is the compression strength of concrete. According to the conservation of energy, the complementary energy of damaged material is equal to that of the elastic material, as long as the stress is converted into an equivalent stress or the elasticity modulus is equal to an equivalent elastic modulus when the material is damaged.

$$W_0^e = W_d^e \tag{3}$$

In Eq. (3), W_0^e is the complementary energy of undamaged material given in Eq. (4):

$$W_0^e = \frac{\sigma^2}{2E_0} \tag{4}$$

And W_d^e is the complementary energy of the damaged material given in Eq. (5):

$$W_d^e = \frac{\sigma'^2}{2E_d} \tag{5}$$

where E_0 and E_d are the undamaged and damaged elasticity moduli, respectively.

The equivalent stress is described using Cauchy's effective stress tensor in Eq. (6), in which D is the damage variable:

$$\sigma' = \frac{\sigma A}{A'} = \sigma/(1-D) \tag{6}$$

where A and A' are the effective bearing area of undamaged and damaged section, respectively; and σ and σ' are the effective stress of undamaged and damaged section. Combining Eqs. (3), (4), (5) and (1) results in Eqs. (7a) and (7b) below:

$$E_d = E_0(1-D)^2 \tag{7a}$$

$$\sigma = E_0(1-D)^2\varepsilon \tag{7b}$$

Adopting the principles above, the uniaxial tensile damage equation can be described in Eqs. (8a) and (8b):

$$D = 0 \quad x \leq 1 \tag{8a}$$

$$D = 1 - \sqrt{\frac{1}{[\alpha_t(x-1)^{1.7}+x]}} \quad x \geq 1 \tag{8b}$$

Similarly, the uniaxial compressive damage equation is described in Eqs. (9a) and (9b):

$$D = 0 \quad x \leq 1 \tag{9a}$$

$$D = 1 - \sqrt{\frac{1}{[\alpha_d(x-1)^2 + x]}} \quad x \geq 1 \tag{9b}$$

3.2 Model Parameters

(1) Dilation angle ψ is a measurement of how much volume increase occurs when the material is sheared. For a Mohr–Coulomb material, dilation is an angle that generally varies between zero (non-associative flow rule) and the friction angle (associative flow rule). Tuo et al. (2008) recommended adopting 30° for concrete material. By comparison with the experimental values, this paper takes a value of 38°.

(2) Flow potential eccentricity ε is a small positive number that defines the rate at which the hyperbolic flow potential approaches its asymptote. This paper takes a value of 0.1. The plastic-damage model assumes non-associated potential flow; $\dot{\varepsilon}^{pl} = \dot{\lambda}\frac{\partial G(\bar{\sigma})}{\partial \bar{\sigma}}$. The flow potential G chosen for this model is the Drucker-Prager hyperbolic function:

$$G = \sqrt{(\varepsilon \sigma_{t0} \tan \psi)^2 + \bar{q}^2} - \bar{p} \tan \psi \tag{10}$$

where Ψ is the dilation angle; σ_{t0} is the uniaxial tensile stress at failure; and ε is a parameter, referred to as the eccentricity, that defines the rate at which the function approaches to the asymptote (the flow potential tends to a straight line as the eccentricity tends to zero). This flow potential, which is continuous and smooth, ensures that the flow direction is defined uniquely.

(3) σ_{b0}/σ_{c0} is the ratio of initial equibiaxial compressive yield stress to initial uniaxial compressive yield stress. This paper uses a value of 1.16. σ_{b0} and σ_{c0} are the equibiaxial compressive yield stress and initial uniaxial compressive yield stress, respectively.

(4) K_c is the ratio of the second stress invariant on the tensile meridian, $q(TM)$, to the compressive meridian, $q(CM)$, at initial yield for any given value of the pressure invariant P such that the maximum principal stress is negative, $\hat{\sigma}_{max} < 0$. It must satisfy the condition $0.5 < K_c < 1.0$. This paper takes a value of 2/3.

(5) Viscosity parameter μ is used for the visco-plastic regularization of the concrete constitutive equations in ABAQUS/Standard analyses.

Material models exhibiting softening behavior and stiffness degradation often lead to severe convergence difficulties in implicit analysis programs. A common technique to overcome some of these convergence difficulties is the use of a viscoplastic regularization of the constitutive equations, which would lead to the consistent tangent stiffness of the softening material to become positive for sufficiently small time increments.

Using the viscoplastic regularization with a small value for the viscosity parameter (this paper takes a value of 0.0005) usually helps to improve the rate of convergence of the model in the softening regime and without compromising results. The lower the coefficient, the higher the calculation accuracy, but the calculation is more time consuming.

3.3 Cohesive Behavior
3.3.1 Basic Principle

Cohesive behavior describes the surface interaction property and is primarily intended for situations in which the interface thickness is negligibly small. It can be used to model delamination at interfaces in terms of traction versus separation. It assumes a linear elastic traction–separation law prior to damage and assumes that failure of the cohesive bond is characterized by progressive degradation of the cohesive stiffness, which is driven by a damage process.

There is no traction–separation model of steel plate-concrete bonded interfaces in the current literature, however, traction–separation models of FRP-concrete bonded interfaces are available in the literature, such as models by Neubauer and Rostasy (1999), Nakaba et al. (2001), Savoia et al. (2003), and Monti et al. (2003). Based on the results from Cicekli et al. (2007) and Fang et al. (2007), this paper uses the bilinear model by Monti et al. (2003) (Fig. 6).

The linear elastic traction–separation behavior is written in terms of an elastic constitutive matrix that relates the normal and shear stresses to the normal and shear separations across the interface. It can be written as:

$$t = \begin{Bmatrix} t_n \\ t_s \\ t_t \end{Bmatrix} = \begin{bmatrix} K_{nn} & K_{ns} & K_{nt} \\ K_{ns} & K_{ss} & K_{st} \\ K_{nt} & K_{st} & K_{tt} \end{bmatrix} \begin{Bmatrix} \delta_n \\ \delta_s \\ \delta_t \end{Bmatrix} = K\delta \tag{11}$$

where t_n, t_s, t_t represent the normal and the two shear tractions nominal traction stress vectors. The corresponding separations are denoted by δ_n, δ_s, δ_t.

Damage modeling allows simulating the degradation and eventual failure of the bond between two cohesive surfaces. The failure mechanism consists of two components: a damage initiation criterion and a damage evolution law. The initial response is assumed to be linear as discussed above. However, once a damage initiation criterion is met, damage can occur according to a user-defined damage evolution law. Damage initiation includes maximum and minimum stress criteria described below.

Maximum stress criterion:

$$\max \left\{ \frac{\langle t_n \rangle}{t_n^0}, \frac{t_s}{t_s^0}, \frac{t_t}{t_t^0} \right\} = 1 \tag{12}$$

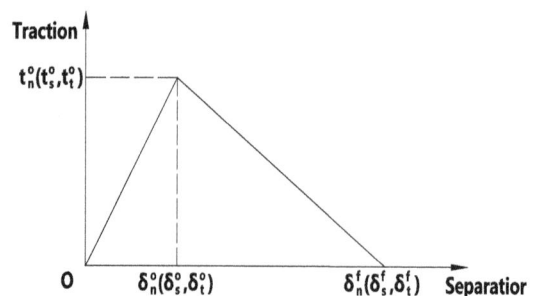

Fig. 6 Bilinear model.

Maximum separation criterion:

$$\max\left\{\frac{\langle\delta_n\rangle}{\delta_n^0},\frac{\delta_s}{\delta_s^0},\frac{\delta_t}{\delta_t^0}\right\}=1 \tag{13}$$

where t_n^0, t_s^0 and t_t^0 represent the peak values of the contact stress when the separation is either purely normal to the interface or purely in the first or the second shear direction, respectively. Likewise, δ_n^0, δ_s^0 and δ_t^0 represent the peak values of the contact separation, when the separation is either purely along the contact normal or purely in the first or the second shear direction, respectively.

For the damage evolution law, a scalar damage variable, D, represents the overall damage at the contact point. D initially has a value of 0. If damage evolution is modeled, D monotonically evolves from 0 to 1 upon further loading after the initiation of damage. The contact stress components are affected by the damage according to:

$$t_n=\begin{cases}(1-D)\bar{t}_n, & \bar{t}_n\geq0\\ \bar{t}_n, & \bar{t}_n\leq0\end{cases} \tag{14a}$$

$$t_n=(1-D)\bar{t}_s \tag{14b}$$

$$t_t=(1-D)\bar{t}_t \tag{14c}$$

where $\bar{t}_n$, $\bar{t}_s$ and $\bar{t}_s$ are the contact stress components predicted by the elastic traction–separation behavior for the current separations without damage.

3.3.2 Basic Parameters

Interfacial stress and separation relationship can be determined by the following:

$$\begin{cases}\tau=\tau_{\max}\frac{\delta}{\delta_0} & \delta\leq\delta_0\\ \tau=\tau_{\max}\frac{\delta_f-\delta}{\delta_f-\delta_0} & \delta_0<\delta\leq\delta_f\\ \tau=0 & \delta>\delta_f\end{cases} \tag{15a}$$

The peak values of the contact stress:

$$\tau_{\max}=1.8\beta_w f_t \tag{15b}$$

The separation corresponding to $\tau_{\max}$ is:

$$\delta_s^0,\delta_t^0=2.5\tau_{\max}\left(\frac{d_a}{E_a}+\frac{50}{E_c}\right) \tag{15c}$$

The ultimate separation is:

$$\delta_f=0.33\beta_w \tag{15d}$$

The width coefficient is:

$$\beta_w=\sqrt{1.125\frac{2-b_f/b_c}{1+b_f/400}} \tag{15e}$$

where E_a represents the cohesive modulus of elasticity, MPa; E_c represents the concrete modulus of elasticity, MPa;

f_t represents tensile strength of concrete, MPa; d_a represents thickness of steel plate, mm; b_f represents width of steel plate, mm; b_c represents width of concrete, mm.

The separation damage constitutive model is defined according to:

$$D=\begin{cases}0 & 0<\delta\leq\delta_0\\ 1-\tau_{\max}\frac{1}{\delta B}\frac{\delta_f-\delta}{\delta_f-\delta_0} & \delta_0<\delta\leq\delta_f\\ 1 & \delta>\delta_f\end{cases} \tag{16}$$

3.4 Proposed Model

In order to accurately simulate the actual behavior of the RC T-beams investigated in this study, a description of the material, model configuration, boundary conditions, and loading are required. For the linear elastic behavior simulations, at least two material constants are required: Young's modulus (E) and Poisson's ratio (v). For nonlinear analysis, the steel and concrete uniaxial behaviors beyond the elastic range must be defined to simulate their behavior at higher stresses. The minimum input parameters required to define the concrete material are the uniaxial compression curve, the ratio of biaxial and uniaxial compressive strength, and the uniaxial tensile strength. The bond between the steel plate and concrete surface was model by cohesive behavior. The anchor was modeled by the node coupled method, where the nodes of the steel reinforcing bar and the steel plate were coupled.

3.4.1 Symmetry

Because the RC T-beams investigated had two axes of symmetry, it is possible to represent the full beam by modeling only one fourth of the beam (Fig. 7). This allowed for reduced analysis time.

3.4.2 Finite Element Type and Mesh

Different element types were evaluated to determine a suitable element type to simulate the behavior of the investigated beams. Because it was of interest to include the response of the concrete under tensile pressure, the concrete elements were modeled as solid elements, which were found to be more efficient both in modeling the behavior and clearly defining the boundaries of their elements. A fine mesh of three-dimensional eight-node solid elements C3D8 (Tuo et al. 2008) was used in this study. The final model contained 5,659 nonlinear concrete elements (C3D8R), 596 three-dimensional linear elastic truss elements (T3D2), and

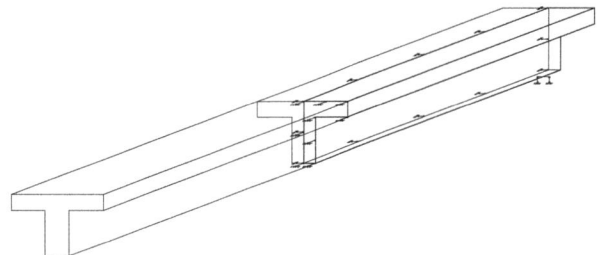

Fig. 7 Schematic drawing of one-fourth of the T-beam.

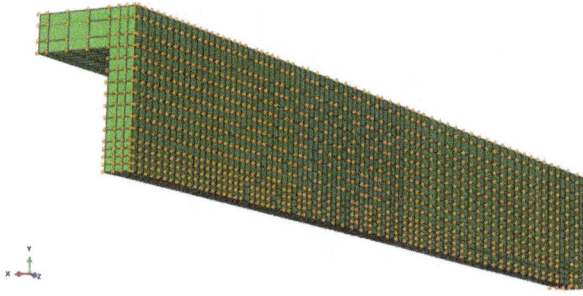

Fig. 8 FE model of one fourth of the T-beam.

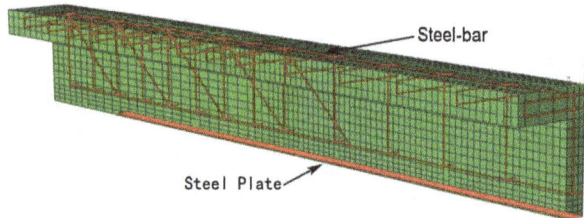

Fig. 9 Reinforcing bar was modeled as embedded regions.

560 shell elements (S4R). The maximum solid elements sizes were $21.3 \times 14 \times 20$ mm at the flange, and the minimum were $10 \times 10 \times 10$ mm at the web. A sketch of the finite element model is shown in Fig. 8.

3.4.3 Embedded Elements

This technique is often used to place embedded nodes at desired locations with the constraints on translational degrees-of-freedom on the embedded element by the host element. The rebar was modeled as embedded regions in the concrete in the interactive module, and assigning the concrete as the host (Fig. 9). Thus, the reinforcing bar elements had translations or rotations equal to those of the host elements surrounding them (i.e., perfect bond) (Garg and Abolmaali 2009).

3.4.4 Boundary Constraints

Due to symmetry, only a quarter of the panel was modeled as shown in Fig. 8. The nodes on the symmetry surfaces were prevented from displacement in X and Y directions, respectively. On the bearing point, only the degree of freedom in the Z direction was constrained.

3.4.5 Loading Method

Two loading methods are commonly used for numerical analysis: (1) application of the force P; or (2) application of the displacement δ. In this study, the loading was applied by incrementally increasing the displacement. This loading method was selected for consistency with the experimental procedure in which the test specimens were loaded under displacement control, as well as for convergence issues. External loads were obtained from the reaction forces, and the peak (failure) load was obtained by plotting the $P - \delta$ response.

3.4.6 Convergence

When solving concrete problems, it is often difficult to obtain convergence. Especially after concrete cracks, strain energy is suddenly released, and the calculation becomes extremely unstable. Through repeated trials, these issues were resolved with the following considerations:

(1) Displacement loading was enforced. At the time of concrete cracking, the initial step was adjusted to 0.005, and the automatic time step was adopted.

(2) Modifications were made to the constitutive equation by introducing the coefficient of viscosity. A high viscosity coefficient will make the structure "hard". Through repeated trials, a viscosity coefficient of 0.0005 was found to give better results.

(3) In concrete nonlinear finite element analysis, the solution is sometimes more important than the calculation accuracy. Therefore, the force and displacement convergence criteria are adjusted to make calculations carried out smoothly. Generally it takes $0.02 \sim 0.03$ load displacement tolerance. Through repeated trials, a value of 0.03 was found to give reasonable results (Jiang et al. 2005).

4. Numerical Analysis and Discussion

4.1 Load–Deflection Relationship

Figure 10a–e shows analytical and experimental load–deflection curves of the five beam specimens. In these figures, indices I–III indicate the loading states. I indicates initiation of flexural cracking, II indicates the yielding of reinforcing bar and/or steel plate, and III indicates the maximum load.

The analytical results for beam B-0 (Fig. 10a) agree well with the experimental results. By comparing the reinforced RC beams shown in Fig. 10b–e, it is confirmed that the stiffness of the RC beam decreases sharply at points I and II due to opening of the flexural cracks and yielding of reinforcing bar and/or steel plate, respectively. After reinforcing bar yielding, the load–deflection curves are almost linear up to the maximum load point. At point III, the applied load decreases suddenly due to concrete crushing.

It can be seen from beam B-5-2 (Fig. 10c) that there is significant difference between analytical and experimental curves after point II. Experimental values of beam B-5-2 are obviously too small after point II by comparison with the other strengthened beams (Fig. 10b, d and e). The reason is that the steel plate was damaged in advance, which happened before the reinforcing bar yielding, because the anchoring holes at the loading point weakened the effective area of the steel plate. The numerical analysis did not consider the debilitating effects of the anchor hole on the steel plate and therefore predicts that beam B-5-2 fails in the sequence of yielding of reinforcing bar, yielding of steel plate, and then crushing of concrete.

These discussions suggest that the load-carrying capacity and failure behavior of RC beams reinforced in flexure with externally bonded steel plate can be simulated with good accuracy using the proposed numerical analysis method.

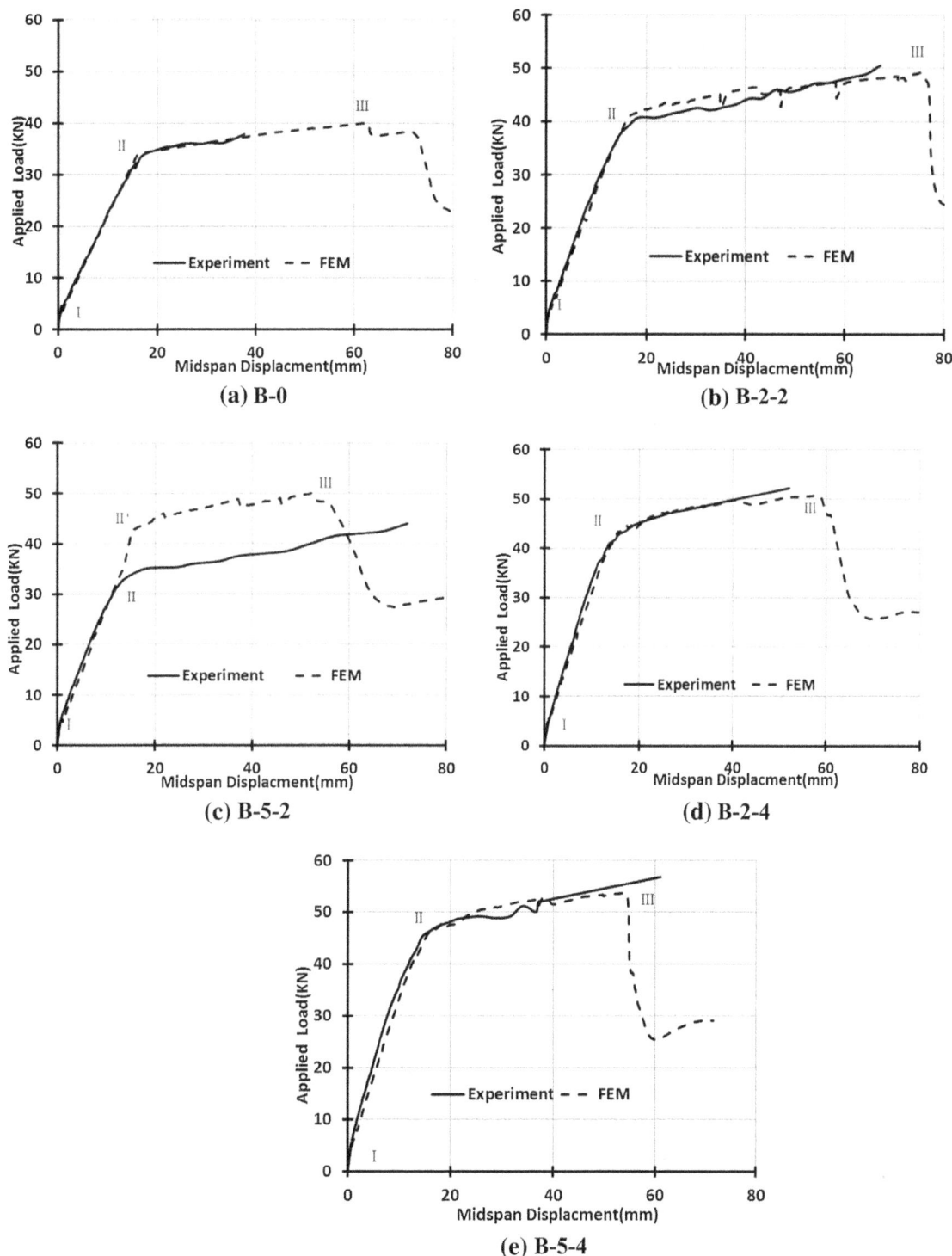

Fig. 10 Comparison of analytical and experimental results—load-midspan displacement responses.

4.2 Strain Distribution of Steel Plate

Figure 11 compares the analytical and experimental strain distributions of the steel plate at the following loading stages: 30, 35, 40, 45, and 50 kN. In this figure, analytical results and experimental results are illustrated using the strains measured over the 1/2 length of the span.

Comparisons for beams B-2-2, B-2-4, B-5-2, and B-5-4 show that analytical results agree roughly with the experimental results at each of the loading states.

Based on the observed analytical results, the strain of steel plate was found to increase suddenly at the section of concrete cracking (plastic strain section). This shows that the cohesive behavior can be used to simulate the stress transfer between concrete and steel effectively. It also can be seen that the strain in the steel plate and local concrete increases rapidly at the end of steel plate. This is consistent with the experimental observation.

In this research, the steel plate was bonded to the concrete by an epoxy adhesive and with mechanical anchors (i.e., bolts). Due to the existence of the bolts, the slip phenomenon

(a) B-2-2

(b) B-5-2

(c) B-2-4

(d) B-5-4

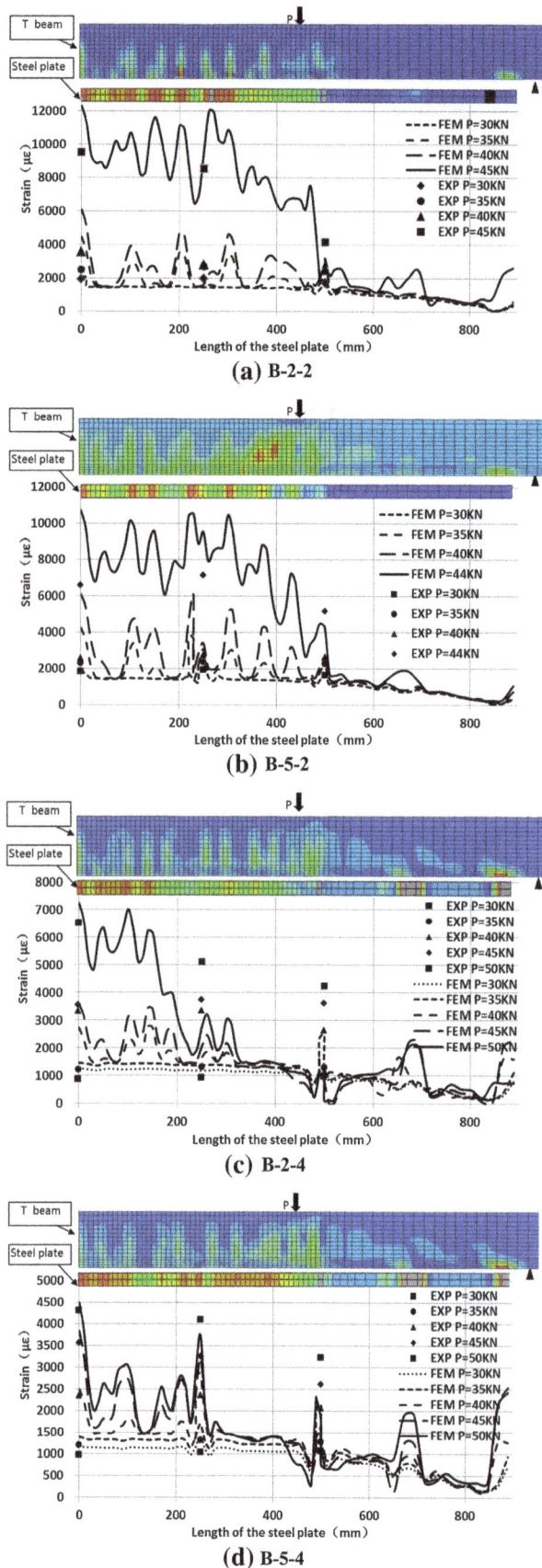

Fig. 11 Comparison of analytical and experimental results—strain distribution of steel plate along the span.

between the concrete and steel plate was been minimized, thus there was a relatively small slip phenomenon between the bolts, which can be observed in Fig. 11.

4.3 The Cracking Load, Yield Load and Failure Loads

The failure load for the numerical analysis was defined as peak load in the $P - \delta$ curves. Table 4 compares the cracking load, yield load, and failure loads from the experiments with those from the numerical analysis, P_{Test} and P_{FE}, respectively. With the exception of beam B-5-2, it can be observed that the correlation is quite good between the numerical results and the experimental data for the cracking load, yield load, and failure loads. The error is within 8.9 % in cracking load, 7.7 % in the yield load, and 9.7 % in failure load.

It can be seen from Table 4 that a relatively large deviation between the experimental and numerical results was found for beam B-5-2. By comparing with the experimental observations, the reason could be attributed to the steel plate damage in advance which influenced the response before the yield load, because the anchor hole at the loading point weakened the effective area of the steel plate (as discussed previously).

4.4 Failure Modes

Figure 12 shows the destruction of the five beams. The failure modes of beams B-0, B-2-2, B-5-2, and B-5-4 were steel plate (rebar) yield/damage and concrete crushing, whereas the failure mode of beam B-2-4 was bolt failure as seen in Fig. 12d. Figures 13, 14 and 15 show the failure modes of the beams predicted by the proposed numerical model. Comparing with the experiment results, it is clearly seen that the proposed FEM model could accurately predict the failure modes and failure loads of the tested beams.

5. Conclusions

In this paper, nonlinear finite element analyses of RC T-beams strengthened in flexure with externally bonded steel plates were performed. Based comparison of the experimental tests and numerical results, the following conclusions could be drawn:

(1) The concrete damaged plasticity model was found to accurately predict the concrete failure characteristics. Cohesive behavior method and the bilinear traction–separation model were used to simulate the load-transferring mechanism between the steel plate and concrete, and satisfactory results were achieved.

(2) The use of bonded steel plate and anchor bolts had a significant impact on the cracking load, yield load, and ultimate load. Increasing the width of steel plate and reducing the spacing of anchor bolts could significantly improve the cracking load, yield load, and ultimate load. However, the cracking load and ultimate load increase was not proportional to the area of steel plate.

(3) Increasing the width of steel plate and decreasing the anchor bolt spacing resulted in change of failure mode from steel yielding to premature failure attributed to the anchor bolt hole.

Table 4 Comparison of cracking load, yield load, and failure loads.

Specimen	Cracking load (kN)			Yield load (kN)			Failure load (kN)		
	P_{Test}	P_{FE}	P_{FE}/P_{Test} (%)	P_{Test}	P_{FE}	P_{FE}/P_{Test} (%)	P_{Test}	P_{FE}	P_{FE}/P_{Test} (%)
B-0	5.08	4.99	98.2	33.60	34.13	101.6	41.05	40.10	97.7
B-2-2	6.04	5.17	85.6	37.64	40.52	107.7	52.95	48.47	91.5
B-5-2	5.55	5.18	93.3	35.06	40.92	116.7	44.75	49.99	111.7
B-2-4	6.04	5.50	91.1	40.90	42.42	103.7	56.10	50.66	90.3
B-5-4	5.89	5.95	101.0	46.29	44.84	96.9	57.23	53.66	93.8

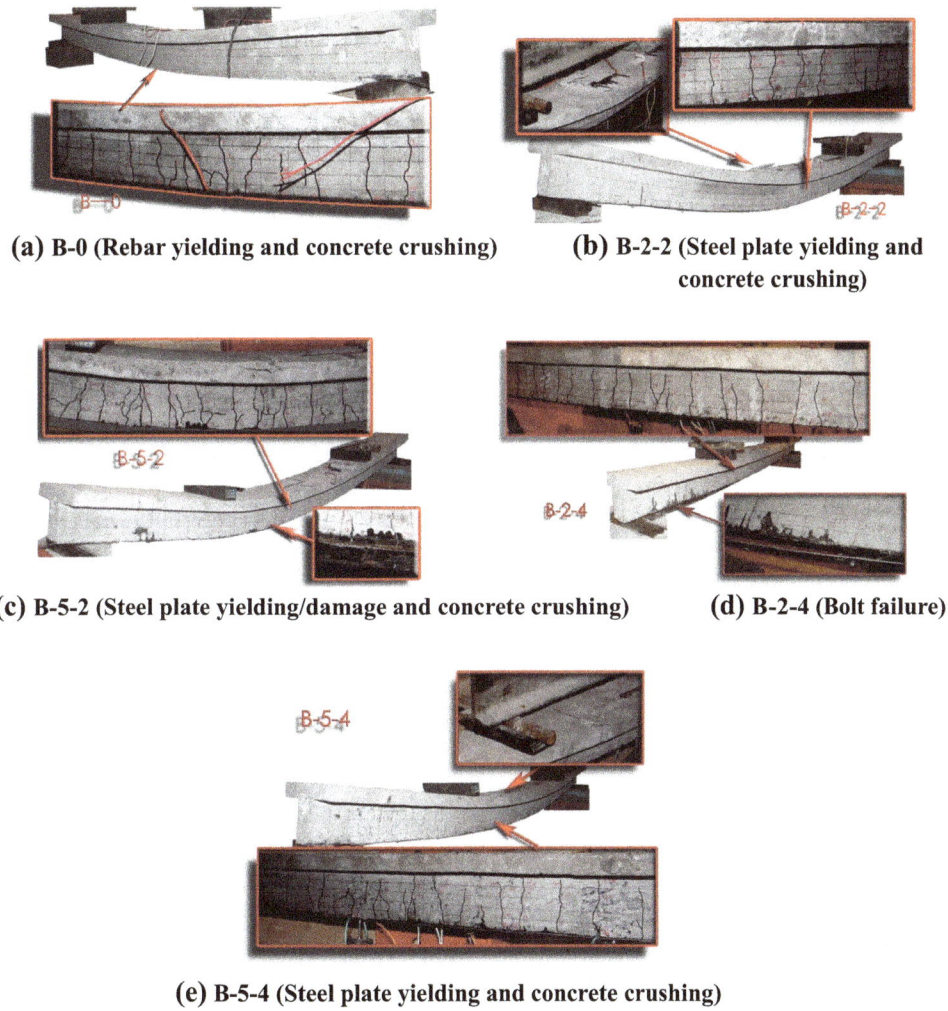

(a) B-0 (Rebar yielding and concrete crushing)

(b) B-2-2 (Steel plate yielding and concrete crushing)

(c) B-5-2 (Steel plate yielding/damage and concrete crushing)

(d) B-2-4 (Bolt failure)

(e) B-5-4 (Steel plate yielding and concrete crushing)

Fig. 12 Experimental observations at failure.

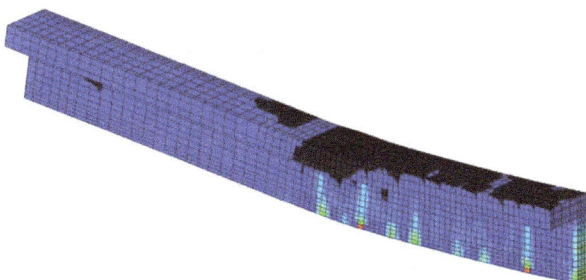

Fig. 13 Failure mode (beam B-0 shown).

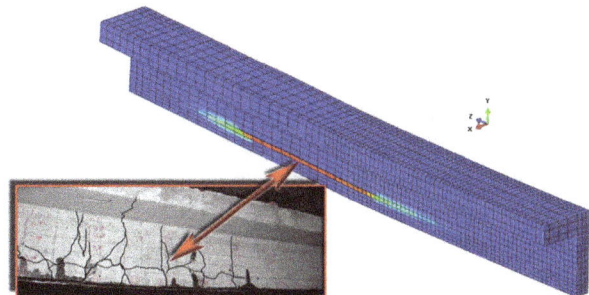

Fig. 14 Cracks along horizontal direction.

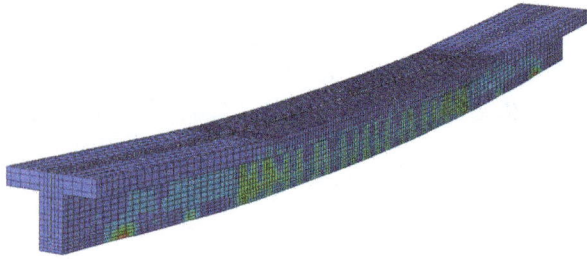

Fig. 15 Failure mode (beam B-5-4 shown).

(4)   As the steel plate area increases, the force distributed in the steel plate increases; therefore the number of anchor bolts should be increased. In addition, the shear resistance of anchor bolts and the tension in the steel plate at the anchor bolt holes should be accounted in the analysis, because the anchor bolt holes could weaken the cross section.

Acknowledgments

This project was funded by Ministry of Transport of the People's Republic of China (2012 319 812 100), the Shaanxi Department of Transportation (14–17 k), and the Guangdong Department of Transportation (2011-02-034).

References

ABAQUS. (2010). ABAQUS, ver. 6.9-3, 2010. Dassault Systemes Simulia Corp., Providence, RI.

Adhikary, B. B., & Mutsuyoshi, H. (2002). Numerical simulation of steel-plate strengthened concrete beam by a nonlinear finite element method model. *Construction and Building Materials, 16*(5), 291–301.

Aprile, A., Spacone, E., & Limkatanyu, S. (2001). Role of bond in RC beams strengthened with steel and FRP plates. *Journal of Structural Engineering, 127*(12), 1445–1452.

China, P. R. (2002). Ministry of construction. GB50010-2002. Code for design of concrete structures GB50010-2002. Beijing, China: China Architecture and Building Press.

Cicekli, U., Voyiadjis, G. Z., & Abu Al-Rub, R. K. (2007). A plasticity and anisotropic damage model for plain concrete. *International Journal of Plasticity, 23*(10), 1874–1900.

Fang, Q., Huan, Y., Zhang, Y. D., & Chen, L. (2007). Investigation into static properties of damaged plasticity model for concrete in ABAQUS. *Journal of PLA University of Science and Technology (Natural Science Edition), 3*, 010. (in Chinese).

Garg, A. K., & Abolmaali, A. (2009). Finite-Element modeling and analysis of reinforced concrete box culverts. *Journal of Transportation Engineering, 135*(33), 121–128.

Guo, Z. H. (2001). *Theory of reinforced concrete*. Beijing, China: Tsinghua University Press. (in Chinese).

Jiang, J. J., Lu, X. Z., & Ye, L. P. (2005). *Finite element analysis of concrete structures*. Beijing, China: Tsinghua University Press. (in Chinese).

Lee, J., & Fenves, G. L. (1998). Plastic-damage model for cyclic loading of concrete structures. *Journal of Engineering Mechanics, 124*(8), 892–900.

Lubliner, J., Oliver, J., Oller, S., & Oñate, E. (1989). A plastic-damage model for concrete. *International Journal of Solids and Structures, 25*(3), 299–326.

Monti, G., Renezelli, M. and Luciana, P. (2003). FRP adhesion in uncracked and cracked concrete zones. In: *Proceedings of the Sixth International Symposium on FRP Reinforcement for Concrete Structures (FRPRCS-6)* (pp. 183–192). Singapore.

Nakaba, K., Kanakubo, T., Furuta, T., & Yoshizawa, H. (2001). Bond behavior between fiber-reinforced polymer laminates and concrete. *ACI Structural Journal, 98*(3), 359–367.

Neubauer, U. and Rostasy, F. S. (1999). Bond failure of concrete fiber reinforced polymer plates at inclined cracks-experiments and fracture mechanics model. In: *Proceedings of 4th International Symposium on Fiber Reinforced Polymer Reinforcement for Reinforced Concrete Structures* (pp. 369–382). MI.

Oehlers, D. J., Ali, M. M., & Luo, W. (1998). Upgrading continuous reinforced concrete beams by gluing steel plates to their tension faces. *Journal of Structural Engineering, 124*(3), 224–232.

Savoia, M., Ferracuti, B. and Mazzotti, C. (2003). Nonlinear bond-slip law for FRP-concrete interface. In: *Proceedings of the Sixth International Symposium on FRP Reinforcement for Concrete Structures (FRPRCS-6)* (pp. 163–172). Singapore.

Transport Planning and Research Institute, Ministry of Transport, China, P. R. (1974). Highway bridge design code. Beijing, China: China Communications Press. (in Chinese).

Tuo, L., Jiang, Q., & Chengqing, L. (2008). Application of damaged plasticity model for concrete. *Structural Engineers, 24*(2), 22–27 (in Chinese).

Xue, Z., Yang, L., and Yang, Z. (2010). A damage model with subsection curve of concrete and its numerical verification based on ABAQUS. In: *2010 International Conference On Computer Design And Applications (ICCDA 2010)* (Vol. 5, pp. 34–37).

Zhao, S. B., Guan, J. F., & Li, X. K. (2011). *Model experiment and optimization design of the reinforced concrete structure*. Beijing, China: China Water Power Press (in Chinese).

11

Sulfate Resistance of Alkali Activated Pozzolans

Dali Bondar[1],*, C. J. Lynsdale[2], N. B. Milestone[3], and N. Hassani[4]

Abstract: The consequence of sulfate attack on geopolymer concrete, made from an alkali activated natural pozzolan (AANP) has been studied in this paper. Changes in the compressive strength, expansion and capillary water absorption of specimens have been investigated combined with phases determination by means of X-ray diffraction. At the end of present investigation which was to evaluate the performance of natural alumina silica based geopolymer concrete in sodium and magnesium sulfate solution, the loss of compressive strength and percentage of expansion of AANP concrete was recorded up to 19.4 % and 0.074, respectively.

Keywords: sulfate attack, geopolymer concrete, alkali activated natural pozzolan (AANP), X-ray diffraction, AANP concrete.

1. Introduction

When concrete is exposed in a solution containing a sufficiently high concentration of dissolved sulfates, the concrete deteriorates due to a series of chemical reactions. This is particularly prevalent in arid regions where naturally occurring sulfate minerals are present in water and ground water. Deterioration due to sulfate attack is generally attributed to reaction of hardened Portland cement hydration products with sulfate ions to form expansive reaction products after hardening, which produce internal stresses and a subsequent disruption of the concrete. It is generally accepted that sulfate ions attack the hydrated cement matrix by reaction of sulfate ions with the hydrated calcium aluminate phases along with $Ca(OH)_2$, forming ettringite and gypsum causing expansion of OPC concrete. Furthermore, with magnesium sulfate attack, brucite ($Mg(OH)_2$), which has low solubility, is produced and assumed to envelop the remainder of the cement gel and protect it against further deterioration. However, Turker reported that this process is only effective in the early stages of attack with deterioration due to brucite becoming dominant as the attack proceeds. Magnesium sulfate will also attack the C–S–H gel forming an M–S–H gel, which is non-cementitious and leads to softening of the cement matrix (Turker et al. 1997). In addition to chemical deterioration, salt crystallization, which usually involves repeated dissolution of the solid salt and re-crystallization, will occur in concrete pores and can be accompanied by large expansions and tension stresses far greater than the concrete bearing capacity (Ganjian and Pouya 2005). Concrete undergoing sulfate attack will suffer from swelling, spalling and cracking. The expansion leading to deterioration usually starts at edges and corners and is followed by progressive cracking with an irregular pattern. The lower the permeability of concrete is, the greater the resistance of concrete to sulfate attack. Thus, factors which reduce the permeability of concrete have a beneficial effect on reducing the vulnerability of concrete to sulfate attack.

Hakkinen evaluated the sulfate resistance of both alkali activated slag cement and Portland cement. The Portland cement samples were destroyed when exposed in 10 % Na_2SO_4 solution for 2 years or immersed in 10 % $MgSO_4$ solution for 1 year, while the alkali activated slag cement samples survived well (Hakkinen 1987, 1986).

Bakharev reported that concrete in which a geopolymer was the binder has a very different durability when exposed in sulfate solutions with the stability of the geopolymeric specimens tested depending on factors such as the type of activator used in specimen preparation, the concentration, and type of cation in the sulfate media. The results of the study are summarized in Table 1. The most significant deterioration reported was for exposure in a sodium sulfate solution where it appeared to be related to the migration of alkalies into solution. In a magnesium sulfate solution, migration of alkalies into the solution and magnesium and calcium diffusion to the subsurface areas was reported for a fly ash geopolymer

[1]Department of Civil, Architectural, and Building, Coventry University, Coventry CV1 5BF, UK.
*Corresponding Author; E-mail: dlbondar@gmail.com;dali.bondar@coventry.ac.uk

[2]Department of Civil and Structural Engineering, University of Sheffield, Sheffield S1 3JD, UK.

[3]Milestone and Associates Ltd, Lower Hutt 5010, New Zealand.

[4]Research Centre of Natural Disasters in Industry, P.W.U.T., Tehran, Iran.

Table 1 Fluctuations of strength of fly ash based geopolymers in sodium and magnesium sulfate solutions, Bakharev (2005).

Immersed solutions	Activators	Fluctuations of strength
Sodium sulfate	Sodium hydroxide	4 % strength increase
	Sodium hydroxide and potassium hydroxide	65 % strength reduction
	Sodium silicate	18 % strength reduction
Magnesium sulfate	Sodium hydroxide	12 % strength increase
	Sodium hydroxide and potassium hydroxide	35 % strength increase
	Sodium silicate	24 % strength reduction

prepared using sodium silicate and a mixture of sodium and potassium hydroxides as activators (Bakharev 2005).

It was shown by Rangan (2008) that heat-cured, low calcium fly ash-based geopolymer concrete exhibited high resistance to sulfate immersion and attack. Specimens exposed in sodium sulfate for up to 1 year showed no visual signs of surface deterioration, cracking or spalling and compressive strength values remained equivalent to those obtained prior to immersion. Moreover, shrinkage was insignificant and was less than 0.015 % of the original dimensions (Rangan 2008).

Chotetanorm studied the resistance of high-calcium lignite bottom ash (BA) geopolymer mortars to sulfate attack. With median particle sizes of 16, 25, and 32 μm the BA was activated with NaOH or sodium silicate and temperature cured to prepare the geopolymer. Relatively high strengths of 40.0–54.5 MPa were obtained for the high-calcium BA geopolymer mortars with the use of fine BA improving the strength and resistance of mortars to sulfate attack. That superior performance was attributed to the high degree of reaction of fine BA giving rise to a low amount of large pores 0.05–100 μm compared with those of a coarse BA. The incorporation of additional water improved the workability of mixes, but the compressive strength, sorptivity, and resistance to sulfate attack decreased due to the increase in the numbers of large pores (Chotetanorm et al. 2013).

It seems that the use of the majority of pozzolans improves the sulfate resistance of mixes where geopolymer binders based on alkali activated natural pozzolans over that of ordinary Portland cement concrete. The simplest explanation for this is the lack of C_3A content, normally present if OPC was used (Bondar 2009). In geopolymer concrete the aluminates, that are prone to attack in OPC concrete along with monosulfate, are held in stable aluminosilicate hydrates which are more resistant to sulfate solutions. Secondly, in comparison with OPC, there is also much less calcium present in natural pozzolans to provide gypsum precipitation. In addition, the type of activator and the curing regime can affect the sulfate attack resistance of alkali activated pozzolan concrete mixture. The use of non-silicate alkaline activators increased the sulfate attack resistance of alkali-activated fly ash and slag cement in a $MgSO_4$ solution (Bakharev 2005; Shi et al. 2006) while steam curing of specimens made with water-glass with a modulus from 1 to 3 decreased the sulfate attack resistance compared to the specimens cured under normal conditions (Shi et al. 2006).

The objective of the present experimental program was to evaluate the performance of natural aluminosilicate based geopolymer concrete in sodium and magnesium sulfate solution. The focus of this paper is on sulfate resistance of alkali activated natural pozzolan concrete mixes and in this regard two types of pozzolanic materials (Taftan and Shahindej) were selected and the effect of two parameters including w/b ratio and different curing temperature and condition were just investigated for Taftan pozzolan. Shahindej pozzolan was studied as another type of natural pozzolan to show the effect of variety of material in raw and calcined form. The performance was assessed on the basis of compressive strengths, expansion and capillary water absorption of the specimens and combined with determination of phases formed discussed in this paper.

2. Experimental Work

2.1 Material and Mixing Procedure

Two natural pozzolans were used, Taftan andesite, the most reactive natural pozzolan in Iran, and Shahindej dacite, both of which are used to produce local Portland pozzolan cement. Optical microscopy showed that the Taftan pozzolan contained feldspar (sodic plagioclase, albite and hornblende), amphibole, quartz, and biotite, whereas Shahindej contained a sodium zeolite, clinoptilolite, albite, quartz and calcite (Ezatian 2002). By calcining Shahindej at 800 °C, the clinoptilolite was converted to opal which rapidly reacts with an alkaline solution (Bondar 2009). Calcination was not considered for Taftan pozzolan since it was applicable in ambient temperature in raw form but shahindej pozzolan needs calcinations to be applicable. Powder X-ray diffraction (XRD) was used to confirm the mineral compositions and the traces are presented in Fig. 1a–c. The chemical compositions were analysed using X-ray fluorescence (XRF) and are presented in Table 2 along with some physical properties.

Potassium hydroxide (KOH) pellets were dissolved in water to produce the alkaline solutions needed for geopolymer concrete production. Sodium silicate solution containing 8.5 % of sodium oxide (Na_2O), 26.5 % of silicon oxide (SiO_2) and 65 % of water; pH = 11.4 was used. The coarse aggregate used in this study was 14 mm and the fine sand 4.75 mm in size. The water absorption coefficient and

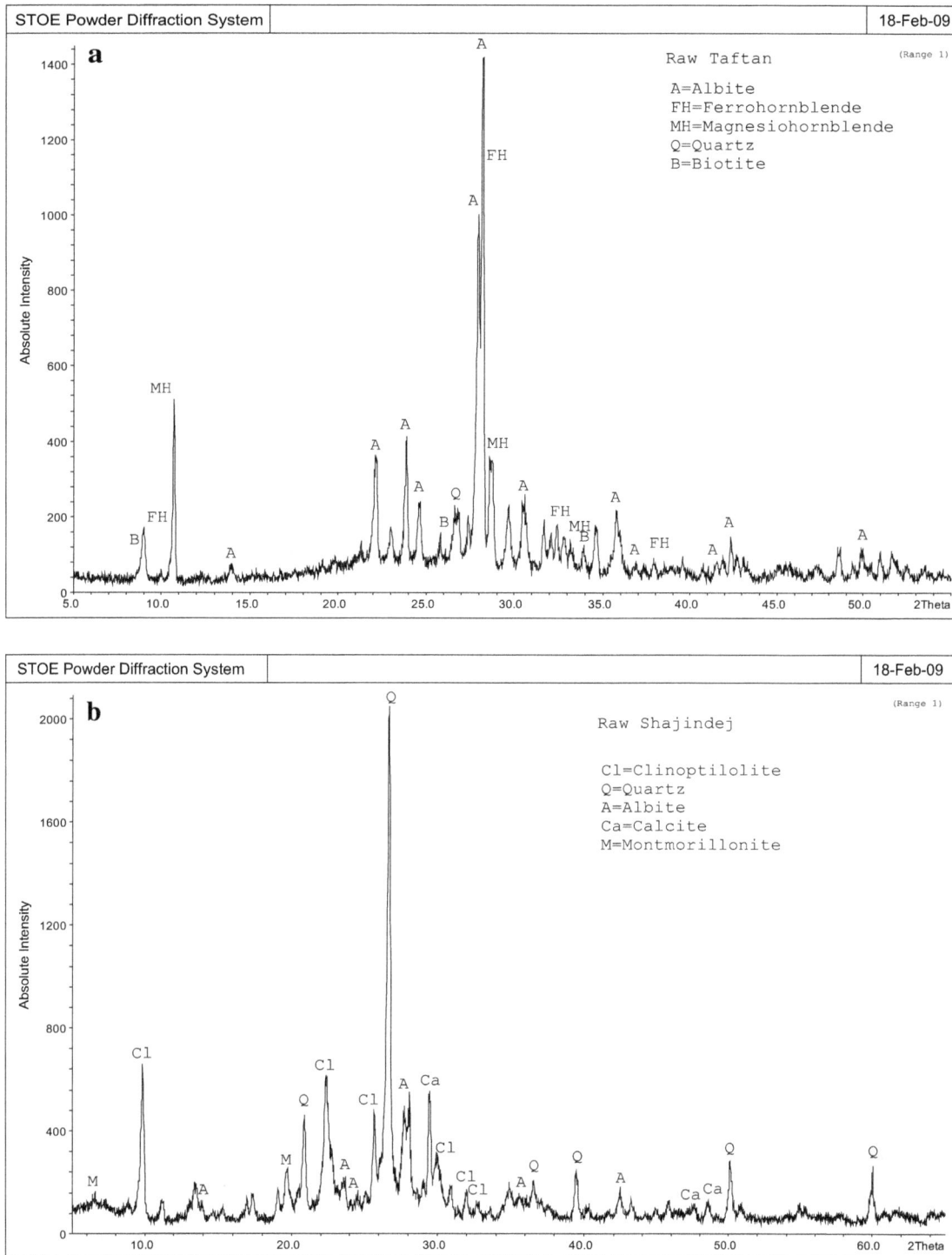

Fig. 1 a Mineralogical composition of Taftan andesite. b Mineralogical composition of Shahindej dacite. c Mineralogical composition of calcined Shahindej dacite.

bulk specific gravity of the saturated surface dry (SSD) aggregates were 0.6 % and 2.62 for the sand while for the gravel they were 0.9 % and 2.6. The fineness modulus of the combined aggregates used for making the concrete mixes was 2.8.

The proportions of the control concrete were based on the BRE (UK) method for design of normal concrete mixtures targeting a 40 MPa (28 days) compressive strength concrete with a slump of 60 mm (Neville 1995). The OPC binder was substituted with an equal quantity (by weight) of the natural pozzolan plus the solids in water

glass and the alkaline solution which was considered as the total binder to calculate the w/b. The mixture calculations including mix water, were made based on the optimum amount of activator needed to activate the pozzolan with no additional water added as the activators were already in solution (Bondar et al. 2011). The details of the different mixtures are presented in Table 3 and the notation for the mixtures is as follows:

ATAF1: Activated raw Taftan pozzolan mixture with w/b = 0.45.

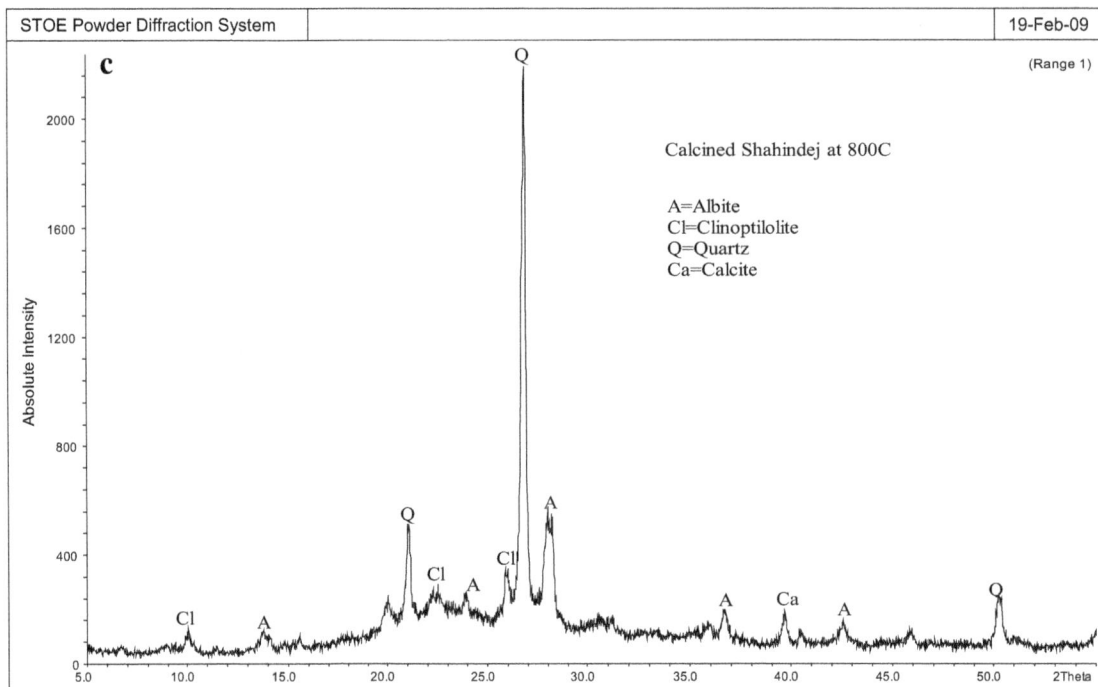

Fig. 1 continued

Table 2 Physical properties and chemical composition (oxide percent) of the materials used in this investigation.

Material	LOI (%)	SiO₂ (%)	Al₂O₃ (%)	Fe₂O₃ (%)	CaO (%)	MgO (%)	TiO₂ (%)	K₂O (%)	Na₂O (%)	Surface area (cm²/g)	Specific gravity
Taftan andesite	1.85	61.67	15.90	4.32	7.99	2.04	0.44	2.12	3.21	3,836	2.22
Shahindej dacite	10.28	70.13	11.11	1.27	2.52	0.92	0.14	2.25	1.01	10,621	2.20
Shahindej dacite-800 °C	5.78	73.44	11.88	1.30	2.55	0.98	0.15	2.30	1.10	5,500	2.25

Table 3 Concrete mix proportion.

Mixture no.	Pozzolan (kg/m³)	KOH (kg/m³)	Na₂SiO₃ (cc/m³)	Water (kg/m³)	Total water (kg/m³)	Total binder (kg/m³)	Fine agg. (kg/m³)	Coarse agg. (kg/m³)	W/B %
ATAF1	325	66	34	158	180	403	578	1,229	0.45
ATAF2	272	72	37	171	195	357	702	1,121	0.55
ARSH	350	67	32	159	180	429	499	1,283	0.42
ACSH	350	67	32	159	180	429	499	1,283	0.42

ATAF2: Activated raw Taftan pozzolan mixture with w/b = 0.55.
ARSH: Activated raw Shahindej pozzolan mixture with w/b ratio = 0.42.
ACSH: Activated calcined Shahindej pozzolan mixture with w/b ratio = 0.42.

The concrete samples were cast in different sized moulds according to the tests described below. The mixtures were cast into pre-oiled moulds in three tamped layers and a vibrating table used to remove any entrapped air. It was observed that geopolymer concrete stuck to the mould and oiling of the moulds was very important to ensure release of the samples. Specimens were left at room temperature, covered by a plastic sheet and de-moulded after 24 h. Taftan specimens were exposed in two curing regimes and three different temperatures:

Table 4 Material type, curing regimes and exposure conditions for different mixes.

Mixture no.	Material type	Curing regime	Exposure conditions
ATAF1*	Raw Taftan	Sealed-20 °C	In a sulfate solution after 28 days curing (in)
			In curing regime till test date (out)
		Sealed-40 °C	In a sulfate solution after 28 days curing (in)
			In curing regime till test date (out)
		Fog-40 °C	In a sulfate solution after 28 days curing (in)
			In curing regime till test date (out)
		Sealed-60 °C	In a sulfate solution after 28 days curing (in)
			In curing regime till test date (out)
ATAF2	Raw Taftan	Sealed-20 °C	In a sulfate solution after 28 days curing (in)
			In curing regime till test date (out)
		Sealed-40 °C	In a sulfate solution after 28 days curing (in)
			In curing regime till test date (out)
		Fog-40 °C	In a sulfate solution after 28 days curing (in)
			In curing regime till test date (out)
		Sealed-60 °C	In a sulfate solution after 28 days curing (in)
			In curing regime till test date (out)
ARSH*	Raw Shahindej	Sealed-60 °C	In a sulfate solution after 28 days curing (in)
			In curing regime till test date (out)
ACSH*	Calcined Shahindej	Sealed-20 °C	In a sulfate solution after 28 days curing (in)
			In curing regime till test date (out)

* For expansion test one set of samples was put in pure water after 28 days curing in the related regime to compare the result with samples ponded in sulfate solution.

(1) *Sealed curing* Three series of specimens were closely wrapped in a special plastic covering shown to be impermeable to water and stored at controlled temperatures of 20 ± 2, 40 ± 2 and 60 ± 2 °C and less than 70 % RH.
(2) Fog curing at 40 ± 2 °C and 98 % RH was used for all of the measurements. To measure the changes of capillary water absorption of the specimens, fog curing at temperatures of 20 ± 2 and 60 ± 2 °C and 98 % RH was considered as well.

Raw and calcined Shahindej specimens both were all cured under sealed curing conditions. It was found that raw Shahindej pozzolan needed at least 60 °C curing to provide moderate to high strength at early stages while the optimum temperature for curing calcined Shahindej was found to be

20 °C. Therefore the ARSH concrete mixtures were cured at 60 °C and ACSH mixtures were cured at 20 °C and less than 70 % RH (Bondar 2009).

2.2 Sample Preparation and Test Method

Comparisons were made between specimens that had been immersed in a sulfate solution after 28 days curing and those that were cured as previously described in sealed or fog conditions till testing date. Material types, curing regimes and exposure conditions for different mixes were summarized in Table 4. Specimens were cast as $100 \times 100 \times 100$ mm cubes (according to BS: 1881: part 116: 1983) for measuring the change of compressive strength and capillary water absorption and as $75 \times 75 \times 285$ mm concrete prisms (according to ASTM C 490) in order to measure the expansion. After de-moulding, the samples were cured as described above using the different curing conditions and temperatures for 28 days. The specimens were then immersed in a solution containing 2.5 % Na_2SO_4 and 2.5 % $MgSO_4$ by weight of water. The samples were placed in containers and left in a temperature controlled room at 20 °C for 6 months with some samples being retained for measuring changes of compressive strength over 2 years. The container solutions were replaced every 2 weeks for the first 3 months and then retained.

2.3 Test Procedure

The methods which test sulfate attack, are designed to assess four properties:

- changes in capillary water absorption (in accordance with the method specified by RILEM-CPC-11.2) (RILEM 1994) and
- compressive strengths of the specimens (according to BS: 1881: part 116: 1983),
- dimensional changes of the specimens which was measured according to ASTM C1012-95a (Astm and 1012–95a 1995), and
- the chemical phases using XRD.

2.4 Water Absorption and Compressive Strength

To assess the variations in capillary water absorption and compressive strength between specimens subjected to sulfate attack and those cured normally in sealed or fog conditions, 100 mm concrete cubes were cast and cured under the different curing conditions. Water absorption measurements were made at different ages up to 180 days for samples exposed in tap water (Na = 105, K = 30, Ca = 120, Mg = 50, Cl = 225, SO_4 = 85 ppm) and to the sulfate solution. In the test procedure specimens were dried to constant weight, the weight noted and the samples were then immersed in tap water or the sulfate solution. At specified times; samples were removed and weighed again. Absorption is expressed as the increase in weight as a percentage of the original weight.

The compressive strength of samples that had been exposed both in and out of the sulfate solution were measured periodically up to 2 years to examine the changes in the strength of cubes exposed in the sulfate solution and those left to cure normally in sealed or fog conditions. For any one mix and age, three cubes were tested.

2.5 Expansion Test

To measure rates of expansion, $75 \times 75 \times 285$ mm concrete prisms were prepared, exposed in sulfate solution and measured (One reference prism was exposed in pure water in order to compare the measurements). The amount of the expansion was calculated by measuring the specimens' length using the length comparator. For each mix three specimens were tested.

2.6 X-ray Diffraction Test

XRD was used to identify the crystalline products present in samples after they had been immersed in sulfate solution for up to 90 days. For analysis, a thin layer 0–5 mm in depth was carefully ground from the sample surface and examined by powder XRD. Its XRD trace was compared to that obtained for a powder prepared from a sample taken from the middle of the sample. This showed there was a difference in phases between the centre and edge of the sample after immersion in sulfate solution.

3. Results and Discussion

The percentage of water absorption for concrete specimens immersed in tap water and sulfate solutions for 180 days are presented in Fig. 2. All specimens recorded an increase in weight over the duration of exposure. The pattern of weight gain is similar in the two series of alkali activated Taftan pozzolan for the various curing regimes. The maximum increase in weight was observed for the ATAF1 specimen (w/b = 0.45) cured at 60 °C and the least gain in weight for the ATAF2 specimen (w/b = 0.55) cured at 20 °C under sealed conditions. The weight gain across all specimens was in the range of 5.1–7.0 %. The increase in weight of ATAF2

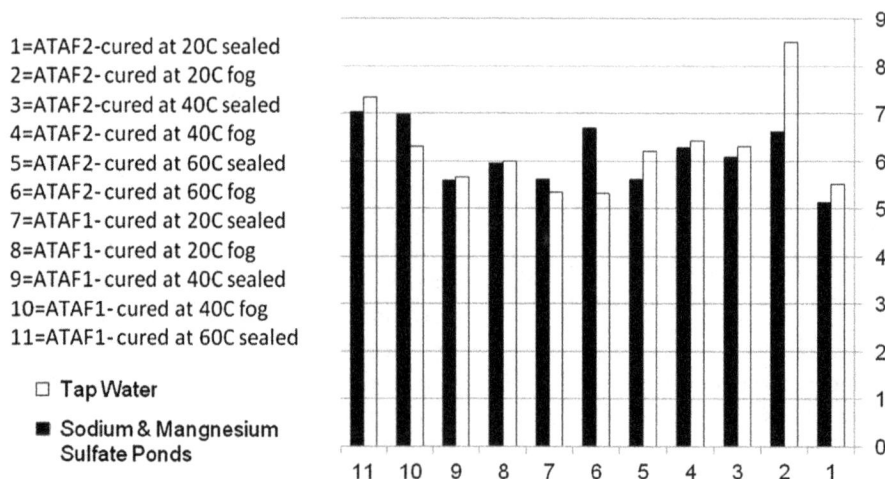

1=ATAF2-cured at 20C sealed
2=ATAF2-cured at 20C fog
3=ATAF2-cured at 40C sealed
4=ATAF2-cured at 40C fog
5=ATAF2-cured at 60C sealed
6=ATAF2-cured at 60C fog
7=ATAF1-cured at 20C sealed
8=ATAF1-cured at 20C fog
9=ATAF1-cured at 40C sealed
10=ATAF1-cured at 40C fog
11=ATAF1-cured at 60C sealed

□ Tap Water

■ Sodium & Mangnesium Sulfate Ponds

Fig. 2 Water absorption (in percent) in AANP concretes exposed in tap water and sodium & magnesium sulfate ponds.

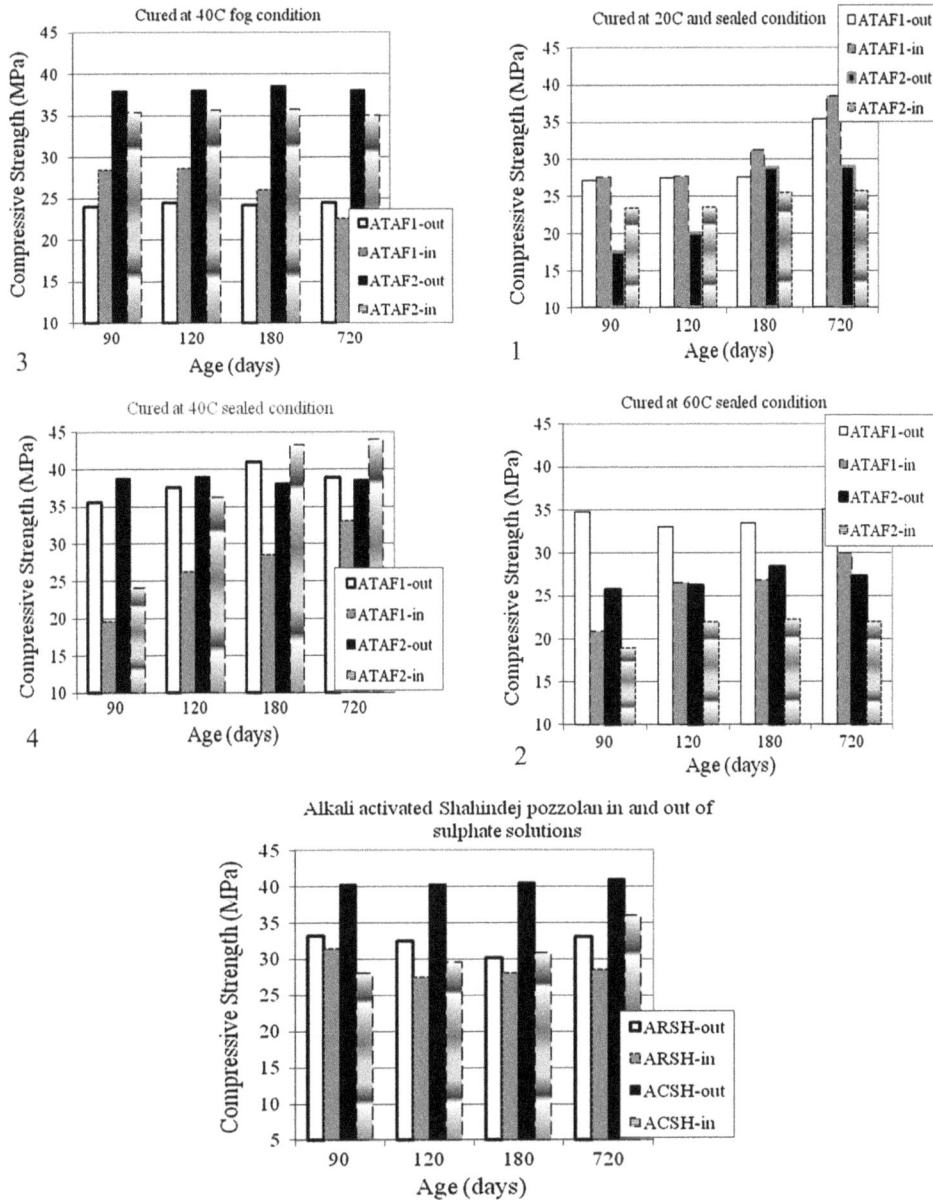

Fig. 3 Compressive strength for ATAF1 and ATAF2 mixes cured under different condition and temperatures, ARSH, and ACSH in and out of the sulfate solution.

specimens in sulfate solution for w/b = 0.55 cured at 60 °C in fog curing condition and ATAF1 specimen (w/b = 0.45) cured at 20 °C in sealed and at 40 °C in fog curing condition was more than the specimens immersed in tap water which might be due to white deposit within the surface pores. Higher amounts of absorption were recorded for all of the fog cured samples which may be due to a more open microstructure. For these samples permeability was higher than for comparable samples cured in a sealed condition (Bondar et al. 2012). Although the water in the geopolymer structure is just a medium that promotes the geopolymerization with further cross linking (Si–O–Al or Si–O–Si bonding) and poly condensation made, the phenomenon is related to more water being retained in the pores, resulting in a more porous microstructure giving rise to higher absorption and follows that found by Zuhua et al. (2009) for calcined kaolin-based geopolymer.

Figure 3 represents the evolution over time of compressive strengths for AANP concrete specimens that are exposed in the sulfate solutions after being cured for 28 days in a specific regime and specimens cured with same curing regime till testing date. The compressive strength results for each case were determined from average of three cubes. It seems there would be the movement of sulfate into the geopolymer. This is related to the absorption capacity of the geopolymer and thus the one absorbed more would have a large amount of sulfate in the matrix and hence affects the strength and the expansion. The strength of all sulfate cured concretes at 180 days was less than that for the samples cured outside of the solution initially, with the exception of the ATAF1 samples cured at 20 °C in sealed and 40 °C fog conditions and the ATAF2 samples cured at 40 °C in sealed conditions. The trend of compressive strength development is to increase, except for Taftan samples cured at 40 °C under fog

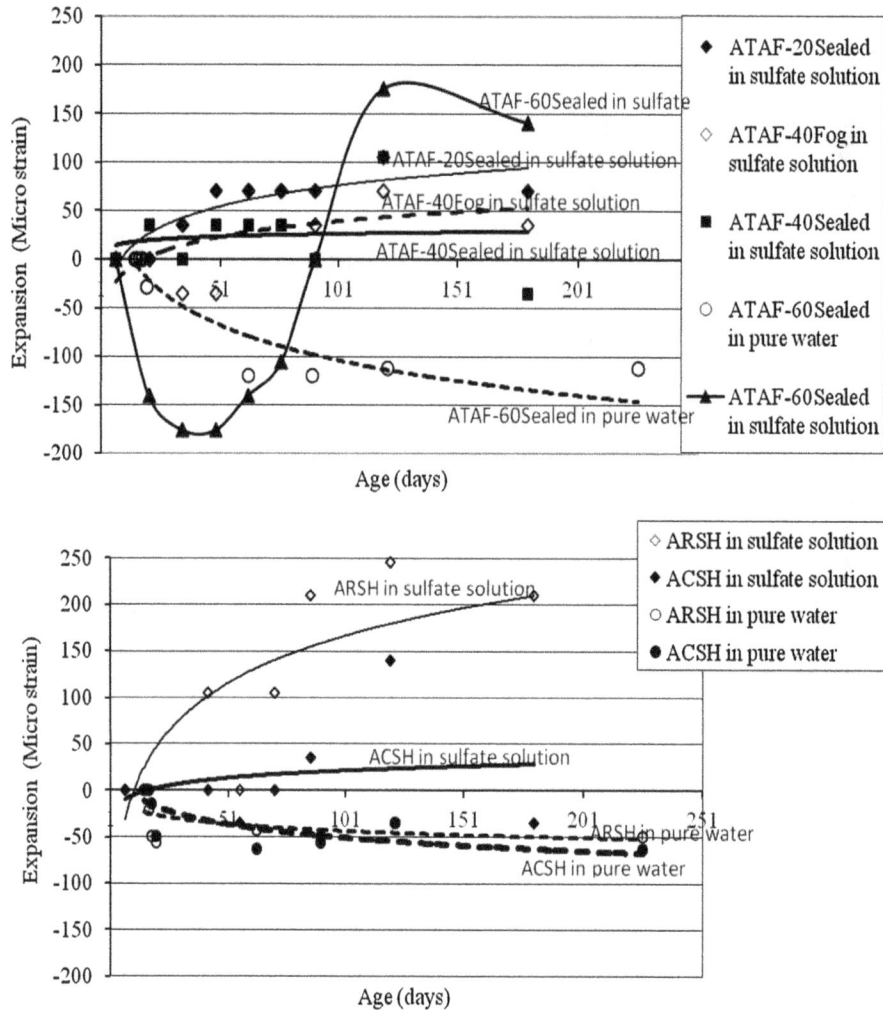

Fig. 4 Expansion at various ages for geopolymer mortar mixes based on alkali activated natural pozzolan in sulfate solution.

conditions and ARSH concrete mixes at this age. The results confirm previous investigations carried out on alkali activated cements (Bakharev 2005; Bakhareva et al. 2002). Lower w/b ratio resulted in higher strength for corresponding specimens that were exposed inside and outside of the sulfate solutions. However, a longer period of exposure to aggressive solutions was conducted to confirm the sulfate resistance of AANP concrete. The strength of all sulfate exposure AANP concrete samples was 8–19.5 % less than those cured outside of the solution by immersing samples for 2 years with the exception of Taftan samples with w/c = 0.45 cured at 20 °C under sealed conditions and Taftan samples with a w/c = 0.55 cure at 40 °C sealed conditions. For these samples, the strength of all sulfate cured concrete was 8.5 and 14 % more than the samples cured outside of the solution, respectively. For 5 months sulfate cured OPC concrete exposed to the same regime, the loss of strength was up to 38 % (Bakharev 2005).

The results of expansion tests against time are shown in Fig. 4. Alkali activated natural pozzolan which has the structure similar to zeolite, partly dehydrates, losing a part of its water molecules at ambient temperatures when the humidity drops below about 60 % and then rehydrates back

form when saturated with water. This reversible hydration associated with wetting and drying is accompanied by a volume change which is sufficient to cause expansion and shrinkage effects in structures which is obvious for Taftan samples cured at 60 °C under sealed conditions. The highest absolute expansion was recorded for the ARSH mortar prism which was most affected by sulfate solution and the lowest amount was recorded for ATAF cured at 40 °C and sealed condition after 6 months immersion. This follows the strength patterns reported previously (Bondar et al. 2011), and may be due to more geopolymer solids in the system causing lower permeability and a decrease of the penetration of sulfate ions so fewer expansive products are formed. The maximum expansion recorded for AANP concrete in this investigation was 0.074 % after 6 months immersion. Swamy reported that very poor durability was shown by OPC samples prepared at w/c = 0.45 and immersed in a sulfate solution. The highest expansion was recorded as 0.25 and 0.35 % for OPC samples immersed in Na and Mg sulfate solutions for 42 days, respectively (Swamy 1998). The maximum percentage of expansion recorded for ARSH mixes was 0.086 % at 6 months and the trend is for this to decrease after this time. Therefore the maximum percentage expansion measured for AANP concrete in this study is less than the 0.1 % which ASTM C1012 Standard (Astm and 1012–95a 1995) suggests as acceptable for OPC concretes in moderate sulfate exposure conditions at 6 months and later.

The crystalline reaction compounds taken from surfaces and middles of cubes were analysed by XRD and the phases present are shown in Fig. 5a–c and summarized in Table 5. It can be seen that the peaks for the reaction products taken from the surface of alkali activated natural pozzolan mortar specimens, are higher than for the same compounds taken from the middle of cubes, after 3 month exposure in the sulfate solutions. For the activated Taftan samples the main crystalline compounds present in both the surface and middle part of specimens were albite (Na-AlSi$_3$O$_8$) and quartz (SiO$_2$) with hornblende and clinoptilolite present for the activated Shahindej samples. The

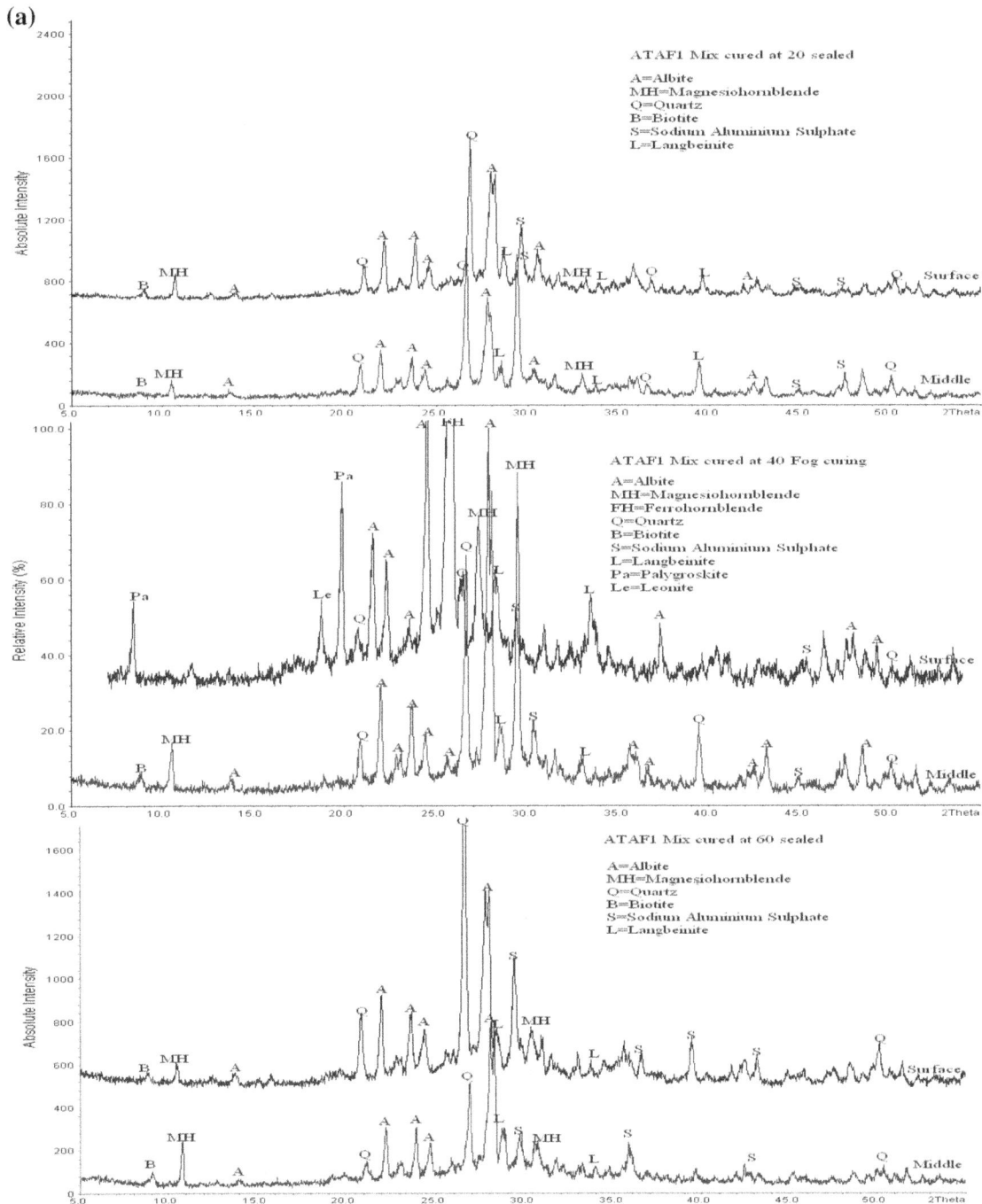

Fig. 5 **a** X-ray diffraction results show the existence of sulfate phases achieved from the powder prepared from the surface and the middle of the samples of ATAF1 mixtures cured at 20 ± 2, 40 ± 2 and 60 ± 2 °C and immersed in sulfate solution. **b** X-ray diffraction results show the existence of sulfate phases and achieved from the powder prepared from the surface and the middle of the samples of ATAF2 mixtures cured at 40 ± 2 and 60 ± 2 °C immersed in sulfate solution. **c** X-ray diffraction results show the existence of sulfate phases and achieved from the powder prepared from the surface and the middle of the samples of ARSH mixtures cured at 60 ± 2 °C and ACSH mixtures cured at 20 ± 2 °C immersed in sulfate solution.

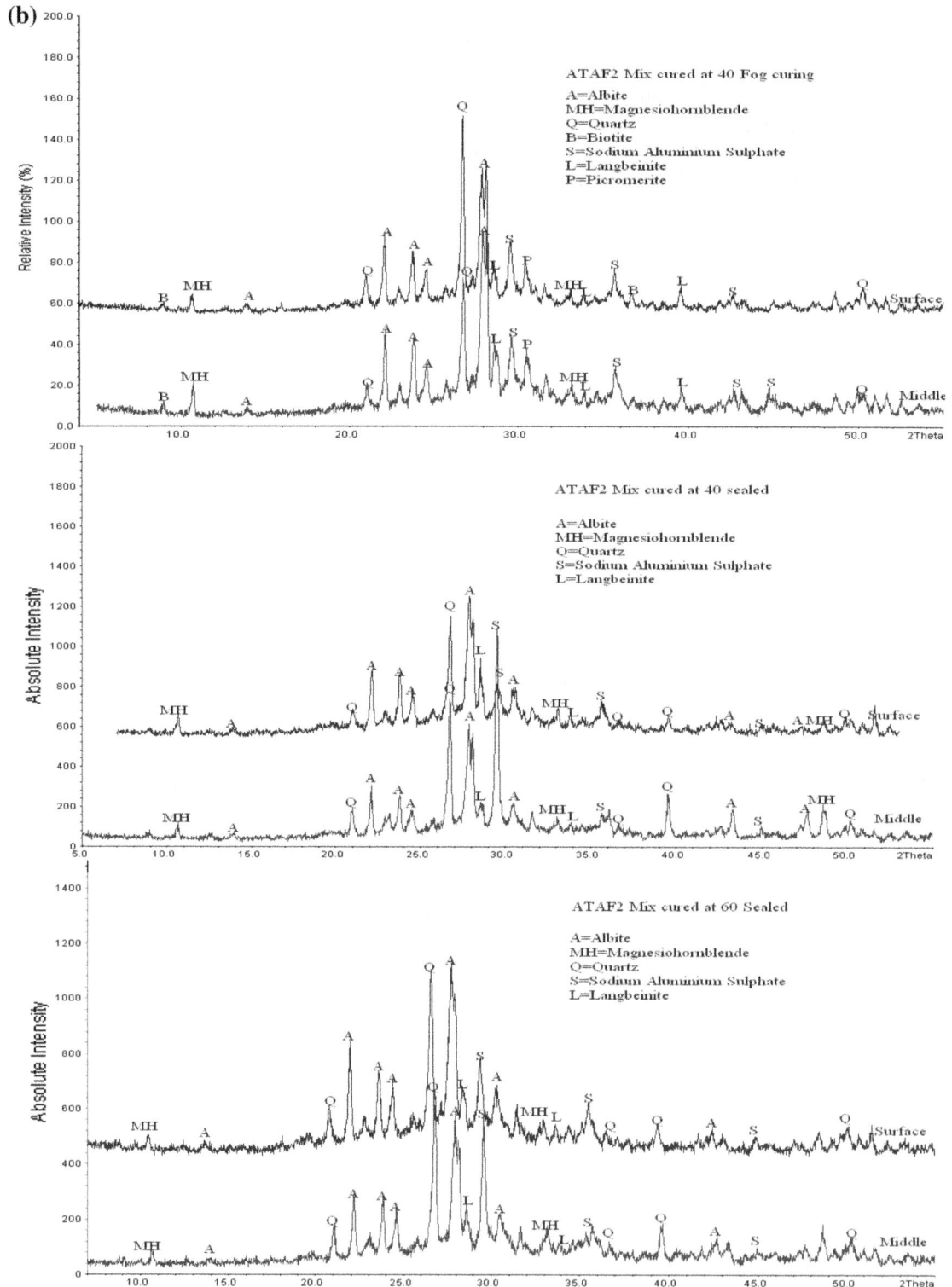

Fig. 5 continued

reaction products consisted mostly of sodium aluminium sulfate [$Na_3Al(SO_4)_3$] and/or langbeinite [$K_2Mg_2(SO_4)_3$] which may cause expansion in this type of concrete particularly at the edge of the samples. By comparing the X-ray traces of samples taken from the surface with those from the middle of the specimen shows that for activated, calcined Shahindej samples, the penetration of sulfate is very limited while for the activated, raw Shahindej, the

peaks only relate to sodium aluminium sulfate. For the ATAF1 mix cured at 40 °C some hydrated sulfate salts such as leonite [$K_2Mg(SO_4)_2.4H_2O$] were detected on the surface of samples. Although the intensity of sulfate crystalline compound peaks in the middle of the specimens is lower than what was found for the surface of the specimens, the results show that in this type of concrete, sulfate could penetrate in the concrete. This can be confirmed

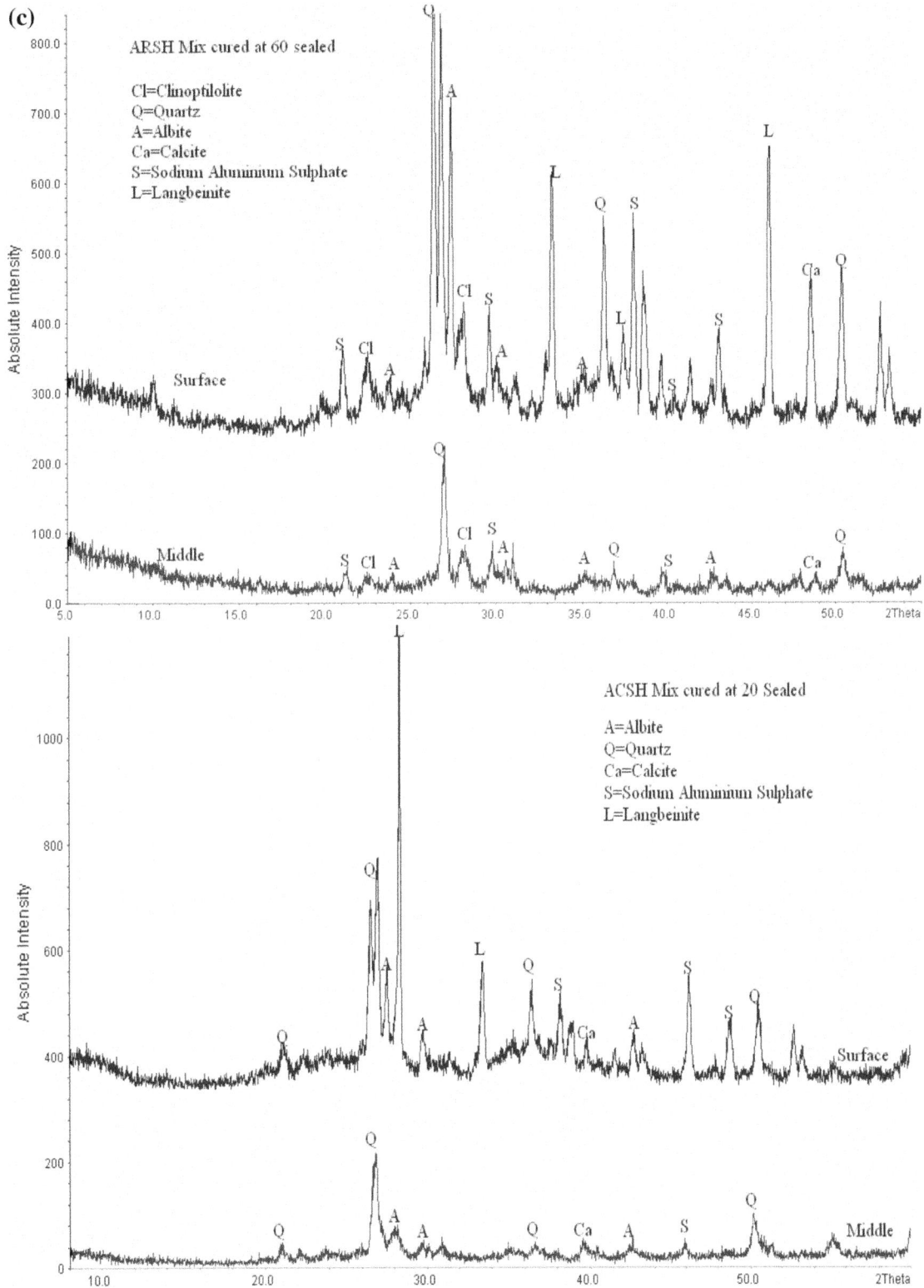

Fig. 5 continued

since the trace of sulfate combination was present in the core of all samples except ACSH. However, further studies exposing samples for 2 years in sulfate solution and strength measurement, has shown that the crystalline products formed in sulfate cured samples do not greatly reduce the resistance of AANP concrete.

Table 5 Summary of X-ray diffraction results show the existence of sulfate phases and achieved from the powder prepared from the surface and the middle of the samples immersed in sulfate solution.

Mix	Water to binder ratio (W/B)	Curing condition and temperature	Position of sampling	Sulfate phases	Mineral name of sulfate phases
ATAF1	0.45	Sealed curing at 20 °C	Surface	$Na_3Al(SO_4)_3$	Sodium aluminium sulfate
				$K_2Mg_2(SO_4)_3$	Langbeinite
ATAF1	0.45	Sealed curing at 20 °C	Middle	$Na_3Al(SO_4)_3$	Sodium aluminium sulfate
				$K_2Mg_2(SO_4)_3$	Langbeinite
ATAF1	0.45	Fog curing at 40 °C	Surface	$Na_3Al(SO_4)_3$	Sodium aluminium sulfate
				$K_2Mg_2(SO_4)_3$	Langbeinite
				$(MgAl)_5(SiAl)_8O_{20}(OH)_2 \cdot 8H_2O$	Palyigorskite
				$K_2Mg(SO_4)_2 \cdot 4H_2O$	Leonite
ATAF1	0.45	Fog curing at 40 °C	Middle	$Na_3Al(SO_4)_3$	Sodium aluminium sulfate
				$K_2Mg_2(SO_4)_3$	Langbeinite
ATAF1	0.45	Sealed curing at 40 °C	Surface	$Na_3Al(SO_4)_3$	Sodium aluminium sulfate
				$K_2Mg_2(SO_4)_3$	Langbeinite
				$(MgAl)_5(SiAl)_8O_{20}(OH)_2 \cdot 8H_2O$	Palyigorskite
				$K_2Mg(SO_4)_2 \cdot 4H_2O$	Leonite
ATAF1	0.45	Sealed curing at 40 °C	Middle	$Na_3Al(SO_4)_3$	Sodium aluminium sulfate
				$K_2Mg_2(SO_4)_3$	Langbeinite
ATAF1	0.45	Sealed curing at 60 °C	Surface	$Na_3Al(SO_4)_3$	Sodium aluminium sulfate
				$K_2Mg_2(SO_4)_3$	Langbeinite
ATAF1	0.45	Sealed curing at 60 °C	Middle	$Na_3Al\ SO_4)_3$	Sodium aluminium sulfate
				$K_2Mg_2(SO_4)_3$	Langbeinite
ATAF2	0.55	Fog curing at 40 °C	Surface	$Na_3Al(SO_4)_3$	Sodium aluminium sulfate
				$K_2Mg(SO_4)_2 \cdot 6H_2O$	Picromerite
ATAF2	0.55	Fog curing at 40 °C	Middle	$Na_3Al(SO_4)$	Sodium
				$K_2Mg_2(SO_4)_3$	Langbeinite
				$K_2Mg(SO4)_2 \cdot 6H_2O$	Picromerite

Table 5 continued

Mix	Water to binder ratio (W/B)	Curing condition and temperature	Position of sampling	Sulfate phases	Mineral name of sulfate phases
ATAF2	0.55	Sealed curing at 40 °C	Surface	$Na_3Al(SO_4)_3$	Sodium aluminium sulfate
				$K_2Mg_2(SO_4)_3$	Langbeinite
ATAF2	0.55	Sealed curing at 40 °C	Middle	$Na_3Al(SO_4)_3$	Sodium aluminium sulfate
				$K_2Mg_2(SO_4)_3$	Langbeinite
ATAF2	0.55	Sealed curing at 60 °C	Surface	$Na_3Al(SO_4)_3$	Sodium aluminium sulfate
				$K_2Mg_2(SO_4)_3$	Langbeinite
ATAF2	0.55	Sealed curing at 60 °C	Middle	$Na_3Al(SO_4)_3$	Sodium aluminium sulfate
				$K_2Mg_2(SO_4)_3$	Langbeinite
ACSH	0.42	Sealed curing at 20 °C	Surface	$Na_3Al(SO_4)_3$	Sodium aluminium sulfate
				$K_2Mg_2(SO_4)_3$	Langbeinite
ACSH	0.42	Sealed curing at 20 °C	Middle	$Na_3Al(SO_4)_3$	Sodium aluminium sulfate
ARSH	0.42	Sealed curing at 60 °C	Surface	$Na_3Al(SO_4)_3$	Sodium aluminium sulfate
				$K_2Mg_2(SO_4)_3$	Langbeinite
ARSH	0.42	Sealed curing at 60 °C	Middle	$Na_3Al(SO_4)_3$	Sodium aluminium sulfate

4. Conclusions

The main results from this investigation on the sulfate resistance of alkali activated, natural pozzolan concrete are summarised as follows:

(1) The absorption percent of AANP concrete after exposure in sulfate solution is greater than 5.1 % but less than 7 % of the weight.

(2) Samples that were fog cured show a higher percentage of absorption. This phenomenon may be related to the retention of water by the geopolymer matrix giving rise to a more open microstructure in this type of concrete.

(3) The maximum expansion was recorded for AANP concrete in this investigation at 0.074 % after 6 months immersion in sulfate solution.

(4) The maximum loss of compressive strength for sulfate cured AANP concrete was 19.5 % following 2 years immersion in sulfate solution.

Acknowledgments

This research described has been lead by the Department of Civil and Structural Engineering, University of Sheffield, Sheffield, UK, with experimental work conducted in the concrete technology laboratory in Civil Department of P.W.U.T., Tehran, Iran. The authors express their gratitude to the Research Centre of Natural Disasters in Industry (RCNDI) in P.W.U.T. for support rendered throughout the research program. X-ray diffraction (XRD) was analysed in the Department of Engineering Materials, The University of Sheffield and X-ray Fluoresence (XRF) analysis was detected in Kansaran Binaloud X-ray laboratory in Tehran, Iran.

References

ASTM C 1012-95a. (1995). standard test method for length change of hydraulic-cement mortars exposed to a sulfate solution, American Society for Testing and Materials, PA.

Bakharev, T. (2005). Durability of geopolymer materials in sodium and magnesium sulfate solutions. *Cement and Concrete Research, 35*, 1233–1246.

Bakhareva, T., Sanjayana, J. G., & Cheng, Y. B. (2002). Sulfate attack on alkali-activated slag concrete. *Cement and Concrete Research, 32*, 211–216.

Bondar, D. (2009). Alkali activation of Iranian natural pozzolans for producing geopolymer cement and concrete. A dissertation submitted to University of Sheffield in fulfilment of the requirements for the degree of Doctor of Philosophy, UK.

Bondar, D., Lynsdale, C. J., Milestone, N. B., & Hassani, N. (2012). Oxygen and chloride permeability of alkali activated natural pozzolan concrete. *ACI Materials, 109*(1), 53–62.

Bondar, D., Lynsdale, C. J., Milestone, N. B., Hassani, N., & Ramezanianpour, A. A. (2011a). Effect of type, form and dosage of activators on strength of alkali-activated natural pozzolans. *Cement and Concrete Composites, 33*(2), 251–260.

Bondar, D., Lynsdale, C. J., Milestone, N. B., Hassani, N., & Ramezanianpour, A. A. (2011b). Engineering properties of alkali-activated natural pozzolan concrete. *ACI Materials, 108*(1), 64–72.

Chotetanorm, C., Chindaprasirt, P., Sata, V., Rukzon, S., & Sathonsaowaphak, A. (2013). High-calcium bottom ash geopolymer: Sorptivity, pore size, and resistance to sodium sulfate attack. *Journal of Materials in Civil Engineering, 25*(1), 105–111.

Ezatian, F. (2002). Atlas of igneous rocks: Classification and nomenclatures. Ministry of Industries and Mines, Geological Survey of Iran (GSI) (in Farsi).

Ganjian, E., & Pouya, H. S. (2005). Effect of magnesium and sulfate ions on durability of silica fume blended mixes exposed to the seawater tidal zone. *Cement and Concrete Research, 35*, 1332–1343.

Hakkinen, T. (1986) Properties of alkali activated slag concrete. VTT Research Notes, Technical Research Centre of Finland (VTT), Finland, No. 540.

Hakkinen, T. (1987). Durability of alkali activated slag concrete. *Nordic Concrete Research, 6*, 81–94.

Neville, A. M. (1995). *Properties of concrete*. Essex, UK: Pearson Educational Limited.

Rangan, B. V. (2008). Fly ash-based geopolymer concrete, available at: http://www.yourbuilding.org/display/yb/Fly+Ash-Based+Geopolymer+Concrete. Accessed 2011.

RILEM CPC-11.2. (1994) Absorption of water by concrete by capillarity QTC14-CPC, 1982, RILEM technical recommendations for the testing and use of construction materials, International Union of Testing and Research Laboratories for Materials and Structures, E and FN Spon, UK.

Shi, C., Krivenko, P. V., & Roy, D. (2006). *Alkali-activated cement and concretes*. London, UK: Taylor & Francis.

Swamy, R. N. (1998). *Blended cement in construction*. London, UK: Taylor and Francis.

Turker, F., Akoz, F., Koral, S., & Yuzer, N. (1997). Effect of magnesium sulfate concentration on the sulfate resistance on mortars with and without silica fume. *Cement and Concrete Research, 27*, 205–214.

Zuhua, Z., Xiao, Y., Huajun, Z., & Yue, C. (2009). Role of water in the synthesis of calcined kaolin-based geopolymer. *Applied Clay Science, 43*, 218–223.

12

Experimental Study on Tensile Creep of Coarse Recycled Aggregate Concrete

Tae-Seok Seo[1], and Moon-Sung Lee[2],*

Abstract: Previous studies have shown that the drying shrinkage of recycled aggregate concrete (RAC) is greater than that of natural aggregate concrete (NAC). Drying shrinkage is the fundamental reason for the cracking of concrete, and tensile creep caused by the restraint of drying shrinkage plays a significant role in the cracking because it can relieve the tensile stress and results in the delay of cracking occurrence. However, up till now, all research has been focusing on the compressive creep of RAC. Therefore, in this study, a uniaxial restrained shrinkage cracking test was executed to investigate the tensile creep properties caused by the restraint of drying shrinkage of RAC. The mechanical properties, such as compressive strength, tensile splitting strength, and Young's modulus of RAC were also investigated in this study. The results confirmed that the tensile creep of RAC caused by the restraint of shrinkage was about 20–30 % larger than that of NAC.

Keywords: recycled aggregate concrete, tensile creep, drying shrinkage, uniaxial restrained shrinkage cracking test.

1. Introduction

As vast amounts of waste materials produced during the demolition of concrete structures create environmental pollution, recycling of construction wastes offers a practical alternative to protect the environment (Enad et al. 2013). Therefore, the use of concrete containing demolished concrete material is an important issue for reducing the environmental load. Because aggregate takes up nearly 70 % of concrete volume, the use of coarse recycled aggregate as a partial replacement for natural aggregate in the manufacturing of concrete has become a common practice. This concrete is called recycled aggregate concrete (RAC).

To encourage the usage of construction waste materials, many researchers have executed research on the mechanical characteristics of recycled aggregate (Soares et al. 2014a; Tavakoli and Soroushian 1996a; Valeria 2010) and the durability of recycled aggregate (Bravo et al. 2015; Ryou and Lee 2014; Sherif et al. 2015; Soares et al. 2014b). According to the studies of Domingo et al. (2009), Gholamreza et al. (2011), Rasiah et al. (2012), Tavakoli and Soroushian (1996b), and Xiao et al. (2014), the drying shrinkage of RAC is larger than that of NAC.

The drying shrinkage is the fundamental reason for the cracking of concrete, and cracks occur when the tensile stress caused by the drying shrinkage exceed the tensile strength of the concrete (Tao et al. 2012). Tensile creep plays a significant role in cracking of concrete because it can relieve the tensile stress and results in the delay of cracking occurrence (Garas et al. 2009). However, up till now, all research has been focusing on the compressive creep of RAC (Domingo et al. 2009; Gholamreza et al. 2011; Xiao et al. 2014).

Therefore, in this study, a uniaxial restrained shrinkage cracking test was executed to investigate the tensile creep properties caused by the restraint of the drying shrinkage of RAC. In a uniaxial restrained shrinkage cracking test, specimens were designed to generate cracks caused by the restraint of drying shrinkage. These specimens were also used to investigate tensile stress, tensile creep, and cracking age. The mechanical properties of RAC, such as compressive strength, tensile splitting strength, and Young's modulus, were also investigated in this study.

2. Experimental Program

2.1 Mixture Proportions and Materials

Two different water-to-cement (w/c) ratios of 0.65 and 0.45, and types of coarse aggregate were used to create four different concrete mixtures. In Korea, when recycled aggregates are used for concrete, the design strength of 21–27 MPa is recommended (KMCT, 2005). However, to encourage the usage of recycled aggregates, the design strength of concrete larger than 27 MPa is included in this research.

[1]Hyundai Engineering and Construction Co., Ltd., Yongin 446-716, Korea.

[2]Division of Architecture and Architectural Engineering, Hanyang University, Ansan 426-791, Korea.

*Corresponding Author; E-mail: moonlee@hanyang.ac.kr

Table 1 Mixture proportions.

Specimen	w/c (%)	s/a (%)	Water (kg/m³)	Unit content (kg/m³)			Aggregate type	
				Cement	Fine aggregate	Coarse aggregate	Fine aggregate	Coarse aggregate
65-NC	65	47.0	175	270	834	957	N	N
65-RC						942		R
45-NC	45	43.1	175	389	724	972		N
45-RC						957		R

w/c Water-to-cement ratio, *s/a* fine aggregate proportion, *N* natural, *R* recycled

Table 2 Properties of aggregate.

Specimens	Recycled aggregate	Natural aggregate	
	Coarse aggregate	Coarse aggregate	Fine aggregate
Density SSD (g/cm³)	2.56	2.60	2.56
Dry density (g/cm³)	2.50	2.58	2.51
Absorption ratio (%)	2.50	1.07	2.25
Fineness modulus	6.69	6.54	2.93

Table 1 shows a list of the concrete mixture proportions. The reference mixture made of 100 % natural aggregate is denoted by NC and the mixture made of 100 % coarse recycled aggregate is denoted by RC. The recycled coarse aggregate was added to the mixture after soaking in water, under conditions comparable to saturated surface-dry conditions. Ordinary Portland cement (specific surface area: 3300 cm²/g, specific gravity: 3.16) was used, and superplasticizer was added to the mixture. The slump and air content of all concrete mixtures satisfied the target values (slump: 18 ± 2.5 cm, air content: 4.5 ± 1.5 %). Table 2 shows the physical properties, such as density of saturated surface dry, dry density, absorption ratio, and fineness modulus, of coarse recycled aggregate and natural aggregate. According to Korean Standards (KMCT 2005), the absorption rate of coarse recycled aggregates should be smaller than 3 %, so the coarse recycled aggregates with the

absorption rate of 2.5 % are used for the mixture. Figure 1 shows the size distribution of coarse aggregate.

2.2 Shape and Kind of Specimens

Figure 2 shows the shape of the uniaxial restrained shrinkage cracking specimen of JIS A 1151 (2002) that was used to investigate shrinkage cracking characteristics. In this specimen, the concrete is restrained by restraint steel (100 × 40 × 40 mm, thickness: 2.3 mm, sectional area: 397 mm²) and is subjected to direct tensile stress due to the restraint of volume change caused by drying shrinkage. Cracking occurs when the tensile stress of steel is larger than that of the concrete. A prism with dimensions of 100 × 100 × 400 mm was made for studying the drying shrinkage. A cylinder with 100 mm diameter and 200 mm height was made for investigating the mechanical properties of concrete, such as, compressive strength, Young's modulus, and splitting tensile strength. Three specimens for each concrete mixture were produced for the uniaxial restrained shrinkage cracking test and these specimens were denoted by URSC 65-NC-1, URSC 65-NC-2, URSC 65-NC-3, URSC 65-RC-1, URSC 65-RC-2, URSC 65-RC-3, URSC 45-NC-1, URSC 45-NC-2, URSC 45-NC-3, URSC 45-RC-1, URSC 45-RC-2, and URSC 45-RC-3, respectively.

2.3 Testing Methods

All specimens were demolded after a day, moist-cured for 7 days, and then exposed to air. They were stored in a controlled environment, temperature of 20 ± 1 °C and relative humidity of 60 ± 5 %. The uniaxial restrained shrinkage cracking tests and drying shrinkage tests were started after 7 days and conducted in accordance with JIS A

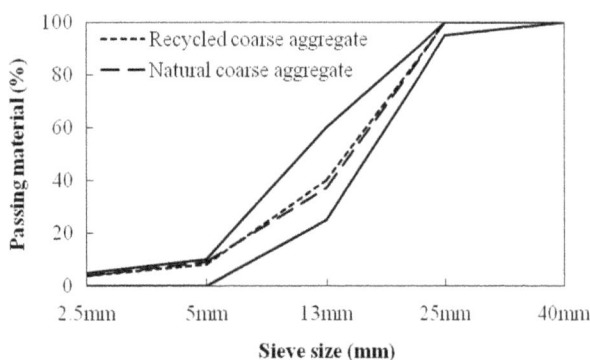

Fig. 1 Size distribution of coarse aggregate.

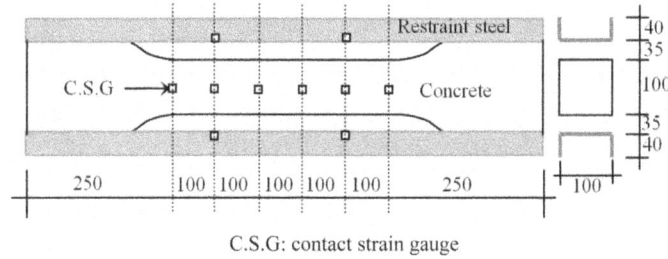

C.S.G: contact strain gauge

Fig. 2 Shape of uniaxial restrained shrinkage cracking specimen.

Table 3 Mechanical properties of concrete.

Specimen	Items	7 days	14 days	28 days
65-NC	f_c (MPa)	26.3	32.5	38.4
	f_{st} (MPa)	2.3	2.6	2.9
	E_c (GPa)	22.9	25.5	27.7
65-RC	f_c (MPa)	22.0	28.6	30.3
	f_{st} (MPa)	2.1	2.3	2.6
	E_c (GPa)	20.9	23.9	24.6
45-NC	f_c (MPa)	39.4	46.1	49.9
	f_{st} (MPa)	2.8	3.1	3.4
	E_c (GPa)	28.0	30.3	31.5
45-RC	f_c (MPa)	33.5	38.7	40.0
	f_{st} (MPa)	2.6	2.8	3.1
	E_c (GPa)	25.8	27.8	28.3

f_c Compressive strength, f_{st} splitting tensile strength, E_c Young's modulus.

1129 (2001) and JIS A 1151 (2002), respectively. A contact strain gauge (CSG) with a precision of 1/1000 was used to measure the strain. Compressive strength and splitting tensile strength tests were conducted in accordance with JIS A 1108 (1999a) and JIS A 1113 (1999b), respectively.

3. Results and Analysis

3.1 Mechanical Properties of Concrete

Table 3 shows the mechanical properties, such as compressive strength, tensile splitting strength, and Young's modulus, of the concrete specimens with 7, 14 and 28 days of curing. Each test results were shown in an average of the results of three specimens. The compressive strength and splitting tensile strength of 65-RC specimens containing coarse recycled aggregate at 28 days showed a decrease of about 20 % and about 10 %, respectively, compared to those of 65-NC specimens containing 100 % natural aggregate. The Young's modulus of 65-RC specimens at 28 days was about 10 % smaller than that of 65-NC specimens. Similar results were also observed for 45-RC and 45-NC specimens at 28 days. Low density and high water absorption and porosity, mainly caused by the heterogeneous nature of recycled aggregate, can lead to low quality concrete (Khaleel and Kypros 2013).

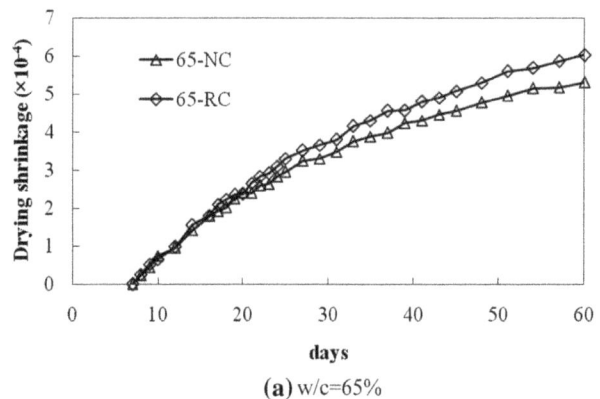

(a) w/c=65%

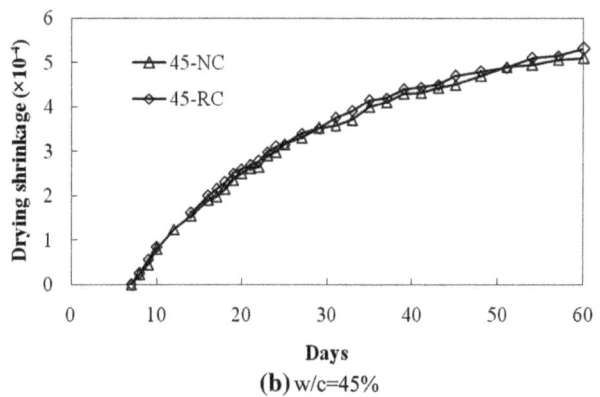

(b) w/c=45%

Fig. 3 Drying shrinkage.

(a) 65-NC

(b) 65-RC

(c) 45-NC

(d) 45-RC

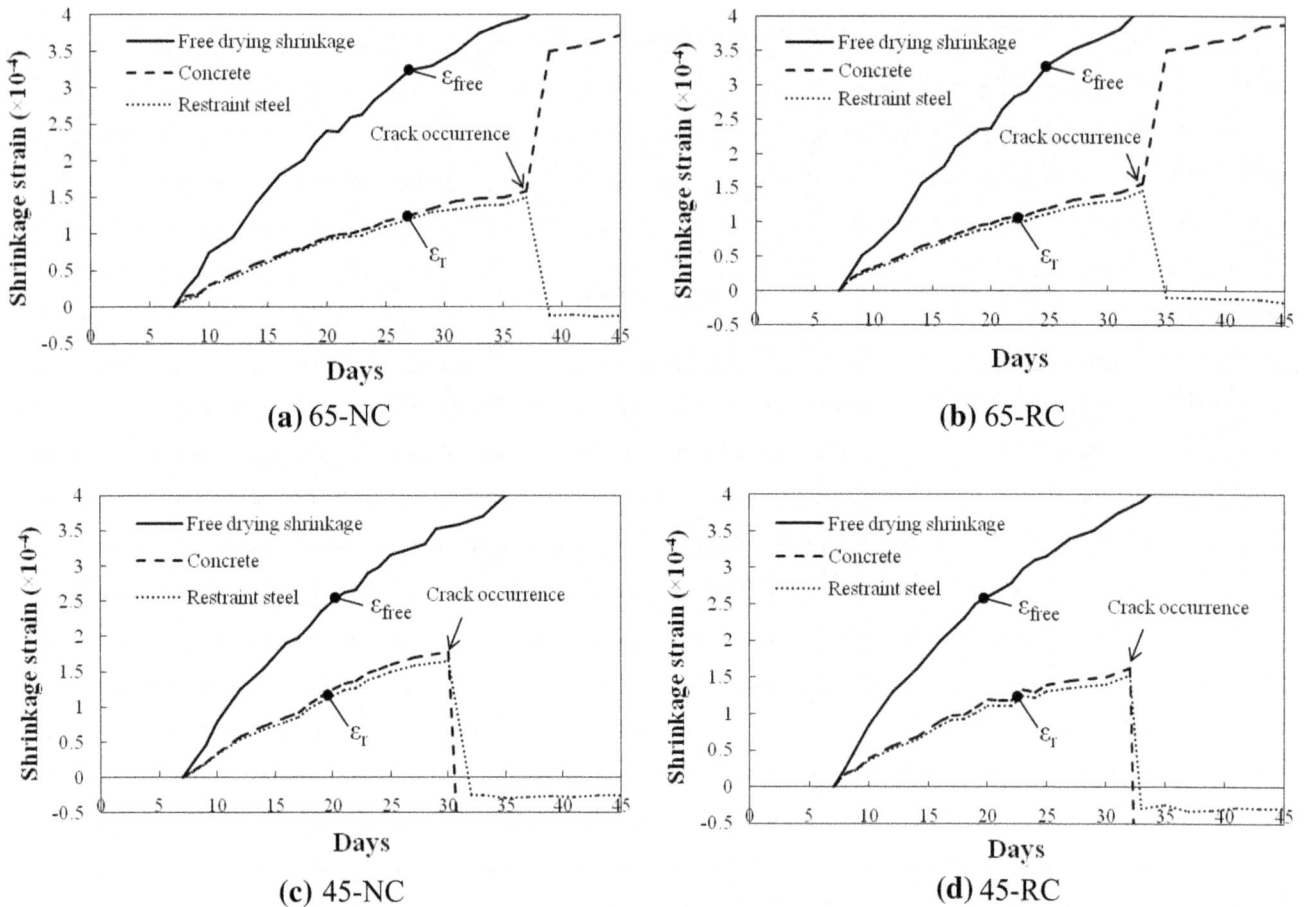

Fig. 4 Shrinkage behavior of uniaxial restrained shrinkage cracking specimens.

Table 4 Cracking age.

Specimen	Cracking age	Average value (cracking age)
URSC 65-NC-1	37 days	34.3 days
URSC 65-NC-2	35 days	
URSC 65-NC-3	31 days	
URSC 65-RC-1	33 days	32.3 days
URSC 65-RC-2	31 days	
URSC 65-RC-3	33 days	
URSC 45-NC-1	30 days	30.0 days
URSC 45-NC-2	29 days	
URSC 45-NC-3	31 days	
URSC 45-RC-1	32 days	30.3 days
URSC 45-RC-2	29 days	
URSC 45-RC-3	30 days	

3.2 Drying Shrinkage

Figure 3 shows the free drying shrinkage strain ε_{free} across the specimens. The results of each drying shrinkage test were shown in an average of the results of three specimens for each concrete mixture. In w/c = 65 % concrete, the drying shrinkage of RC specimens at 60 days was about 20 % larger than that of NC specimens, and in w/c = 45 % concrete, the drying shrinkages of NC specimens and RC specimens were almost the same. In AIJ (2003), to ensure the quality of drying shrinkage the maximum shrinkage should be smaller than 800 µε. All the specimens tested satisfied this quality control regulation.

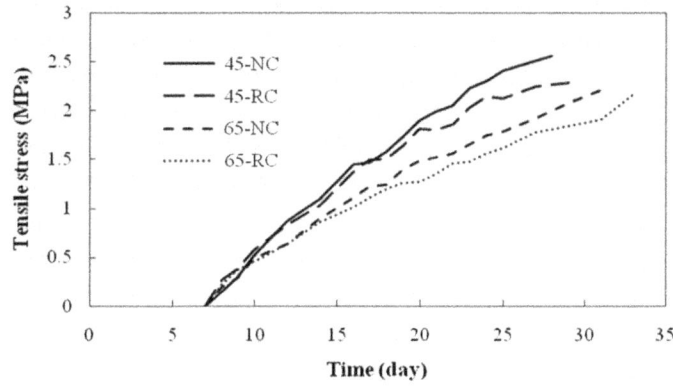

Fig. 5 Behavior of tensile stress.

3.3 Uniaxial Restrained Shrinkage Cracking Test

3.3.1 Cracking Age

Figure 4 shows the restrained shrinkage strain, ε_r, of the uniaxial restrained shrinkage cracking specimens (the first specimen out of the three specimens for each concrete mixture, URSC 65-NC-1, URSC 65-RC-1, URSC 45-NC-1, and URSC 45-RC-1) and the free drying shrinkage, ε_{free} (see Fig. 3), of concrete. The restrained shrinkage strains, ε_r, of concrete and restraint steel were almost identical. The cracking age of each specimen was determined based on the point where the sudden change in restrained shrinkage occurred. Table 4 presents the cracking age. Within the specimens with the same w/c ratio, there was no significant difference in cracking age between NC specimens and RC specimens.

3.3.2 Histories of Tensile Stress

The tensile stress of concrete was obtained using Eq. (1) based on the force equilibrium between the restraint steel and concrete. Figure 5 shows the behavior of tensile stress of concrete from the uniaxial restrained shrinkage cracking test. These tensile stress results were shown in an average of the results of three specimens for each concrete mixture. For example, the results of 65-NC were an average of the results of the specimens URSC 65-NC-1, URSC 65-NC-2 and URSC 65-NC-3. The tensile stress behaviors between these three results were similar to each other. The results of the specimens with other concrete mixtures, 65-RC, 45-NC, and

45-RC, were also shown in Fig. 5. The tensile stress of all concrete mixtures almost linearly increased with time. Regardless of the w/c ratio, the tensile stresses of RC specimens showed about 10 % reduction with time compared to those of NC specimens. Behavior like this was observed since the stress relaxation of RAC is expected to be larger than that of NAC.

$$\sigma_t = \frac{E_s \cdot \varepsilon_r \cdot A_s}{A_c} \qquad (1)$$

where E_s is the elastic modulus of restraint steel, ε_r is the restrained shrinkage strain, A_s is the sectional area of the restraint steel, and A_c is the sectional area of concrete.

3.3.3 Histories of Tensile Creep

Figure 6 shows the conceptual diagram of tensile strain of concrete of the restrained shrinkage specimen. Since the restrained tensile strain, ε_t, is defined as the sum of the elastic strain, ε_e, and the tensile creep strain, ε_c, according to previous studies (Kanda 2005; Shima and Ichikawa 2009), the tensile creep strain can be obtained from Eq. (2). The elastic strain, ε_e, for Eq. (2) was obtained by using the tensile stress of concrete from Eq. (1) and the Young's modulus from the approximate regression equation shown in Fig. 7 based on measured Young's modulus shown in Table 3. Figure 8 shows the behavior of tensile creep, found from Eq. (2), of concrete occurred in the restrained shrinkage specimens. The tensile creep of specimens with w/c = 65 % was larger than

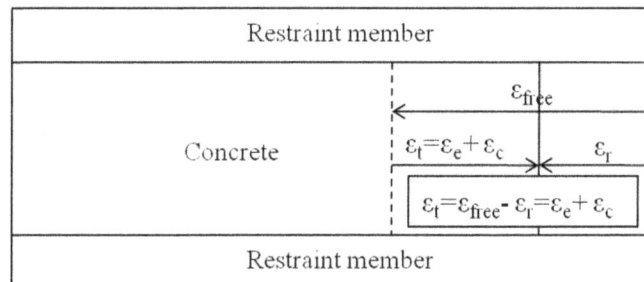

ε_c : tensile creep strain, ε_{free} : free drying shrinkage, ε_e : elastic strain ε_r :

restrained shrinkage strain, ε_t : restrained tensile strain

Fig. 6 Conceptual diagram of restrained tensile strain.

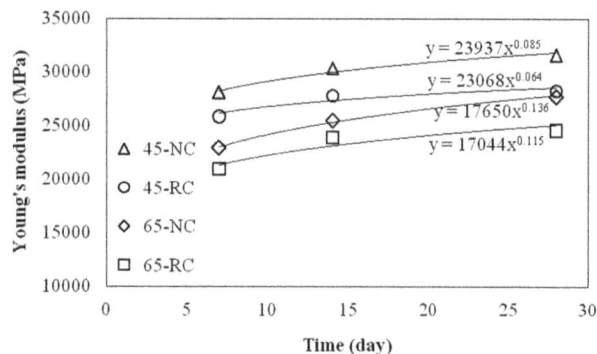

Fig. 7 Concrete Young's modulus.

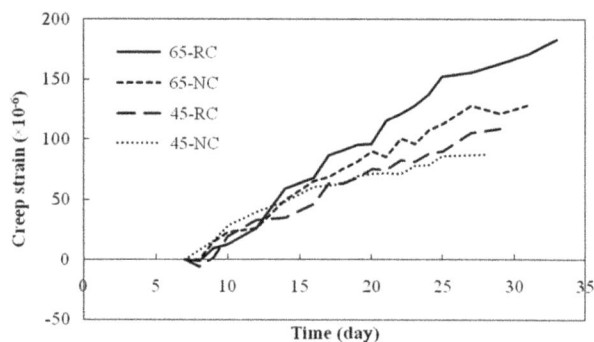

Fig. 8 Tensile creep strain.

that of specimens with w/c = 45 %. For specimens with w/c = 65 %, the tensile creep of concrete of RC was about 30 % larger than that of NC. For specimens with w/c = 45 %, after 17 days, the tensile creep of concrete of RC was about 20 % larger than that of NC. According to previous studies (Domingo et al. 2009; Gholamreza et al. 2011; Xiao et al. 2014), for compressive creep, the creep deformation of RAC is 20–60 % larger than that of NAC. Based on the results of this study, for tensile creep due to restraint of shrinkage, the creep deformation of RAC is 20–30 % larger than that of NAC. Although the tensile strength of the concrete with recycled aggregates is smaller than that of the concrete with natural aggregates, the cracking age was similar to each other, since the stress relaxation of the concrete with recycled aggregates caused by tensile creep is larger than that of the concrete with natural aggregates.

$$\varepsilon_c = \varepsilon_{free} - \varepsilon_e - \varepsilon_r \tag{2}$$

where ε_c is the tensile creep strain, ε_{free} is the free drying shrinkage, ε_e is the elastic strain and ε_r is the restrained shrinkage strain.

4. Conclusion

In this study, a uniaxial restrained shrinkage cracking test was carried out to investigate the tensile creep properties caused by the restraint of drying shrinkage of RAC. The mechanical properties of RAC, such as compressive strength, tensile splitting strength, and Young's modulus, were also investigated. All the conclusions drawn from this research are limited to coarse recycled aggregates with the absorption rate of 2.5 % and should be used with caution due to the limited number of mixes tested. The study results are summarized as follows:

(1) For both w/c = 65 % and w/c = 45 % concrete, the compressive strength and splitting tensile strength of RC specimens containing coarse recycled aggregate at 28 days showed a decrease of about 20 % and about 10 %, respectively, compared to those of NC specimens containing 100 % natural aggregate at 28 days.

(2) In w/c = 65 % concrete, the drying shrinkage at 60 days of RC specimens was about 20 % larger than that of NC specimens. In w/c = 45 % concrete, the drying shrinkages of NC specimens and RC specimens were almost the same.

(3) The tensile stress of all concrete mixtures almost linearly increased with time, and the tensile stress of RC specimens showed about 10 % reduction compared to that of NC specimens.

(4) According to previous studies, for compressive creep, creep deformation of RAC is 20–60 % larger than that of NAC. Based on the results of this study, for tensile creep due to restraint of shrinkage, the creep deformation of RAC is 20–30 % larger than that of NAC.

(5) For cracking age, if the specimens have the same w/c ratio, the difference between NC specimens and RC specimens was very small. Although the tensile strength of the concrete with recycled aggregates is smaller than that of the concrete with natural aggregates, the cracking age was similar to each other, since the stress relaxation of the concrete with recycled aggregates caused by tensile creep is larger than that of the concrete with natural aggregates.

References

AIJ. (2003). Recommendations for practice of crack control in reinforced concrete structures (design and construction), Architectural Institute of Japan (in Japanese).

Bravo, M., Brito, J., Pontes, J., & Evangelista, L. (2015). Durability performance of concrete with recycled aggregates from construction and demolition waste plants. *Construction and Building Materials, 77*, 357–369.

Domingo, A., Lázaro, C., López, F., Serrano, M., Serna, P., & Castaño, J. (2009). Creep and shrinkage of recycled aggregate concrete. *Construction and Building Materials, 23*(7), 2545–2553.

Enad, M., Ahmad, I., Hassan, E., & Varun, C. P. (2013). Self consolidating concrete incorporating high volume of fly ash, slag, and recycled asphalt pavement. *International Journal of Concrete Structures and Materials, 7*(2), 155–163.

Garas, V., Kahn, L., & Kurtis, K. (2009). Short-term tensile creep and shrinkage of ultra high performance concrete. *Cement and Concrete Composite, 30*(3), 147–152.

Gholamreza, F., Ghani, R., Burkan, I., Abdelgadir, A., Benoit, F., & Simon, F. (2011). Creep and drying shrinkage characteristics of concrete produced with coarse recycled concrete aggregate. *Cement & Concrete Composites, 33*(10), 1026–1037.

JCI Committee. (1999a). *JIS A 1108: Method of test for compressive strength of concrete.* Tokyo, Japan: Japanese Standards Association (in Japanese).

JCI Committee. (1999b). *JIS A 1113: Method of test for splitting tensile strength of concrete.* Tokyo, Japan: Japanese Standards Association (in Japanese).

JCI Committee. (2001). *JIS A 1129: Method of test for length change of mortar and concrete.* Tokyo, Japan: Japanese Standards Association (in Japanese).

JCI Committee. (2002). *JIS A 1151: Method of test for drying shrinkage cracking of restrained concrete.* Tokyo, Japan: Japanese Standards Association (in Japanese).

Kanda, T. (2005). Quantitative evaluation of shrinkage cracking initiation. *Concrete Journal of the Japan Concrete Institute, 43*(5), 60–66 (in Japanese).

Khaleel, H., & Kypros, P. (2013). Strength prediction model and methods for improving recycled aggregate concrete. *Construction and Building Materials, 49*, 688–701.

KMCT. (2005). Quality standard of recycled aggregate, Korea Ministry of Construction and Transportation (in Korean).

Rasiah, S., Neo, D. H. W., & Lai, J. W. E. (2012). Mix design for recycled aggregate concrete. *International Journal of Concrete Structures and Materials, 6*(4), 239–246.

Ryou, I. S., & Lee, Y. S. (2014). Characterization of recycled coarse aggregate (RCA) via a surface coating method. *International Journal of Concrete Structures and Materials, 8*(2), 165–172.

Sherif, Y., Kareem, H., Anaam, A., Amani, Z., & Hiba, I. (2015). Strength and durability evaluation of recycled aggregate concrete. *International Journal of Concrete Structures and Materials, 9*(2), 219–239.

Shima, H., & Ichikawa, D. (2009). Measurement of tensile creep of concrete under restrained drying shrinkage conditions using ring specimens. *Japan Society of Civil Engineering, 65*(4), 477–489 (in Japanese).

Soares, D., Brito, J., Ferreira, J., & Pacheco, J. (2014a). In situ materials characterization of full-scale recycled aggregates concrete structures. *Construction and Building Materials, 71*, 237–245.

Soares, D., Brito, J., Ferreira, J., & Pacheco, J. (2014b). Use of coarse recycled aggregates from precast concrete rejects: mechanical and durability performance. *Construction and Building Materials, 71*, 263–272.

Tao, J., Chen, C., Zhuang, Y., & Lin, X. (2012). Effect of degree of ceramsite prewetting on the cracking behavior of LWAC. *Magazine of Concrete Research, 61*(1), 1–9.

Tavakoli, M., & Soroushian, P. (1996a). Strength of recycled aggregate concrete made using field-demolished concrete as aggregate. *ACI Materials Journal, 93*(2), 178–181.

Tavakoli, M., & Soroushian, P. (1996b). Drying shrinkage behavior of recycled aggregate concrete. *Concrete International, 18*(11), 58–61.

Valeria, C. (2010). Mechanical and elastic behavior of concretes made of recycled-concrete coarse aggregates. *Construction and Building Materials, 24*(9), 1616–1620.

Xiao, J., Li, L., Tam, V. W. Y., & Li, H. (2014). The state of the art regarding the long-term properties of recycled aggregate concrete. *Structural Concrete, 15*(1), 3–12.

13

Strength and Durability Properties of Concrete with Starch Admixture

A. A. Akindahunsi*◉, and H. C. Uzoegbo

Abstract: This paper examines some properties of concrete, such as strength, oxygen permeability and sorptivity using starch [cassava (CA) and maize (MS)] as admixtures. Concrete cubes containing different percentages of the CA and MS by weight of cement (0, 0.5, 1.0, 1.5 and 2.0 %) were cast. Compressive strength tests were carried out after 3, 7, 14, 21, 28, 56, 90, 180, 270 and 365 days of curing. Oxygen permeability and sorptivity tests were carried out on another set of concrete specimens with the same percentages of starch at 7, 28, 90, 180, 270 and 365 days. Oxygen permeability and sorptivity tests data obtained were subjected to Kruskal–Wallis one-way analysis of variance by ranks. The strength increase after 1 year over the control for CA 0.5 and CA 1.0 are 2.7 and 3.8 % respectively, while MS 0.5 and MS 1.0 gave 1.5 % increase over control. These results showed a decrease in oxygen permeability and rates of sorptivity, with concretes containing starch as admixtures giving better performance than the control concretes.

Keywords: starch, concrete, strength, sorptivity, oxygen permeability, statistical analysis.

1. Introduction

Concrete structures according to Baroghel-Bouny et al. (2009) should be able to serve the purposes for which they were built throughout their service life. Baroghel-Bouny et al. (2009) noted that safety, economy and environmental factors are major issues in the long-term durability of structures. Achieving durability in concrete therefore should be a very significant factor in the design and construction of new structures and in the evaluation of the condition of existing structures (Merretz et al. 2009). According to Folić and Zenunović (2010), the mode of interaction of concrete with its environment will influence the likely mechanisms of deterioration. The ability of concrete to resist chemical attack, abrasion, weathering action and other deterioration effects is very important during the service life of the structure. Materials used in construction play a major role in the durability of concrete. However, Gjørv (2011) pointed out that design, materials used and workmanship are very important factors in achieving good quality construction which will enhance durability of concrete. Deterioration mechanisms in concrete structures are influenced by interaction with the environment. The system design of concrete and civil engineering structures that involve materials selection, structural shape, construction work and maintenance should be carried out in a manner that is environmentally friendly because this will contribute to environmental sustainability (Folić 2009). The durability of concrete may be affected by physical, chemical and biological factors. These factors may be due to weathering conditions (temperature, and moisture changes), abrasion, attack by natural or industrial effluents and gases, or biological agents (Nagesh 2012). Increased knowledge of materials properties is vital in durability considerations for concrete structures. According to Chidiac (2009) durability of concrete depends on the qualities of the materials, construction, design and exposure conditions. The importance of materials quality cannot be over-emphasized in concrete durability. Elahi et al. (2010) examined the mechanical and durability properties of high performance concretes containing supplementary cementitious materials and concluded that the combination of different cementitious materials and the precise choice combinations should be on the basis of the physical properties relevant to the durability and performance expected from the concrete, as well as the exposure conditions.

Chemical admixtures are used in the production of concrete in order to achieve various durability properties. Khayat (1998) reviewed the use of viscosity-enhancing admixtures such as water soluble synthetic and natural organic polymers. Polymers used as admixtures are said to enhance the joining of the mixing constituents as a result of intertwining polymer film which, according to Chung (2004), produces concrete of better mechanical and durability characteristics. Chemical admixtures are used as high range water reducer admixtures (HRWRA) and have impacted on the rheological and mechanical behaviour of cement-based systems. This allows for a latent time that permits casting of concrete in excellent condition. These chemical admixtures are oil based, non-renewable products such as polynaphthalene sulphonate (PNS),

School of Civil and Environmental Engineering,
University of the Witwatersrand (WITS),
Johannesburg 2050, Gauteng, South Africa.
*Corresponding Author;
E-mail: akindehinde.akindahunsi@students.wits.ac.za

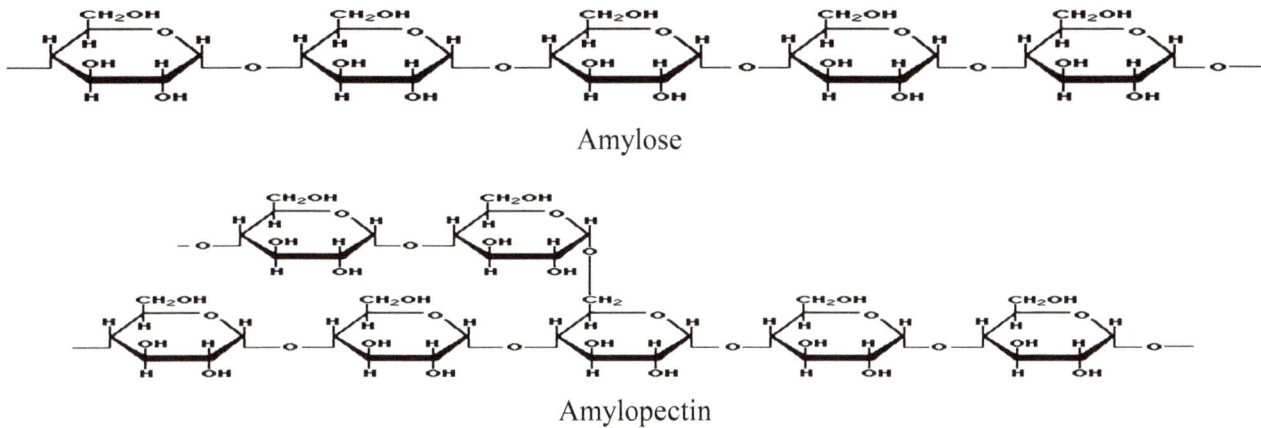

Fig. 1 Arrangement of glucose units in Amylose and Amylopectin macromolecules (Source: www.intechopen.com).

polycarboxylate (PC) and polyacrylate (PA). They contain formaldehyde which when accidentally or intentionally released into the environment may result in undesirable environmental toxic effects (Crépy et al. 2011; Akindahunsi et al. 2013). Recently, interest has been developing in the use of organic admixtures to modify various properties in concrete because they are available in abundance; their preparation is not so sophisticated. They are renewable materials, therefore contributing to sustainable green construction. The use of organic admixtures including starch and its derivatives to modify different properties of cement and concrete has been reported by various authors such as: Luke and Luke (2000); Peschard et al. (2004); Crépy et al. (2011); Akindahunsi et al. (2011), (2012); Lasheras-Zubiate et al. (2012) among others.

Cassava (CA) and maize (MS) are abundant in sub-Saharan Africa and good sources of starch which is generally known to have a wide variety of industrial applications. Izaguirre et al. (2010) noted that the thickening action of starches and their derivatives can be studied by considering the relationship between amylose and amylopectin molecules, units of starch consisting of the two macromolecule types. The chemical structure of the two macromolecules of starch is shown in Fig. 1. The amylose builds a helix while the amylopectin exhibits a tree structure consisting of numerous connected linear glucose units. Amylose properties are these of thickener, water binder, emulsion stabilizer, and gelling agent while Amylopectin is what makes up a larger percentage of starch and is highly soluble.

Starch is used for different purposes such as a thickener/stabilizer and gelling agent. Starch pastes and gels are used to control the consistency of some manufactured products. It is also used as starting material in the production of sweeteners and polygons (BeMiller and Hubber 2011). Starch is equally used in the plastics industry to produce biodegradable plastics which require starches that have small granules (Wang et al. 1998). Furthermore, it is used in the construction industry as concrete block binder, asbestos, clay and limestone binder, fire-resistant wallboard, plywood/chipboard adhesive, gypsum board binder and paint filler (Satin 1998). One of the fears exhibited in the use of organic admixtures is that it is biodegradable and its long term effect on concrete might be negative. This paper therefore

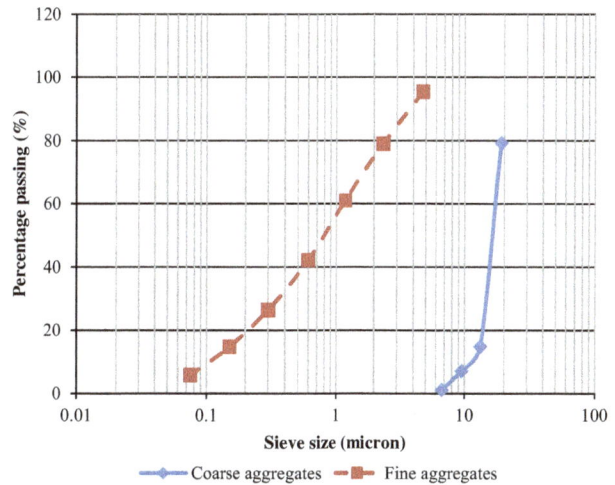

Fig. 2 Aggregates grading curve.

examines the use of CA and MS starches as admixtures in concrete and this long term durability characteristics.

2. Materials and Methods

2.1 Materials

Crushed granite used as (coarse and fine) aggregates were obtained from AfriSam aggregates in Johannesburg, South Africa in compliance with South African National standards (SANS 1083:2006). The maximum coarse aggregate size used is 22 mm (compacted bulk density of 1750 kg/m^3 and specific gravity of 2.67). Figure 2 gives the particle size distribution. The fine aggregates grading used conforms to the recommended grading of South Africa as contained in Fulton's concrete technology (2009). CEM1 52.5N cement used in the preparation of different concrete mixes was supplied by Pretoria Portland Cement, Johannesburg, South Africa. CA and MS starches were obtained from Nigeria.

2.2 Methods
2.2.1 Determination of Particle Sizes of Cement and Starches Used

The particle size distributions of the cement and the starches (CA and MS) used for this investigation were

Table 1 Mix proportions (kg/m³) of the various concrete used.

Mix no	Cement (CEM1 52.5 N)	Coarse aggregates	Fine aggregates	w/c	Starch %	Slump (mm)
			Content (kg/m³)			
Control	380	1024	738	0.54	0	85
CA 0.5	380	1024	738	0.54	0.5	65
CA 1.0	380	1024	738	0.54	1.0	47
CA 1.5	380	1024	738	0.54	1.5	32
CA 2.0	380	1024	738	0.54	2.0	23
MS 0.5	380	1024	738	0.54	0.5	68
MS 1.0	380	1024	738	0.54	1.0	46
MS 1.5	380	1024	738	0.54	1.5	35
MS 2.0	380	1024	738	0.54	2.0	26

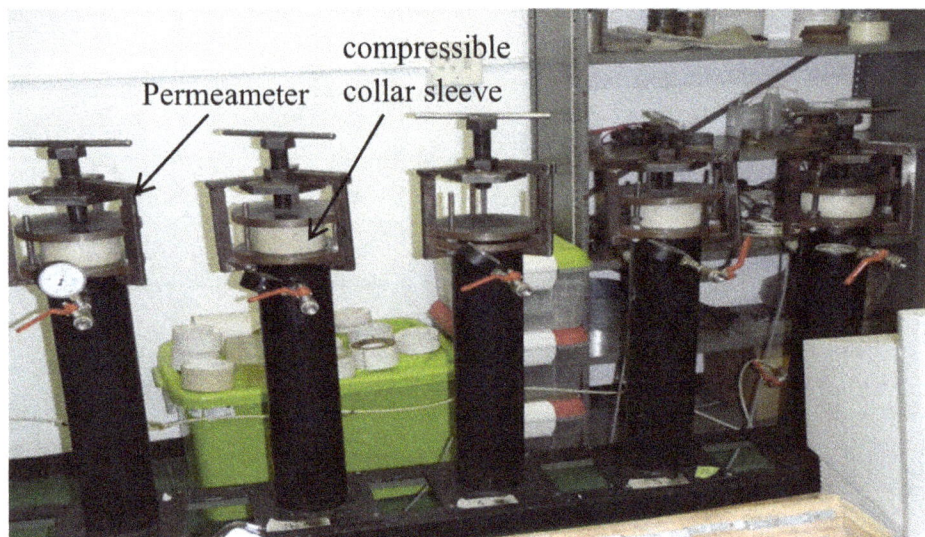

Fig. 3 Permeameters used for oxygen permeability tests.

determined by means of a Malvern (2005) particle size analyzer. It is an automated light scattering instrument, the laser particle size analyser measures the size of particles, powders and suspensions or emulsions using diffraction and diffusion of a laser beam. The sample particles to be measured are passed through concentrated laser beam and the particles scatter light at an angle that is inversely proportional to their size (Malvern 2005). The laser diffraction result generated for the particle size distribution of the CA and MS starches used are volume based. The laser diffraction computed median (D50) is used for the point specification (Malvern 2005). The two most common points used to describe the dispersal of the particles are the finest (D10) and the coarsest (D90) distribution. D50, therefore, is the average particle size. The D10 is the particle size that has ten percent smaller and ninety percent larger. The D90 refers to ninety percent of the particle size distribution having smaller particle size and ten percent having the larger particle size.

2.2.2 Starch Activation

The MS starch powder is factory pre-treated and the starch properties can be activated in water at ambient room temperature and the required dosage (in powder form) added directly to the mix. CA starch however, has to be activated with hot water at a temperature between 70 to 90 °C. Therefore, the required dosage has to be prepared separately and allowed to cool down in order not to contribute to temperature rise in the mix in which it is going to be used. The quantity of water used in the starch activation was deducted from that required in a mix.

2.2.3 Starch Morphology

Field emission Scanning Electron Microscope (JSM-7600F model) was used to examine the morphology of the starch (MS and CA) materials used in this study. The JSM-7600F operates with the use of a T-FE electron gun and sem-in-lens objective lens in its electron optics system and a robust structure is able to operate in a broad range of

Table 2 Chemical composition of cement.

Oxides	Oxides composition
Chemical tests	(%)
SiO_2	20.0
Al_2O_3	4.60
Fe_2O_3	1.68
Mn_2O_3	0.90
TiO_2	0.27
CaO	59.9
MgO	4.50
P_2O_5	0.13
SO_3	2.61
Cl	0.004
K_2O	0.65
Na_2O	0.18
LOI	3.89
Total	99.3
Insoluble residue	1.11
Compounds calculated from Bogue's equation	
C_3S	53.59
C_2S	16.91
C_3A	9.35
C_4AF	5.11

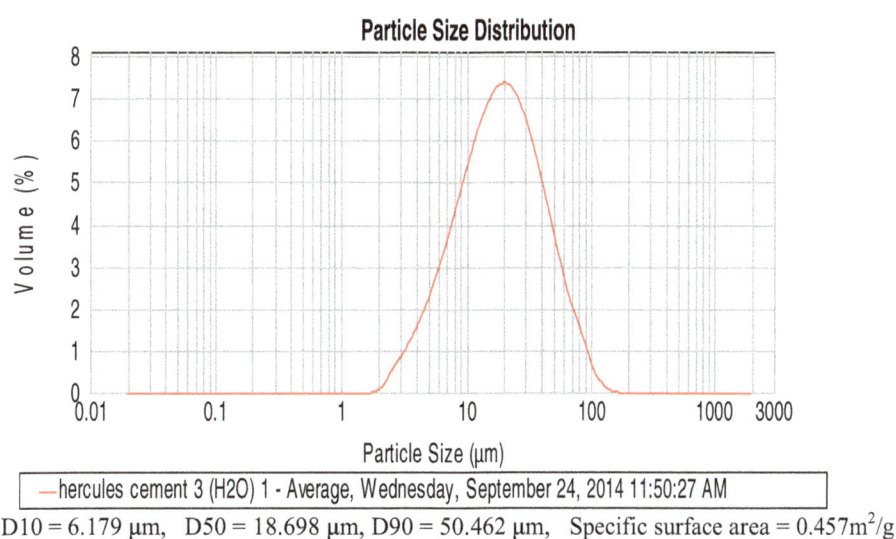

D10 = 6.179 μm,   D50 = 18.698 μm, D90 = 50.462 μm,   Specific surface area = 0.457m^2/g

Fig. 4 Particle size of cement used for this investigation.

installation environments. This enables the achievement of high resolution and high quality images. The scanning electron microscope incorporates an energy filter (r-filter) for the secondary and backscattered electrons, and this increases image resolution on nonconductive specimens and semi-conductor devices.

2.2.4 Setting Times Tests

Setting times (initial and final) for the different percentages of CA and MS starch additions in cement were determined under controlled humidity and temperature. The percentages of starch added to the cement are 0, 0.5, 1.0, 1.5, 2.0 respectively. The tests complied with the South African

Particle Size Distribution

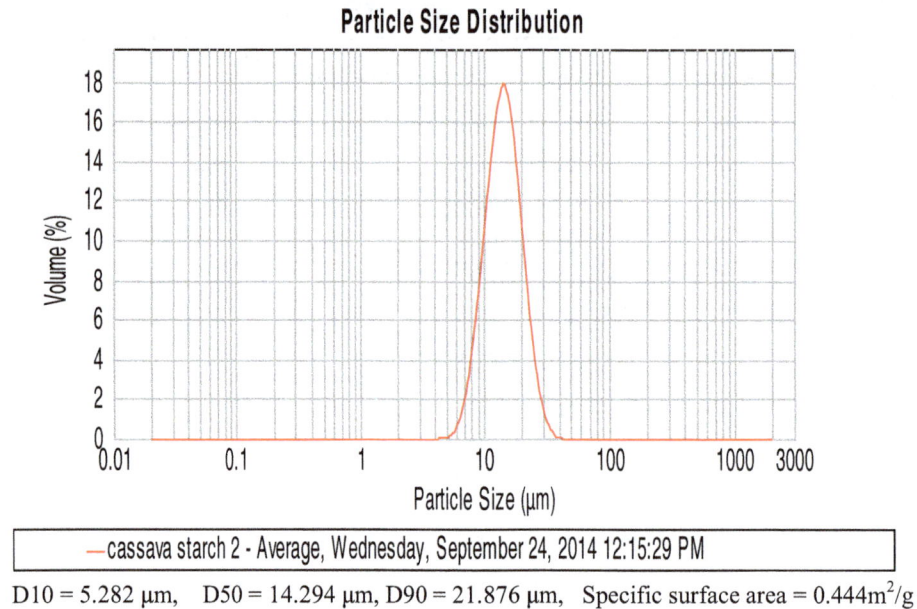

D10 = 5.282 µm,   D50 = 14.294 µm, D90 = 21.876 µm,   Specific surface area = 0.444m^2/g

Fig. 5 Particle size of the cassava starch powder.

Particle Size Distribution

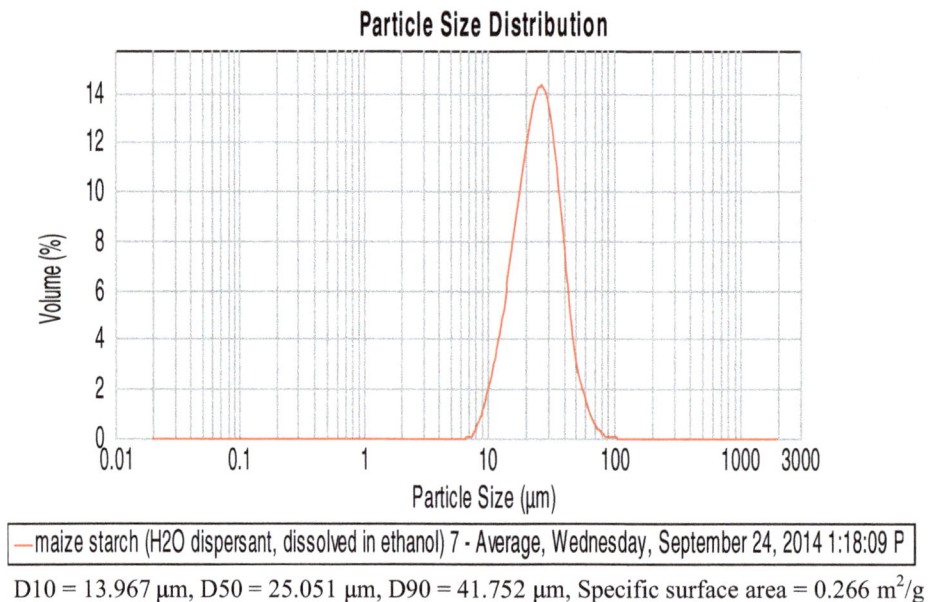

D10 = 13.967 µm, D50 = 25.051 µm, D90 = 41.752 µm, Specific surface area = 0.266 m^2/g

Fig. 6 Particle size of the maize starch powder.

standard SANS 50196-3:2006 and EN 196-3:2005. Standard mortar mixer set to EN mixing standard was used to carry out the mixing of the different cement pastes with various concentrations of starches. The setting times were determined using automatic Vicat needle apparatus ToniSet Expert model 7320 manufactured by Toni Tecknik, Germany.

2.2.5 Concrete Mixes

The concrete used for the study was prepared according to ACI (1999). A pan mixer was used for making the concrete mixes. The concrete mixes are as presented in Table 1, showing different concrete mixes containing different percentages of CA and MS starches by weight of cement cast

(0, 0.5, 1.0, 1.5 and 2.0 %) using the same mix proportions. The air content used in the mix proportion was 1 %. For each mix listed in Table 1 three samples were cast in 100 mm cube moulds for compressive testing at 3, 7, 14, 21, 28, 56, 90, 180, 270 and 365 days. The cast cubes were covered with plastic sheets in a controlled laboratory environment at a temperature of 23 ± 2 °C. The concrete cubes were demoulded after 24 h and kept in a water curing tank with temperature maintained at 23 ± 2 °C. The cubes for each mix were taken out of the curing tank, weighed and tested for compressive strength at each of the stated curing days.

The compressive strengths of the concrete cubes were determined in accordance with South African codes SANS

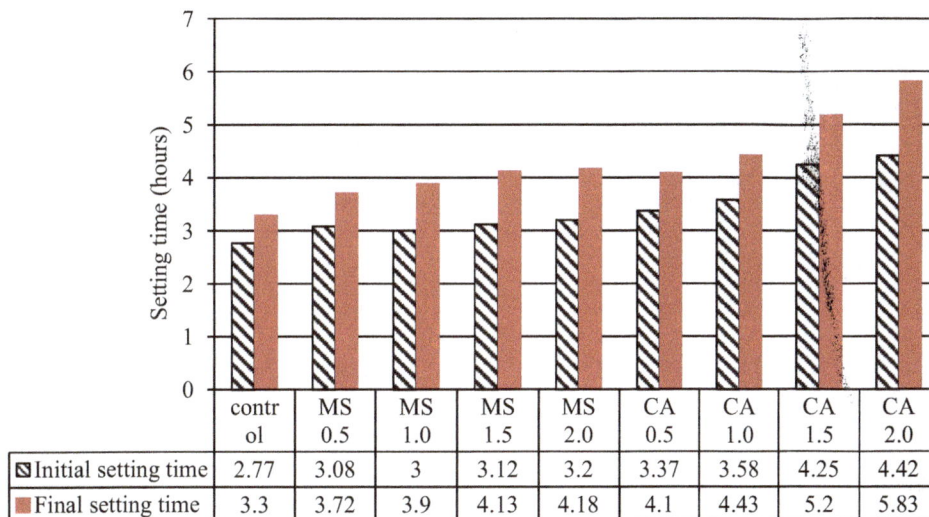

Fig. 7 Setting time of cement paste with different concentrations of starches.

	contr ol	MS 0.5	MS 1.0	MS 1.5	MS 2.0	CA 0.5	CA 1.0	CA 1.5	CA 2.0
Initial setting time	2.77	3.08	3	3.12	3.2	3.37	3.58	4.25	4.42
Final setting time	3.3	3.72	3.9	4.13	4.18	4.1	4.43	5.2	5.83

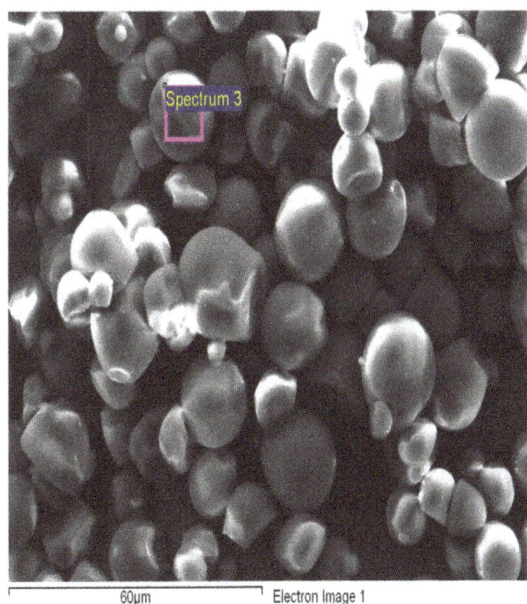

Fig. 8 SEM micrograph of cassava starch.

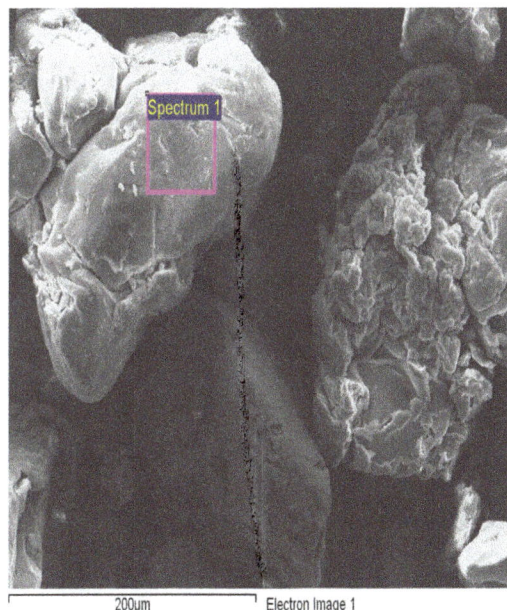

Fig. 9 SEM micrograph of maize starch.

5860 (2006), 5861-2 (2006) and 5861-3 (2006) which are used as benchmark for concrete quality and as an index of the strength of concrete. The cubes were crushed in saturated moist condition in compliance with the standard as stated in Fulton's concrete technology (2009). An Amsler compression machine with a capacity of 2000 kN was used to carry out the compression tests.

2.2.6 Oxygen Permeability Test

A set of concrete cubes were cast for durability tests (sorptivity, oxygen permeability) at 7, 28, 90, 180, 270 and 365 days. The permeability coefficient based on the derivation of the D'Arcy coefficient of permeability was obtained by monitoring a falling head pressure carried out on the concrete discs. The cast concrete cubes moist cured at the stated cured age were removed from the curing tank and cored with a coring machine. Oxygen permeability

tests were carried out according to the durability index testing procedure manual of the Cement and Concrete Institute (2010) of South Africa. The study carried out by Alexander et al. (2008) showed that this method can be used for the assessment of performance-based design and specifications in concrete structures. The set-up of the permeameters is as shown in Fig. 3. The permeameters have both inlet and outlet valves, the inlet valves connected to an oxygen cylinder allows oxygen gas inflow into the permeameters. Pressure gauges having a range from 0 to 150 kPa with an accuracy of 0.5 kPa are connected to the permeameters. Pressure was allowed to build up to 100 ± 5 kPa inside the permeameter and oxygen permeability test was then commenced and it was for a duration of 2 h. The equation governing the operation of oxygen permeability test is shown in Eq. (1). The test data were captured by a data logger connected to all the

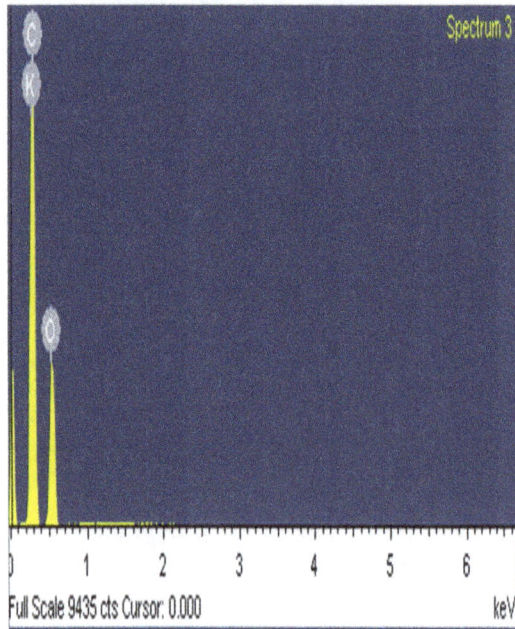

Fig. 10 EDXA of cassava starch.

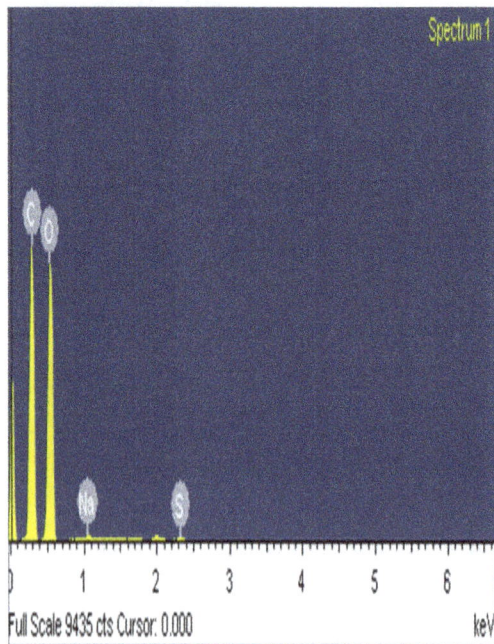

Fig. 11 EDXA of maize starch.

permeameters. The slope of the linear regression line forced through the line (0, 0) point is calculated from the equation.

$$z = \frac{\sum [In(P_0/P_t)]^2}{\sum [In(P_0/P_t)t]} \tag{1}$$

where P_0 is the initial pressure at start of test (at time t_0) to the nearest 0.5 kPa, P_t is the subsequent readings in pressure to the nearest 0.5 kPa at times t, measured from t_0. The D'Arcy coefficient of permeability is given by

$$k = \frac{wVgdz}{RA\theta} \tag{2}$$

where k is the coefficient of permeability of test specimen (m/s), w is the molecular mass of oxygen (0.032 kg/mol), V is the volume of oxygen under pressure in permeameter recorded to the nearest 0.01 litre, g is the acceleration due to gravity (9.81 m/s), R is the universal gas constant (8.313 Nm/K mol), d is the average specimen thickness (m), A is the cross sectional area of the specimen (m2), θ is the absolute temperature (K), z is the slope of the line determined in the regression analysis.

The coefficient of permeability was calculated for each of the test specimens, and the oxygen permeability index (OPI) is given as the negative log of the average coefficients of the specimens.

$$OPI = -\log_{10}\left[^1/_4(k_1 + k_2 + k_3 + k_4)\right] \tag{3}$$

2.2.7 Water Sorptivity Test

Fluid flow into a porous, unsaturated substance under the action of capillary forces is referred to as absorption. The capillary pressure is dependent on the pore geometry and the saturation level of concrete. The water sorptivity test was carried out according to the durability index testing procedure manual of the Cement and Concrete Institute (2010) of South Africa. This test quantifies the rate of absorption of the concrete material tested. Two concrete cubes per specimen type were cored and two concrete discs of 70 ± 2 mm diameter, 30 ± 2 mm thickness cut from each cube. The circular edges of the core are sealed with tape to ensure unidirectional absorption. Dry masses of the samples were taken and the thickness of each sample measured in four different directions and diameters taken at two distinct points. The sealed circular edges of the concrete disc were placed in the tray with saturated calcium hydroxide. The specimens were weighed at 3, 5, 7, 12, 16, 20 and 25 min after patting them once on a piece of absorbent paper. Thereafter, the specimens were placed standing on their curved edges in a vacuum saturation tank. The tank was

Table 3 Chemical composition of CA starch.

Element	Weight %	Atomic %
C K	49.99	57.18
O K	49.76	42.73
K K	0.25	0.09
Total	100.00	

Table 4 Chemical composition of MS starch.

Element	Weight %	Atomic %
C K	43.47	50.68
O K	56.00	49.02
Na K	0.38	0.23
S K	0.14	0.06
Total	100.00	

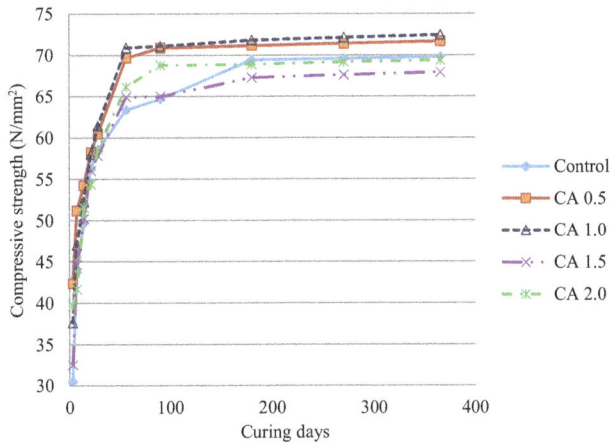

Fig. 12 Compressive strength of concrete with cassava starch addition.

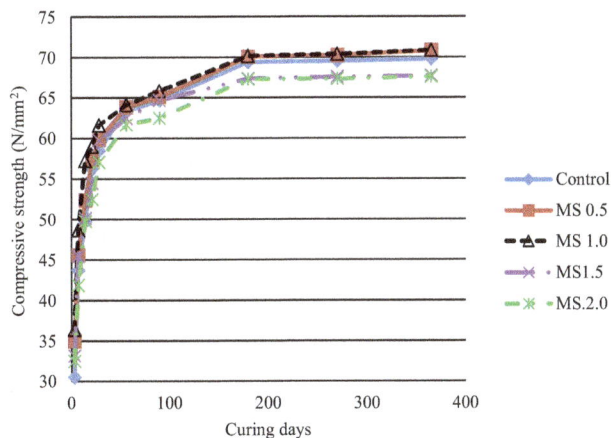

Fig. 13 Compressive strength of concrete with maize starch addition.

evacuated to maintain the specimens between −75 and −80 kPa under vacuum for 3 h. After 3 h, the tank was isolated and water was allowed into the chamber until it was filled to approximately 40 mm above the specimens. Thereafter, vacuum was re-established to between −75 and −80 kPa and maintained for 1 h after which the vacuum was released and air was allowed to enter into the tank and left for 18 h. Thereafter, they were surface dried and the weights were immediately measured. The sorptivity (S) is given by:

$$S = \frac{\Delta M_t}{t^{1/2}} \cdot \frac{d}{M_{sat} - M_0} \qquad (4)$$

where ΔM_t is the change of mass with respect to the dry mass (g), M_{sat} is the saturated mass of concrete (g), M_0 is the dry mass of concrete (g), d is the sample thickness (mm), t is the period of absorption (hr).

3. Results and Discussion

3.1 Materials Characterization

The chemical composition of the cement used for this investigation is shown in Table 2. Observation from the result shows that C_3S has the highest constituent percentage of the cement compounds. This therefore implies that high early strength development should be expected when the cement is used.

3.2 Particle Sizes

The average particle size of cement used for the tests as illustrated in Fig. 4 is 18.698 μm. The specific surface area of the cement as determined by the Malvern particle size analyzer is 0.457 m²/g. The average particle sizes of the CA and MS starch powders as presented in Figs. 5 and 6 are 14.294 μm and 25.051 μm respectively. The specific surface areas of the CA and MS starch granules are 0.444 m²/g and 0.266 m²/g respectively. Observations from Figs. 5 and 6 show that CA starch powder is finer than MS starch powder and has larger specific surface area. The obtained particle size distribution of the CA starch granules is consistent with the findings of Swinkels (1985) and Moorthy (2002). However, the average particle size obtained for MS starch granules is higher than that obtained by Swinkels (1985), but consistent with Moorthy (2002).

3.3 Setting Times Tests

The results of the setting times of cement paste with starches (CA and MS) is presented in Fig. 7. It can be observed from Fig. 7 that control cement exhibited the least initial and final setting times. The results showed that with increasing concentrations of starches, the setting times, whether initial or final, increases irrespective of the type of starch used. It also showed that control has the least initial and final setting times followed by MS 0.5, MS 1.0, MS 1.5, MS 2.0, CA 0.5, CA 1.0, CA 1.5 and CA 2.0 respectively. Observation from MS pastes showed continuous small increments in both initial and final setting times as the concentrations of MS starch increases. However, CA pastes of similar concentration to MS pastes have longer initial and final setting times.

The prolonged setting times of the cement pastes with starch additions may be attributed to retardation of hydration of some cement compounds such as tricalcium silicate (C_3S) and tricalcium aluminate (C_3A) (Ramachandran 1995). According to Brooks et al. (2000) it may also be due to adsorption of the admixture onto the cement grain surfaces. It should however, be pointed out that though admixtures may delay setting times in concrete, mechanical properties of the concrete is not negatively affected. Bishop and Barron (2006) found that though tartaric acid retarded hydration of cementitious systems yet the strength was enhanced. The small smooth and spherical nature of CA starch (Fig. 8) granules coupled with the larger specific surface area (Figs. 5) in comparison to that of MS starch (Fig. 6) would result in adsorption of more CA starch onto the surfaces of cement grains and covering of the cement grains.

3.4 Morphology of Starches

The micrographs of the starches are presented in Figs. 8 and 9. It can be observed that characteristic features of CA starch granules (Fig. 8) are smooth and spherical in nature while MS starch granules (Fig. 9) are also smooth but polygonal in shape, similar to what Swinkels (1985) reported. The specific surface area of the CA starch (Fig. 5) being larger than that of MS starch granules (Fig. 6) would make CA starch more available for adsorption onto the cement grains surfaces. The gel formed from CA starch will cover the surfaces of cement grains very easily because of the particle size, shape and the specific surface area. However, because of the size and polygonal shape of MS starch granules the gel would not cover as much surface of the cement grains as that of CA starch. One of the effects of starch addition to concrete is that it increases the viscosity of the material and this is seen from the slump measurements of different concrete mixes used in this investigation (see Table 1). Further comparison between CA concretes and MS concretes shows that CA concretes of the same concentrations with MS concretes have lower slump, which is due to adsorption of more CA starch onto the surfaces of the cement grains. This results in higher viscosity and also confirms one of the findings of Swinkels (1985) that CA starch has higher congealing power and thus leads to higher viscosity.

Energy dispersive X-ray analysis (EDXA) that gives the chemical composition of the CA and the MS starches were carried out as shown in Figs. 10 and 11 respectively, with the results presented in Tables 3 and 4 respectively. The major elements in those starches are carbon, oxygen and hydrogen, though EDXA cannot give the percentage of hydrogen in any compound. The general structural arrangement of starch shown in Fig. 1 confirms the presence of hydrogen. CA starch has Potassium (K) as a trace element while MS starch has Sodium (Na) and Sulphur (S) as trace elements.

3.5 Compressive Strength Tests

The compressive strength results are presented in Figs. 12 and 13. The strength development pattern generally consistent with the recommendations in Table 11 of BS 8110 (1985). Generally observed high early strength development of the concretes can be attributed to high percentages of C_3S and C_3A in the cement composition as shown in Table 2. These compounds are responsible for the early strength development in cement. Neville and Brooks (1990) reported that that the strength ratio of 365-day to 28-day concrete without admixtures should be about 1.25. The result obtained from this investigation is 1.20, which is reasonably close to the stated ratio.

The compressive strength results of concrete mixes containing different percentages of CA by weight of cement (0, 0.5, 1.0, 1.5 and 2.0) after 28 days of curing are: 58.53, 60.5, 61.43, 57.83 and 58.63 N/mm^2 respectively. This represents 3.4, 4.9, 0.5 and 0.2 % increase in compressive strength of CA concretes over the control. For MS starch additions (0.5, 1.0, 1.5 and 2.0) the percentage increase in strength are 59.85, 61.6, 59.77 and 57.1 N/mm^2 respectively. This represents 2.3, 5.2, 2.1 and −2.4 % increase over the control. After 1 year of curing, percentage strength increase over the control for CA 0.5 and CA 1.0 are 2.7 and 3.8 % respectively while MS 0.5 and MS 1.0 gave same value of 1.5 %. Drop in strength experienced in CA 1.5, CA 2.0 MS 1.5 and MS 2.0 are 2.7, 0.6, 2.9 and 3.1 % respectively. From Figs. 12 and 13, CA 0.5, CA 1.0 MS 0.5 and MS 1.0 attained strength of more than 70 N/mm^2, however, CA 0.5 and CA 1.0 attained it at 90 days while MS 0.5 and MS 1.0 attained at 180 days. The obtained compressive strength of more than 70 MPa is reasonable with the use of high strength cement of 52.5 N at water-cement ratio of 0.54. The long-term strength of the concretes is seen to slow down after 180 days because much of the strength has been gained in the early days (of strength development) as observed from the high percentage of C_3S in the constituents of cement used (Table 1). The gain in compressive strength after 100 days as observed from Figs. 12 and 13 is limited by the availability of hydration sites which is reduced as the cement matures.

A comparison of CA concretes (Fig. 12) and MS concretes (Fig. 13) with the control shows that CA concretes attained high strength earlier than MS concretes and the control. This may be because CA starch is known to have higher degree of polymerisation than MS starch (Swinkels 1985), resulting in greater binding force. This explains why CA starch concretes have lower slump when compared to the control concretes or MS starch concretes. Swinkels (1985) showed that the extent of polymerisation of starch molecule is influenced by the source of starch.

3.6 Oxygen Permeability Test Results

The results of oxygen permeability test are presented in Figs. 14 and 15. From the result of coefficients of permeability k (m/s) presented in Fig. 14, there is progressive decrease (based on time) in the coefficients of permeability

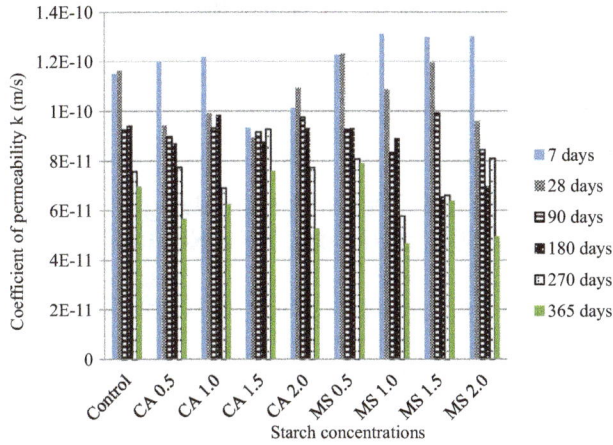

Fig. 14 Coefficient of permeability of different starch concentrations.

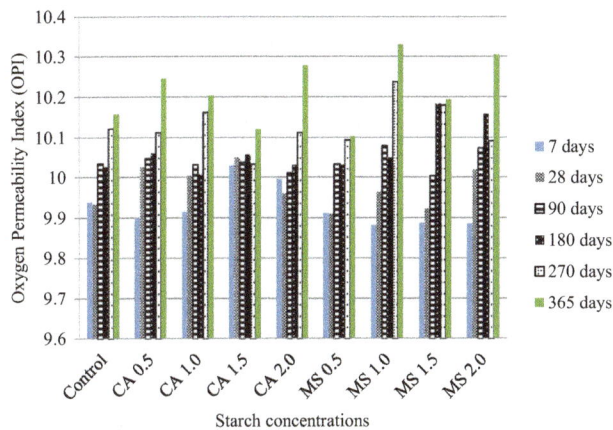

Fig. 15 Oxygen Permeability Indices for different starch concentrations.

for all the starch concentrations except for spikes experienced at 6 months for control, CA 1.0 and MS 1.0, at 90 and 270 days for CA 1.5, at 28 days for CA 2.0 and at 270 days for MS 2.0. The illustrated results of the oxygen permeability indices (OPI) in Fig. 15 show that the indices increased over time except for the drops shown at 180 days for control, CA 1.0 and MS 1.0, at 3 and 9 months for CA 1.5, at 28 days for CA 2.0 and at 270 days for MS 2.0. The results show that the permeability characteristics of the concrete specimens improved with time. For concrete specimens with CA and MS starches, their permeability characteristics improved significantly as can be seen in Fig. 14 from 7 to 365 days. The consistent decrease in coefficients of permeability which result increase in OPI of starch modified concretes over time implies improved durability characteristics. This is in agreement with the findings of Akindahunsi et al. (2012) and shows that in general starch improves durability of concretes over the monitored period.

The relationship between coefficients of permeability and oxygen permeability indices for concrete samples with various concentrations of starch are presented in Figs. 16 and 17. This indicates positive correlation between the coefficients of permeability and oxygen permeability indices for

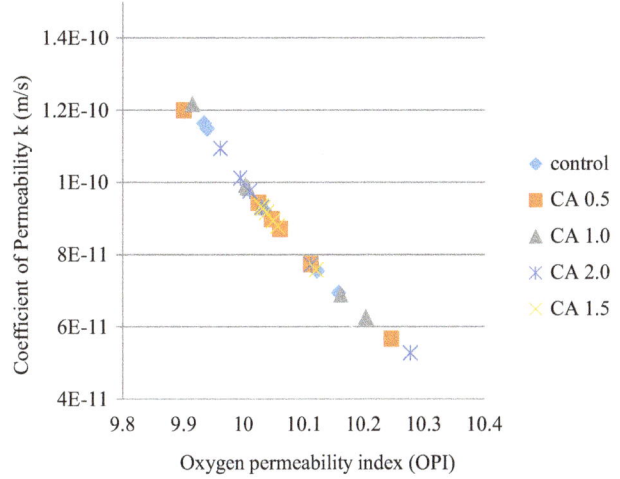

Fig. 16 Relationship between coefficients of permeability and oxygen permeability indices with different concentrations of cassava starch in concrete.

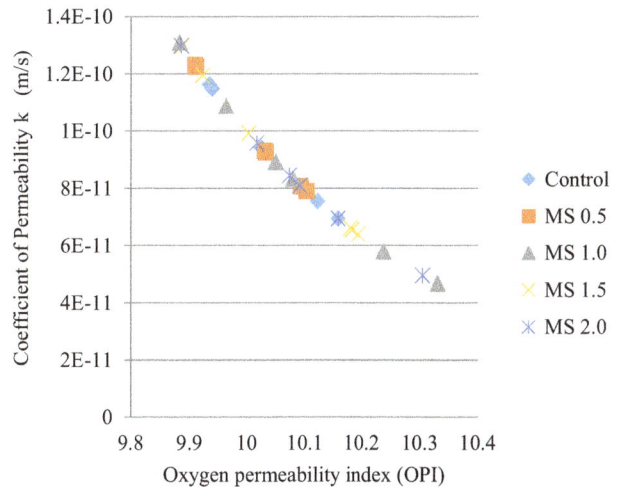

Fig. 17 Relationship between coefficients of permeability and oxygen permeability indices with different concentrations of maize starch in concrete.

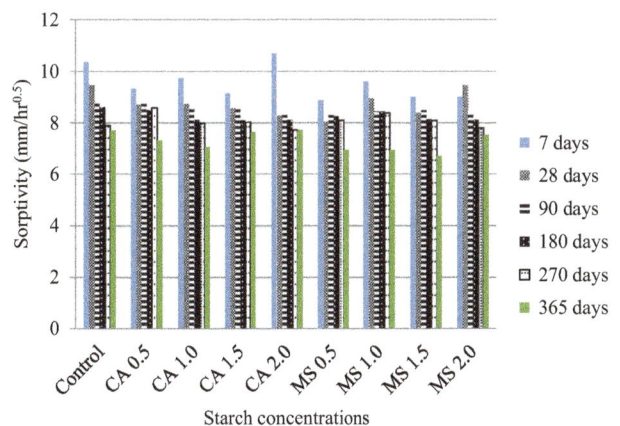

Fig. 18 Sorptivity for different starch concentrations.

the various concrete samples. The linear correlations of control, CA 0.5, CA 1.0, CA 1.5, CA 2.0, MS 0.5, MS 1.0, MS 1.5 and MS 2.0 are 0.9947, 0.9922, 0.9901, 0.9994, 0.9902, 0.9971, 0.9771, 0.9953 and 0.97 respectively. The

OPI indicates the rate at which oxygen can permeate through the concrete. A low value of coefficient of permeability will imply a high OPI as seen in Figs. 16 and 17, which is an indication of a low permeable concrete material. Beushausen and Alexander (2008) pointed out that oxygen permeability is a dependable method of assessing the durability of concrete characteristics and can be used in ascertaining material properties of concrete.

3.7 Sorptivity

The result of the sorptivity tests is presented in Fig. 18. The illustrated result generally showed a decrease in sorptivity of the concrete samples over time, in agreement with Akindahunsi et al. (2012). Generally, sorptivity at 7 days curing was the highest for all the concretes except for CA 1.0 where 28-day sorptivity was higher than that of 7 days. The sorptivity results of the concrete specimens at one year testing showed that the control has a sorptivity rate of 7.7 $mm/hr^{0.5}$ while concrete specimens with CA starch shows that CA 1.0 has a sorptivity of 7.1 $mm/hr^{0.5}$ followed by CA 0.5 (7.5 $mm/hr^{0.5}$). Concrete with MS starch additions (MS 0.5, MS 1.0, MS 1.5 and 2.0) show lower sorptivity rates (6.96, 6.95, 6.7 and 7.6 $mm/hr^{0.5}$ respectively) for concrete specimens tested at 365 days. Alexander et al. (2008) explained the importance of the sorptivity test in construction as a property that is sensitive to the cover provided for reinforcement in the concrete. The authors noted that sorptivity can be used on site as a control measure for durability requirements.

4. Statistical Evaluation of Oxygen Permeability and sorptivity Results

The results of the oxygen permeability and sorptivity tests were subjected to some statistical analyses to determine their normality. That is, statistical tests were conducted to examine if there was any significant difference in the results obtained for the different concentrations of starch additions in concrete at each of the test periods (i.e. 7, 28, 90, 180, 270 and 365 days) and if there was any significant difference between control specimens and CA and MS concretes. STATA® statistical software was used for the analyses. Sktest and swilk tests were used to check the normality of the oxygen permeability and sorptivity data and the results showed skewness. The histogram and the qnorm plots are illustrated in Figs. 19 and 20.

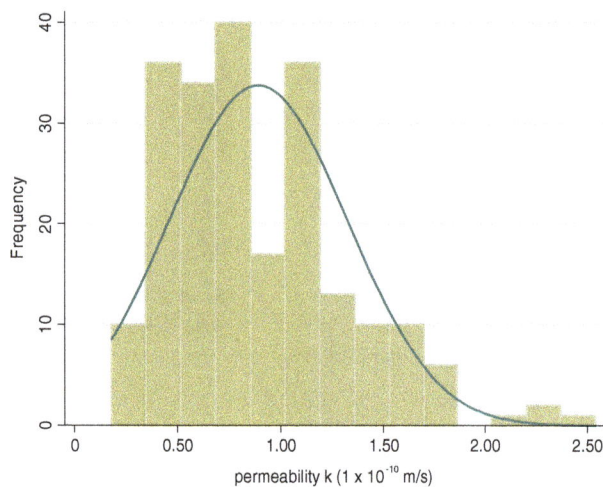

(a) Histogram for oxygen permeability

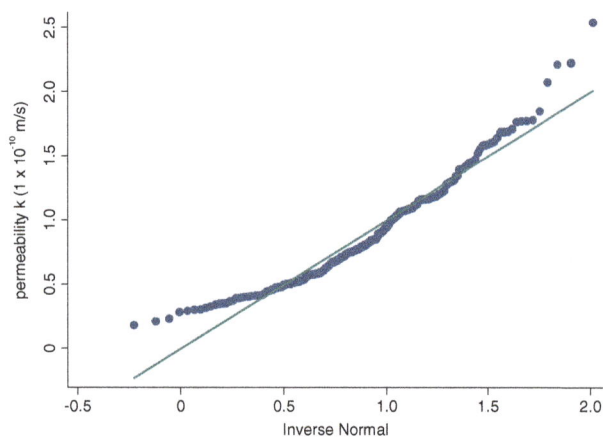

(b) Qnorm for oxygen permeability

Fig. 19 Skewness of permeability data.

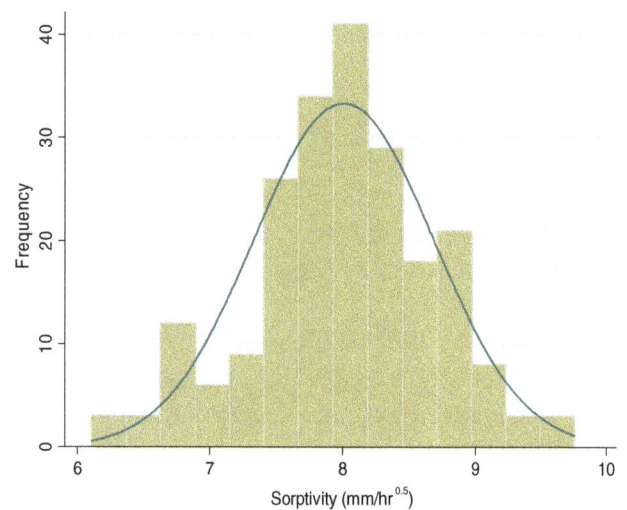

(a) Histogram for sorptivity

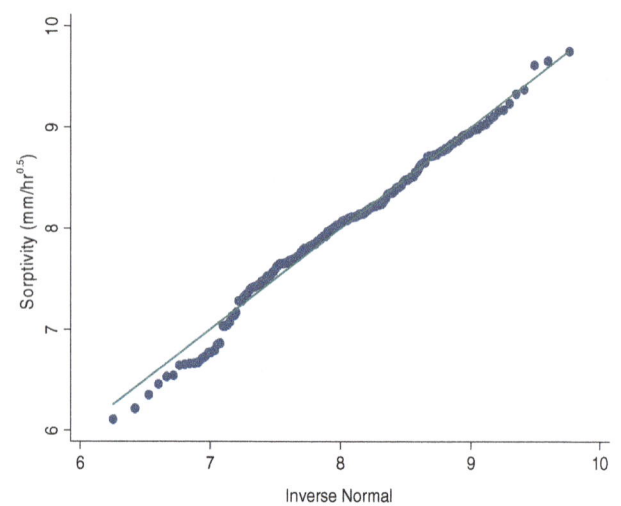

(b) Qnorm for sorptivity

Fig. 20 Skewness of sorptivity data.

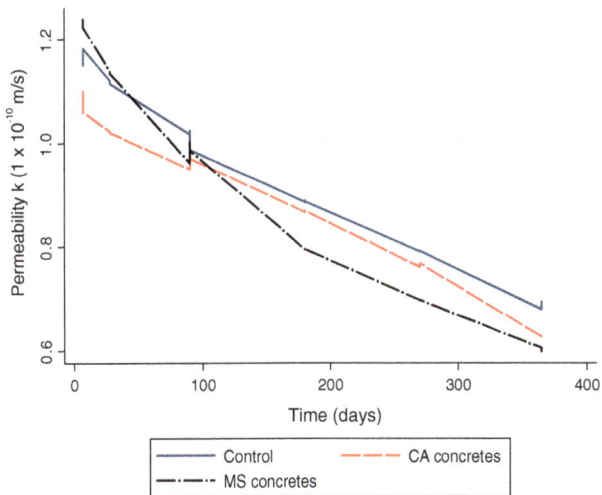

Fig. 21 Effect of different materials on oxygen permeability test.

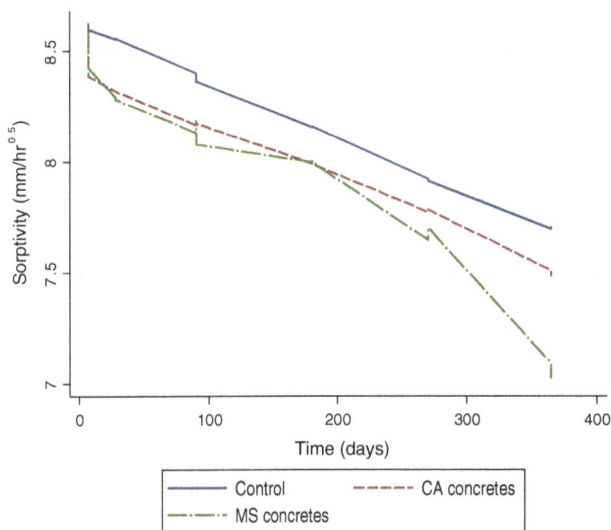

Fig. 22 Effect of different materials on sorptivity test.

Since the data are skewed, a non-parametric statistic was used to examine if there was any statistically significant difference in oxygen permeability and sorptivity at the different test days for concrete specimens containing different concentrations of starches. Kruskal–Wallis one-way analysis of variance by ranks was used compare the mean permeability of the concretes and rates of sorptivity at different test periods. By combining different concentrations of the same starch material, a test was conducted to check if there were significant differences between control, CA and MS concretes over the monitored test period. The result indicated that there are significant differences over time especially from after 100 to 365 days and they are illustrated in Figs. 21 and 22. This shows that the addition of starch to the concretes improves their durability properties.

5. Conclusion

The effects of different starch concentrations as an admixture in concrete were examined and the following inferences can be made:

1. Starches used in this investigation generally delay the setting time of cement which may be an advantage for use where a longer period of time is required for casting the concrete.
2. The morphology, particle size and specific surface areas of the starches gave an indication of the form of adhesion the starches would have on the cement grains. CA starch with a smaller average particle size thus adheres more strongly onto cement grains. Hence, it will lead to a higher viscosity and less slump in concretes when compared to MS starch.
3. The compressive strength tests result at 28 days showed that CA 0.5, CA MS 0.5 and MS 0.5 have 1.0 3.4, 4.9, 2.3 and 5.2 % strength gain over the control respectively. After a year the strength increase for CA 0.5 and CA 1.0 are 2.7 and 3.8 % respectively over the control, while MS 0.5 and MS 1.0 gained 1.5 % strength over control.
4. Incorporation of CA and MS starches into concrete improves the durability properties of the concrete.
5. The statistical analyses of concrete mixed with different starches when compared with control indicated that generally concrete with starches performed better than control. However, overall the concretes with addition of different concentrations of MS starch performed better than concretes with different concentrations of CA starch.

Further investigations may be required to determine the suitability of the use of starch under different environmental conditions in order to assess the full impact of durability properties of the material in concrete.

References

ACI 211. (1999). *Standard practice for selecting proportions for normal, heavyweight, and mass concrete.* Farmington Hills, MI: American Concrete Institute.

Akindahunsi, A. A., Schmidt, W., Uzoegbo, H. C., & Iyuke, E. S. (2011). Technological Advances in the Potential Uses of Cassava Starch in Concrete. In *Proceedings of the 6th International Conference of Africa Materials Research Society.* Victoria Falls, Zimbabwe.

Akindahunsi, A. A., Schmidt, W., Uzoegbo, H. C. & Iyuke, S. E (2013). The influence of starches on some properties of concrete. In *Proceedings of the 1st ACCTA International*

Conference on Advances in Cement and Concrete Technology in Africa. Johannesburg, South Africa.

Akindahunsi, A. A., Uzoegbo, H. C., & Iyuke, S. E. (2012). Use of starch modified concrete as a repair material. In *Proceedings of the 3rd International Conference on Concrete Repair, Rehabilitation and Retrofitting.* Cape Town, South Africa.

Alexander, M. G., Ballim, Y., & Stanish, K. (2008). A framework for the use of durability indexes in performance and specifications of reinforced concrete structures. *Materials and Structures, 41*, 921–936.

Baroghel-Bouny, V., Nguyen, T. Q., & Dangla, P. (2009). Assessment and prediction of RC structure service life by means of durability indicators and physical/chemical models. *Cement & Concrete Composites, 31*, 522–534.

BeMiller, J. N., & Hubber, K. C. (2011). *Starch. Ullmann's Encyclopaedia of industrial chemistry.* Co Weinheim, Germany: Wiley-VCH Verlag GmbH.

Beushausen, H., & Alexander, M. G. (2008). The South African index tests in an international comparison. *Journal of South African Institution of Civil Engineering, 50*(1), 25–31.

Bishop, M., & Barron, A. R. (2006). Cement hydration inhibition with sucrose, tartaric acid, and Lignosulfonate: Analytical and Spectroscopic Study. *Industrial and Engineering Chemistry Research, 45*, 7042–7049.

Brooks, J. J., Megat Johari, M. A., & Mazloom, M. (2000). Effect of admixtures on the setting times of high-strength concrete. *Cement & Concrete Composites, 22*, 293–301.

C&CI. (2010). *Durability index testing procedure manual.* Midrand: Published by Cement and Concrete Institute.

Chidiac, S. E. (2009). Sustainability of civil engineering structures – Durability of concrete. *Cement & Concrete Composites, 31*, 513–514.

Chung, D. D. L. (2004). Review use of polymers for cement-based structural materials. *Journal of Materials Science., 39*, 2973–2978.

Crépy, L., Petit, J. Y., Joly, N., Wirquin, E., & Martin. P. (2011). First step towards bio-superplasticizers. *In Proceedings of the 13th international Congress on the Chemistry of Cement.* Madrid, Spain.

Elahi, A., Basheer, P. A. M., Nanukuttan, S. V., & Khan, Q. U. Z. (2010). Mechanical and durability properties of high performance concretes containing supplementary cementitious materials. *Construction and Building Materials, 24*, 292–299.

Folić, R. (2009). Durability design of concrete structures—Part 1: Analysis fundamentals. *Architecture and Civil Engineering, 7*(1), 1–18.

Folić, R., & Zenunović, D. (2010). Durability design of concrete structures—part 2: modelling and structural assessment. *Facta Universitatis Series: Architecture and Civil Engineering, 8*(1), 45–66.

Gjørv, O. E. (2011). Durability of concrete. *Arabian Journal Science Engineering, 36*, 151–172.

Izaguirre, A., Lanas, J., & Álvarez, J. I. (2010). Behaviour of starch as a viscosity modifier for aerial lime-based mortars. *Carbohydrate Polymers, 80*, 222–228.

Khayat, K. H. (1998). Viscosity-enhancing admixtures for cement-based materials—An overview. *Cement & Concrete Composites, 20*, 171–188.

Lasheras-Zubiate, M., Navarro-Blasco, I., Fernández, J. M., & Álverez, J. I. (2012). Effect of the addition of chitosan ethers on the fresh state properties of cement mortars. *Cement & Concrete Composites, 34*, 964–973.

Luke, K., & Luke, G. (2000). Effect of sucrose on the retardation of Portland cement. *Advances in Cement Research. 12*(1), 9–18.

Malvern. (2005). *Malvern mastersizer 2000.* Malvern, UK. www.malvern.com. Accessed 5 Jan 2013.

Merretz, W., Smith, G., & Borgert, J. (2009). Achieving durability in construction. *CIA Concrete Durability Workshop.* www.sciaust.com.au. Accessed 27 May 2013.

Moorthy, S. N. (2002). Physicochemical and functional properties of tropical tubers starches: A review. *Starch/Stärke, 54*, 559–592.

Nagesh, M. (2012). *Concrete Technology.* Ramanagara, India: Government Engineering College.

Neville, A. M., & Brooks, J. J. (1990). *Concrete technology* (2nd ed.). Harlow, UK: Longman Scientific & Technical.

Owens, G. (2009). *Fulton's concrete technology* (9th ed.). Midrand: Cement & Concrete Institute.

Peschard, A., Govin, A., Grosseau, P., Guilhot, B., & Guyonnet, R. (2004). Effect of polysaccharides on the hydration of cement paste at early ages. *Cement and Concrete Research, 34*, 2153–2158.

Ramachandran, V. S. (1995). *Concrete admixtures handbook properties, Science, and Technology.* Park Ridge, NJ: by Noyes Publications.

SANS 1083. (2006). *Aggregate from natural sources. Aggregates for concrete.* Pretoria, South Africa: South African Bureau of Standards.

SANS 5860. (2006). *Concrete tests—dimensions, tolerances and uses of cast test specimens.* Pretoria, South Africa: South African Bureau of Standards.

SANS 5861–2. (2006). *Concrete tests—sampling of freshly mixed concrete.* Pretoria, South Africa: South African Bureau of Standards.

SANS 5861-3. (2006). *Concrete tests—making and curing of test specimens.* Pretoria, South Africa: South African Bureau of Standards.

Satin, M. (1998). *Functional properties of starches. In spotlight tropical starch misses market.* AGSI report, Agriculture 21. FAO-Magazine, p. 11.

Swinkels, J. J. M. (1985). Composition and Properties of Commercial Native Starches *Starch/starke.* 37 Nr. 1, S. 1–5.

Wang, T. L., Bogracheva, T. Y., & Hedley, C. L. (1998). Starch: as simple as A, B, C? *Journal of Experimental Botany, 49*(320), 481–502.

14

Strength Prediction of Corbels Using Strut-and-Tie Model Analysis

Wael Kassem*

Abstract: A strut-and-tie based method intended for determining the load-carrying capacity of reinforced concrete (RC) corbels is presented in this paper. In addition to the normal strut-and-tie force equilibrium requirements, the proposed model is based on secant stiffness formulation, incorporating strain compatibility and constitutive laws of cracked RC. The proposed method evaluates the load-carrying capacity as limited by the failure modes associated with nodal crushing, yielding of the longitudinal principal reinforcement, as well as crushing or splitting of the diagonal strut. Load-carrying capacity predictions obtained from the proposed analysis method are in a better agreement with corbel test results of a comprehensive database, comprising 455 test results, compiled from the available literature, than other existing models for corbels. This method is illustrated to provide more accurate estimates of behaviour and capacity than the shear-friction based approach implemented by the ACI 318-11, the strut-and-tie provisions in different codes (American, Australian, Canadian, Eurocode and New Zealand).

Keywords: corbels, load-carrying capacity, shear strength, strut-and-tie model.

1. Introduction

Reinforced concrete (RC) corbels, defined as short cantilevers jutting out from walls or columns having a shear span-to-depth ratio, a_v/d, normally less than 1, are commonly used to support prefabricated beams or floors at building joints, allowing, at the same time, the force transmission to the vertical structural members in precast concrete construction. Corbels are primarily designed to resist vertical loads and horizontal actions owing to restrained shrinkage, thermal deformation and creep of the supported beam and/or breaking of a bridge crane. They are becoming a common feature in building construction with the increasing use of precast concrete. Owing to their geometric proportions, corbels are commonly classified as a discontinuity region (D-region), where the strain distribution over their cross-section depth is nonlinear, even in the elastic stage (MacGregor and Wight 2009), and their strength is predominantly controlled by shear rather than flexure (Yang and Ashour 2012).

The ACI 318-11 code (ACI Committee 318 2011) requires corbels having shear span-to-depth ratio, a_v/d, less than 2 to be designed using the strut-and-tie method and those with shear-to-span ratio less than 1 to be designed either using strut-and-tie method, or by the closely related traditional ACI design method based on shear-friction approach. However, the shear-friction hypothesis has little correlation with the observed failure phenomenon of concrete crushing in the diagonal strut (Hwang et al. 2000b).

Strut-and-tie models (STM) have been generally recognized as an acceptable rational design approach for D-region members including deep beams and corbels (Schlaich et al. 1987). In addition, most current design codes [ACI Committee 318; Australian code AS 3600 (2009); Canadian code (CSA A23.3-04); Eurocode 2 (2004) and New Zealand code (NZS 3101-1)] have recommended the STM approach as a design tool for RC corbels. However, shear capacity of corbels evaluated from STMs and available formulae and computing procedures showed substantial scatter when compared to experimental results (Hwang et al. 2000b; Ali and White 2001; Russo et al. 2006). A rational design procedure to produce safe and economic corbels is therefore required.

In the current paper, a strut-and-tie based method intended for determining the load-carrying capacity of corbels is presented. In addition to the normal strut-and-tie force equilibrium requirements, the proposed model accounts for strain compatibility and constitutive laws of cracked reinforced concrete, and uses a secant stiffness formulation. A similar approach was used previously to calculate the shear capacity of of squat walls (Hwang et al. 2001), deep beams (Hwang and Lee 2000), beam-column joints (Hwang and Lee 1999, 2000, 2002), dapped-end beams (Lu et al. 2003), and corbels (Hwang et al. 2000a), while using a statically indeterminate truss for modeling the flow of forces and an approximate estimation of members stiffness in evaluating the capacity.

Division of Construction Engineering, College of Engineering at Al-Qunfudah, Umm Al-Qura University, P.O. BOX 288, Al-Qunfudah 21912, Saudi Arabia.
*Corresponding Author; E-mail: wakassem@uqu.edu.sa

2. Research Significance

In the present study, a strut-and-tie based method is developed for calculating the load-carrying capacity of reinforced concrete corbels. The proposed method is based on an

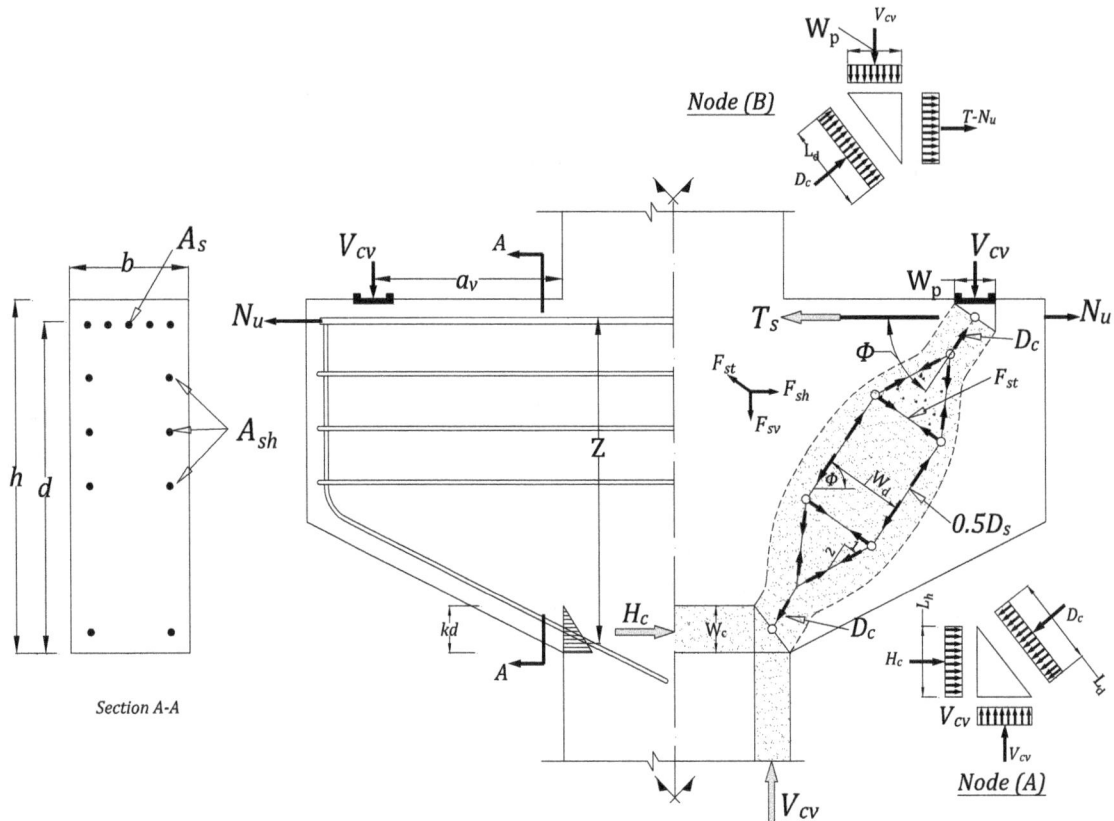

Fig. 1 Geometry and strut-and-tie model with forces acting on corbel.

iterative, secant stiffness formulation and employs constitutive laws for cracked reinforced concrete, while considering strain compatibility. The secant stiffness formulation approach has previously been implemented in nonlinear finite element procedures to predict the nonlinear response of reinforced concrete membrane elements (Vecchio 1989), as well as to estimate the load-carrying capacity of deep beams (Park and Kuchma 2007). The method accounts for the failure modes due to crushing of the nodal compression zone at the top of the diagonal strut, yielding of the longitudinal reinforcement, as well as that of strut crushing or splitting. This method is used successfully to predict the load-carrying capacity of 455 corbels that have been tested experimentally. The findings illustrate that the strut-and-tie model proposed by different code provisions provide conservative and scattered estimates of the strength of corbels, which should be expected since these provisions were developed for the design of all forms of discontinuity regions and not explicitly for corbels.

3. Compatibility-Based Strut-and-Tie Model Approach for Corbels

Strut-and-tie modelling is a generalisation of the truss analogy in which a structural continuum is transformed into a discrete truss with compressive forces being resisted by concrete and tensile forces by reinforcement. The method is based on the lower bound theorem of plasticity. Consequently, there are an unlimited number of possible solutions

with only some having sufficient ductility for the assumed stress distribution to develop. In the proposed approach, a simple and statically determinate strut-and-tie load path is proposed to model the force transferring within the corbel as shown in Fig. 1. Statically determinant model requires no knowledge of the member stiffness which makes it simple to calculate member forces using simple statics rules. The proposed strut-and tie models assumes that the corbel resists the loads by compressive struts feeding directly into the column, and a tension tie is required to resist the out-of-balance forces at the loading point.

4. Equilibrium Conditions

Figure 1 presents loads acting on a corbel and the proposed force transferring mechanisms in view of the proposed strut-and-tie model. For corbels with short span-to-depth ratios, a large portion of the applied vertical shear force is directly transferred to the supporting columns or walls through inclined strut, with the formation of a full-length horizontal tie to balance the thrust of the inclined struts (Fig. 1). The corbel is loaded by the vertical force V_{cv} applied at the distance a_v from the column face and it is assumed that the horizontal outward load, N_u, is directly applied at the centroid of the principal tensile reinforcement and the effect of shifting is neglected for simplicity (Hwang et al. 2000b). The angle between the compressed diagonal concrete strut and the horizontal direction ϕ, can be defined as (Russo et al. 2006):

$$\phi = \tan^{-1}\left(\frac{Z}{a_v}\right) \tag{1}$$

where Z is the distance of the lever arm from the centroid of the principal tension steel to the resultant compressive force and a_v is the shear span. According to linear bending theory, the lever arm Z of a singly reinforced rectangular section can be estimated as (Hwang et al. 2000a):

$$Z = d - \frac{kd}{3} \tag{2}$$

where d is the effective depth of the corbel; kd is the depth of the neutral axis of the cross section; and coefficient k can be defined as (Hwang et al. 2000b):

$$k = \sqrt{(n\rho_f)^2 + 2n\rho_f} - n\rho_f \tag{3}$$

in which n is the ratio of the elastic moduli of steel and concrete, $n = E_s/E_c$, and the flexural reinforcement ratio ρ_f is assumed to be given by:

$$\rho_f = \frac{A_s + \Omega A_{sh} - A_n}{bd} \tag{4}$$

where A_n is the cross-sectional area of principal reinforcement used to resist the applied outward load, taken as $A_n = N_u/f_{ys}$, where f_{ys} is the yielding strength of the principal reinforcement, A_s and A_{sh} are the cross-sectional areas of the principal tensile reinforcement and horizontal web reinforcement, respectively, and Ω is an efficiency factor representing the contribution of the web horizontal reinforcement, assumed equal to 0.2 (He et al. 2012). The value of n is obtained by assuming, from ACI 318-11, that $E_s = 200$ GPa and $E_c = 4700\sqrt{f_c'}$ (MPa), it follows that:

$$n = 42.6/\sqrt{f_c'}. \tag{5}$$

The diagonal strut is assumed to have a bottle-shaped form. That is, it spreads laterally along its length. The lateral spreading of the bottle-shaped strut introduces tensile force transverse to the strut, F_{st}. The tensile force could potentially cause cracking along the length of the strut resulting in a premature failure. Hence, transverse skin reinforcement should be provided in order to control the cracking. The strut compressive force is assumed to spread at a 2:1 slope (longitudinal: transverse direction) as suggested by the ACI 318-11. The considered strut-and-tie model leads to the following equilibrium equations:

$$D_c = \frac{V_{cv}}{\sin \phi} \tag{6}$$

$$H_c = \frac{V_{cv}}{\tan \phi} \tag{7}$$

$$F_{st} = \frac{V_{cv}}{4 \sin \phi} \tag{8}$$

where D_c, H_c are the compressive forces in the diagonal and horizontal concrete struts, respectively; and F_{st} is the

bursting tensile force in the tie of the strut-and-tie model. Because the bursting force F_{st} represents a quarter of the compressive force of the diagonal strut, D_c, the horizontal and vertical components, F_{sh} and F_{sv}, of the tie force can be obtained from equilibrium as follows:

$$F_{sh} = \frac{V_{cv}}{4} = \frac{D_s \sin \phi}{4} \tag{9}$$

$$F_{sv} = \frac{V_{cv}}{4 \tan \phi} = \frac{D_s \cos \phi}{4}. \tag{10}$$

4.1 Secant Stiffness Formulation

The proposed procedure is based on a compatibility—based iterative, secant stiffness formulation and employs constitutive relations for cracked concrete and reinforcement. The secant stiffness approach was used to calculate the normal strains in the horizontal concrete strut, diagonal concrete strut, horizontal web steel, vertical web steel, and the longitudinal steel tie according to the following equations:

$$\varepsilon_d = \frac{D_c}{\overline{E}_d A_d} \tag{11}$$

$$\varepsilon_c = \frac{H_c}{\overline{E}_c A_c} \tag{12}$$

$$\varepsilon_h = \frac{2F_{sh}}{\overline{E}_{sh} A_{sh}} \tag{13}$$

$$\varepsilon_v = \frac{2F_{sv}}{\overline{E}_{sv} A_{sv}} \tag{14}$$

$$\varepsilon_s = \frac{T_s}{\overline{E}_s A_s} \tag{15}$$

where A_d, A_c are the cross-sectional areas of the diagonal and horizontal concrete struts; A_{sh}, A_{sv} and A_s are the cross-sectional areas of horizontal, vertical and longitudinal steel ties; $\overline{E}_d$, $\overline{E}_c$, $\overline{E}_{sh}$, $\overline{E}_{sv}$ and $\overline{E}_s$, are the corresponding secant moduli. Given compatible stress and strain fields, secant moduli can be defined for the concrete and reinforcement (shown in Fig. 2). Secant moduli can be estimated by (Park and Kuchma 2007):

$$\overline{E}_d = \frac{\sigma_{2d}}{\varepsilon_d} \tag{16}$$

$$\overline{E}_c = \frac{\sigma_{2c}}{\varepsilon_c} \tag{17}$$

$$\overline{E}_{sh} = \frac{f_{sh}}{\varepsilon_h} \tag{18}$$

$$\overline{E}_{sv} = \frac{f_{sv}}{\varepsilon_v} \tag{19}$$

$$\overline{E}_s = \frac{f_s}{\varepsilon_s} \tag{20}$$

where σ_{2d}, σ_{2c}, f_{sh}, f_{sv} and f_s are the uniaxial stresses that are obtained from the constitutive relations of each member.

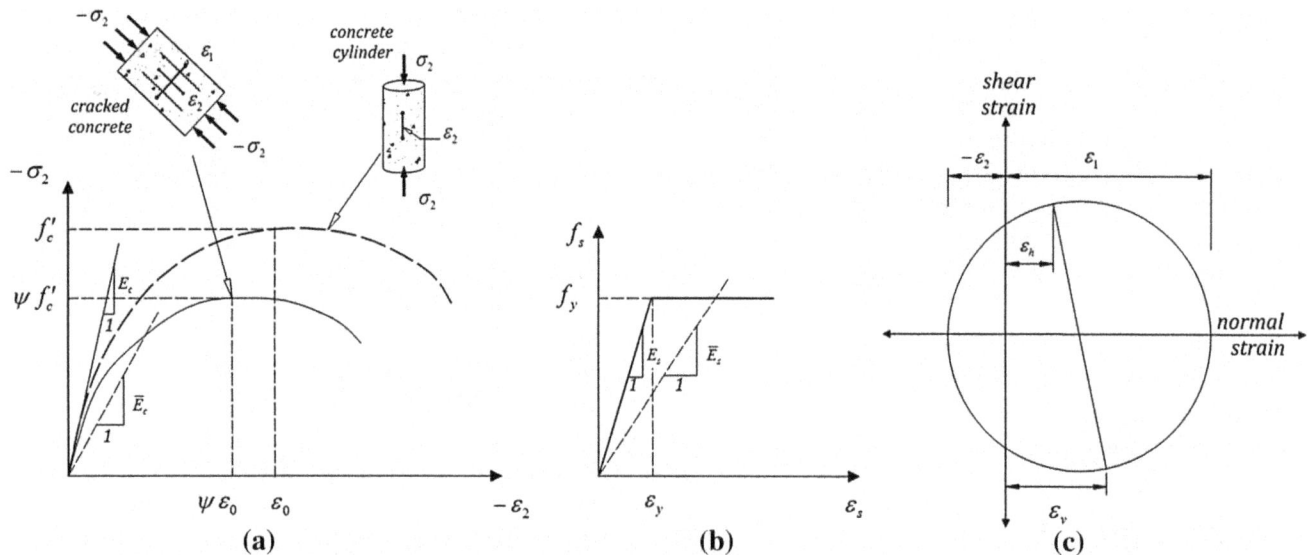

Fig. 2 Constitutive relations and secant moduli used in analysis procedures for **a** concrete in compression, **b** reinforcing steel; and **c** compatibility conditions for diagonally cracked concrete.

5. Constitutive Relationships of Concrete and Steel

5.1 Softened Concrete in Compression

Cracked reinforced concrete in compression has been observed to exhibit lower strength and stiffness compared with uniaxially compressed concrete, see Fig. 2a. This phenomenon of strength and stiffness reduction is commonly referred to as compression softening. Applying this softening effect to the strut-and-tie model, it is recognized that the tensile straining perpendicular to the strut will reduce the capacity of the concrete strut to resist compressive stresses. The stress in the concrete is determined from the strains according to the following equations (Vecchio and Collins 1993):

- The ascending branch

$$\sigma_2 = \Psi f_c' \left[2 \left(\frac{\varepsilon_2}{\Psi \varepsilon_0} \right) - \left(\frac{\varepsilon_2}{\Psi \varepsilon_0} \right)^2 \right] \left(\frac{\varepsilon_2}{\Psi \varepsilon_0} \right) \leq 1 \quad (21a)$$

- The descending branch

$$\sigma_2 = \Psi f_c' \left[1 - \left(\frac{(\varepsilon_2/\psi \varepsilon_0) - 1}{4/\psi - 1} \right)^2 \right] \left(\frac{\varepsilon_2}{\Psi} \varepsilon_0 \right) \geq 1 \quad (21b)$$

$$\Psi = \frac{1}{1 + k_c k_f} \quad k_c = 0.35 \left(\frac{-\varepsilon_1}{\varepsilon_2} - 0.28 \right)^{0.8} \geq 1.0$$

$$k_f = 0.1825 \sqrt{f_c'} \geq 1.0$$

where σ_2 is the average principal stress of concrete in the 2 direction; ψ is the softening coefficient; f_c' is the compressive strength of a standard concrete cylinder in unit of MPa; ε_2 and ε_1 are the average principal strains in the 2 and 1 directions, respectively; and ε_0 is the concrete cylinder strain corresponding to the cylinder strength f_c', which can be defined approximately as:

$$\varepsilon_0 = 0.002 + 0.001 \left(\frac{f_c' - 20}{80} \right) \quad 20 \leq f_c' \leq 100 \, \text{MPa}. \quad (22)$$

6. Reinforcing Steel

The stress–strain relationship of steel is assumed to be linear up to yielding, followed by a yield plateau (Fig. 2b). This elastic–perfectly plastic type of stress–strain relationship is represented mathematically by:

$$f_s = E_s \varepsilon_s \quad \text{for} \quad \varepsilon_s \leq \varepsilon_y \quad (23a)$$

$$f_s = f_y \quad \text{for} \quad \varepsilon_s > \varepsilon_y \quad (23b)$$

where E_s is the elastic modulus of the steel bars; f_s and ε_s are the average tensile stress and strain of the reinforcing bars, respectively; and f_y and ε_y are the yield stress and strain of the bars, respectively.

7. Compatibility Condition

In the proposed approach, the normal tensile strains in the horizontal and vertical web steel, ε_h and ε_v and the principal compressive and tensile strain in concrete strut, ε_2 and ε_1, have a simple relationship that satisfies the compatibility condition of Mohr's circle (Fig. 2c):

$$\varepsilon_h + \varepsilon_v = \varepsilon_2 + \varepsilon_1. \quad (24)$$

The compatibility equation employed in this paper is the first strain invariant. Equation (24) is used to estimate the value of the principal tensile strain, ε_1, which is directly related to the extent of softening of the concrete, as per Eq. (24). Hwang and Lee (2000) pointed out that the used concrete softening model tended to overestimate the softening effect in situations where behaviour was

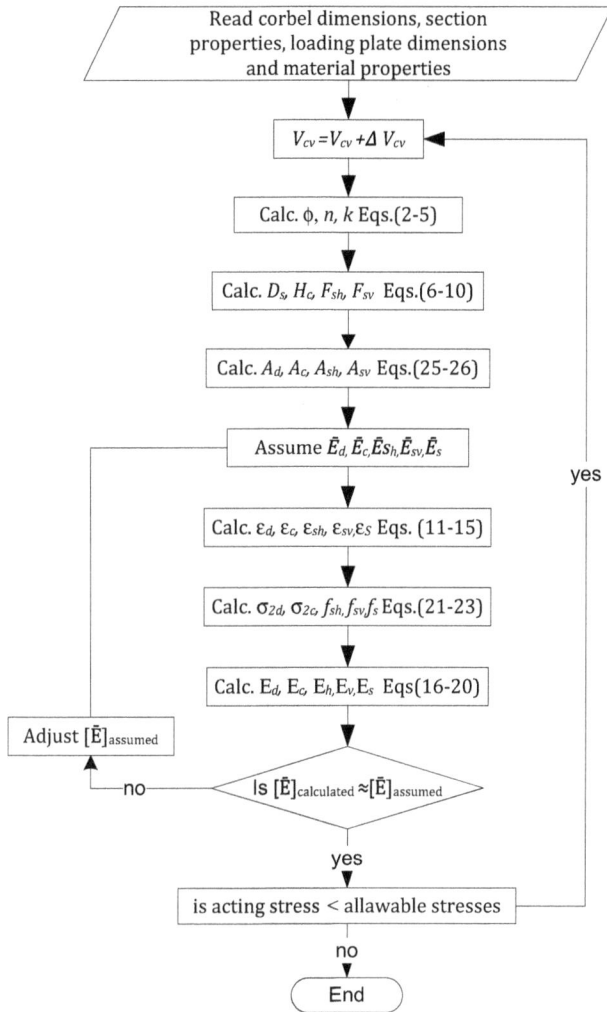

Fig. 3 Flow chart showing solution algorithm.

governed by yielding of all reinforcement crossing the crack direction. To guard against this, a limiting value of the principal tensile strain, ε_1, was proposed. Thus, the value of tensile strain, ε_h, in Eq. (24) is limited by the yielding strain, ε_{yh}, after yielding, or the value of ε_h is set to a yielding strain of 0.002 for the corbels not detailed with a horizontal shear reinforcement. Since all the corbels considered in the current study were not provided with vertical shear reinforcement, the tensile strain ε_v is conservatively taken as 0.002 in Eq. (24).

8. Effective Depth of Concrete Struts

Diagonal struts frequently are wider at mid-length than at their ends because strut stresses is greater at mid-length than at the ends of the strut. The curved, dashed outlines of the strut in Fig. 1 represent the effective boundaries of the diagonal strut. In the proposed model, the bottle-shaped strut is idealized as the prismatic struts shown by the straight, solid-line boundaries of the struts in Fig. 1. The effective depth of the diagonal strut, W_d, was assumed equal to (Park and Kuchma 2007):

$$W_d = \frac{a_v \sin \phi}{2} kd \cos \phi \qquad (25)$$

where $a_v/2$ should not be less than the loading plate width, W_p, and kd is the depth of the compression zone at the section. The horizontal bottom strut was assumed to have a uniform prismatic cross section over its length with effective depth W_c which is presumed equal to the depth of the neutral axis (He et al. 2012):

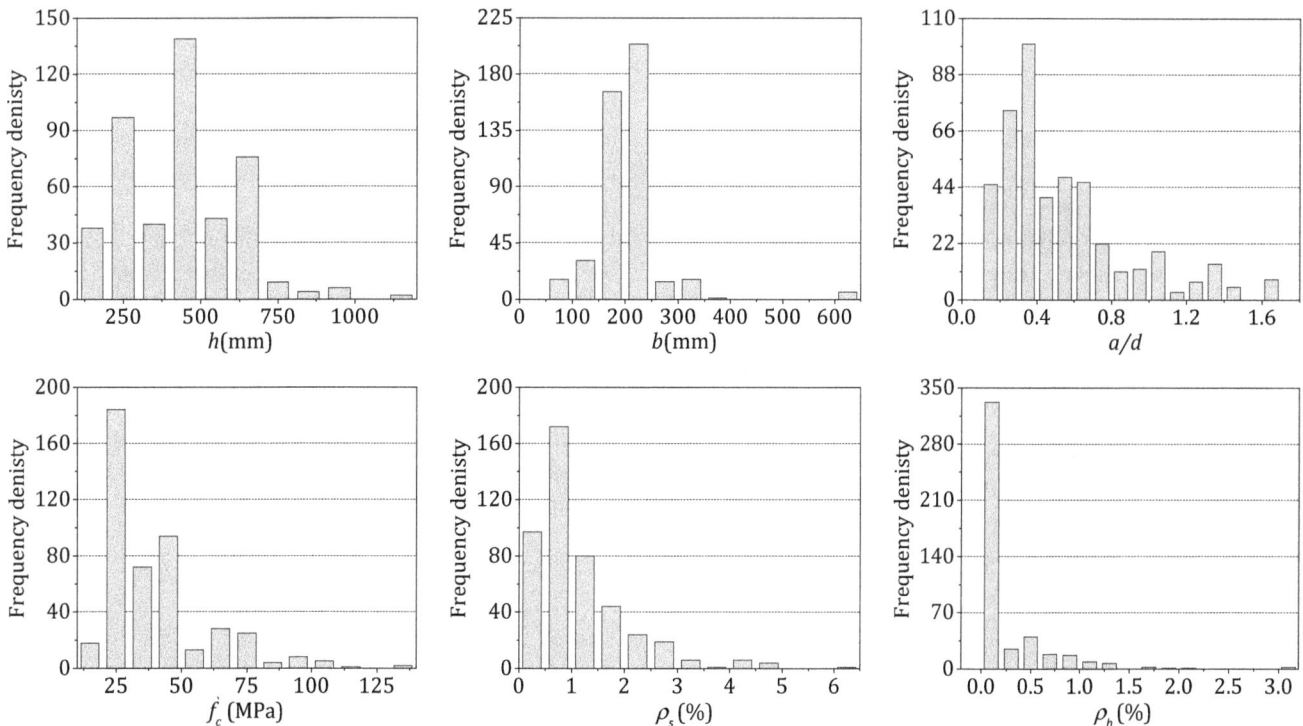

Fig. 4 Histograms of geometric and material properties of 455 reinforced concrete corbels.

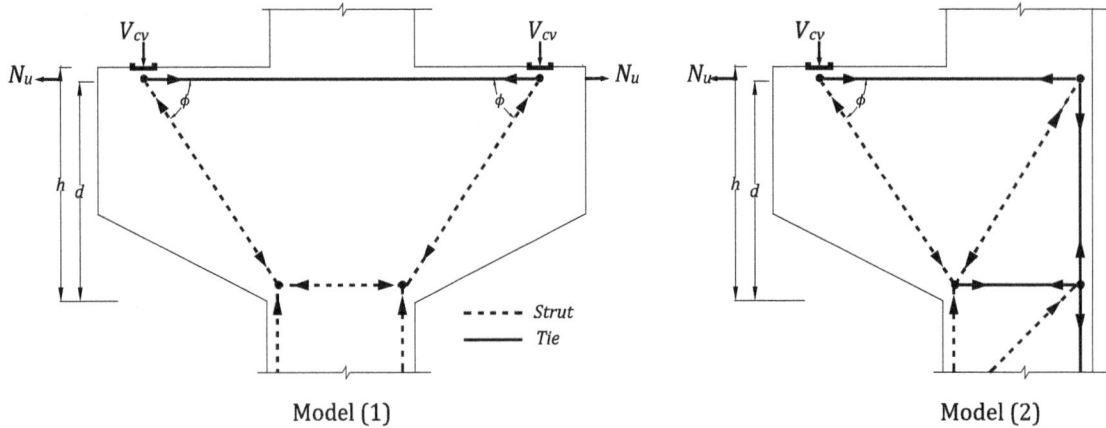

Fig. 5 Selected strut and tie models.

$$W_c = kd \tag{26}$$

9. Dimensions of Nodal Zone

Following the suggestion of Paulay and Priestley (1992), the effective width of the bottom node of the horizontal concrete strut was approximated by the depth of the flexural compression zone of the elastic column as:

$$L_h = \left(0.25 + \frac{0.85N_u}{A_c f_c'}\right) h \tag{27}$$

where N_u is the applied horizontal tension load (negative for tension), A_c is the gross sectional area of corbel, and h is the corbel overall depth, see Fig. 1. The effective width of the top and bottom nodes in the face of the diagonal concrete strut was taken as:

$$L_d = W_p \sin \phi + L_h \cos \phi \tag{28}$$

10. Proposed Solution Procedures

The failure modes associated with nodal crushing, yielding of the principal tensile reinforcement, and crushing or splitting of the diagonal strut were used to evaluate the ultimate load-carrying capacity of the corbels. The algorithm in Fig. 3 starts with a selection of the vertical corbel shear force V_{cv} and can be proceeded as outlined in following major steps:

1. According to the member forces D_c, H_c, F_{sh}, and F_{sv}, calculated from Eqs. (6) to (10), the values of the strains in concrete struts and steel reinforcements are estimated for the selected V_{cv} using Eqs. (11) through (15). In initiating the analysis, an initial estimate of the material secant stiffness can be made by assuming linear elastic values. Alternatively, the stiffness determined in a previous analysis can be used as the starting values;

2. Using the state of strain in each member, the normal stresses are determined from the stress–strain relations of Eqs. (21a, 21b) through (23a, 23b);

3. The secant moduli for each member are then calculated by Eqs. (16) through (20) using the strain and stresses values calculated in the previous step;

4. If the differences between the secant moduli in step 3 and those assumed in Eqs. (11) through (15) are larger than the specified tolerance, then the assumed secant moduli are considered incorrect and must be revised until convergence;

5. The stresses in the diagonal and horizontal struts, σ_{2d} and σ_{2c}, are compared to their capacity. The capacity of the diagonal strut can be estimated from $v_{cv1} = 0.85\,\beta_s f_c'$, where $\beta_s = 0.6$ as suggested by the ACI 318-11(American Concrete Institute 2011) for bottle-shaped strut with web reinforcements not satisfying the minimum reinforcement requirements, while the capacity of the horizontal strut is taken as $v_{cv2} = 0.85\,f_c'$.

6. The stresses on the on nodes' vertical back face and node-to-strut interface are compared to nominal strengths due to crushing, assumed equal to $V_{cv4} = 0.85 f_c'$ and $V_{cvs5} = 0.68 f_c'$ for nodal zones bounded by compressive struts (node A) and nodal zones crossed by tension tie reinforcement in one direction (node B) respectively, refer to Fig. 1;

7. If the acting stress determined in Steps 5 and 6 is less than the allowable stress, iteration continues from Step 1 by increasing the value of V_{cv}; and

8. The predicted strength employed in the proposed analysis method is the minimum value of the nominal strengths computed from the different failure modes, which are crushing of the horizontal and diagonal concrete strut, crushing of the compression zone, and yielding of principal tensile reinforcement.

11. Experimental Verification

11.1 Experimental Results Database

Combining the results of wide-ranging research into a single database provides the ability to examine code provisions as well as develop new models for use in design.

Table 1 Comparison of experimental and calculated load-carrying capacities of corbels in the database using different methods.

Strength ratio (v_{test}/v_{calc})			Avg	Std	CoV (%)
Current study			1.39	0.29	21
Strut-and-tie based models	ACI 318-11	Model 1	1.52	1.21	79
		Model 2	1.52	1.21	79
	CSA A23.3-04	Model 1	1.54	1.21	78
		Model 2	1.5	1.21	81
	Eurocode 2	Model 1	1.57	1.21	77
		Model 2	1.6	1.22	76
	NZS 3101-1	Model 1	1.68	1.12	67
		Model 2	1.68	1.12	67
	CSA A23.3-04	Model 1	1.52	1.21	79
		Model 2	1.52	1.21	79
Shear-friction based model			1.68	0.87	52

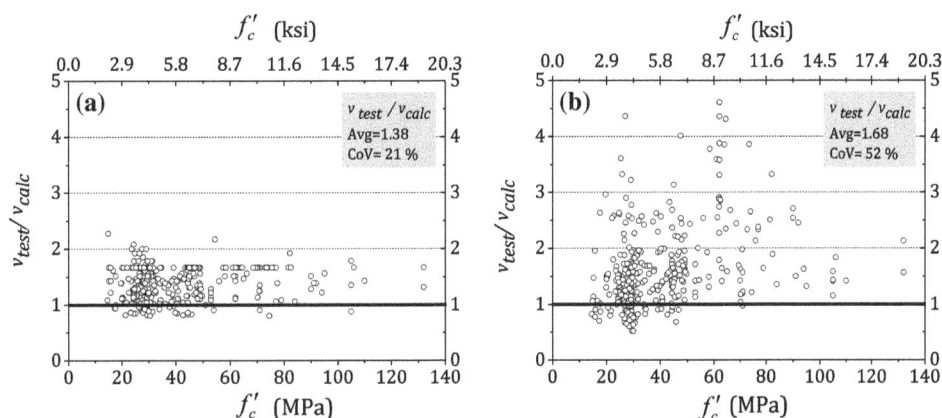

Fig. 6 Effect of concrete strength, f_c, on shear strength predictions by means of: **a** proposed strut-and-tie model approach; and **b** the shear-friction model used by the ACI 318-11 for 357 reinforced concrete corbels.

Aimed at verifying the accuracy of the proposed compatibility-based strut-and-tie method and assessing the performance of code provision that are used in concrete corbels design, a database with relevant information from tests was constructed. The database contains the results of tests of 550 reinforced concrete corbels collected from (in chronological: Abdul-Wahab (1989); Alameer (2004); Bourget et al. (2001); Chakrabarti et al. (1989); Clottey (1977); Fattuhi (1987); Fattuhi (1994); Fattuhi (1990); Foster et al. (1996); Hermansen and Cowan (1974); Kriz and Raths (1965); Lu et al. (2009); (Mattock 1976); Yong and Balaguru (1994) and Yong et al. (1985).

Several possible failure modes of corbels have been identified from past experimental testing, including shearing along the interface between the column and the corbel, yielding of the principal reinforcement and crushing or splitting of the compression strut (Russo et al. 2006). Premature failure modes, such as anchorage failure of principal reinforcement and bearing failure under loading plate, would

be avoided by correctly designing the corbel details (ACI Committee 318 2011). The results of corbels that were reported to have failed prematurely and those with insufficient information on the test setup and material properties were excluded from the database, leaving only 455 results in the database. Fig 4 presents summary information associated with different parameters, in the form of histograms, on the 455 RC corbels considered in this study. The test specimens in the database were made of plain and fibrous concrete having a relatively low compressive strength of 14.5 MPa and very high compressive strength of 132 MPa. The shear span-to-overall depth ratio of corbels ranged from 0.11 to 1.69. The primary tension reinforcements were anchored using a structural weld to transverse bars, bending to form a horizontal loop, or using headed bars. The main longitudinal reinforcement ratio varied between 0.1 and 6.5 %, whereas the horizontal shear reinforcement ratio varied from 0 to 3.05 %. All the corbel specimens included in the database had no vertical shear reinforcement. The horizontal load to

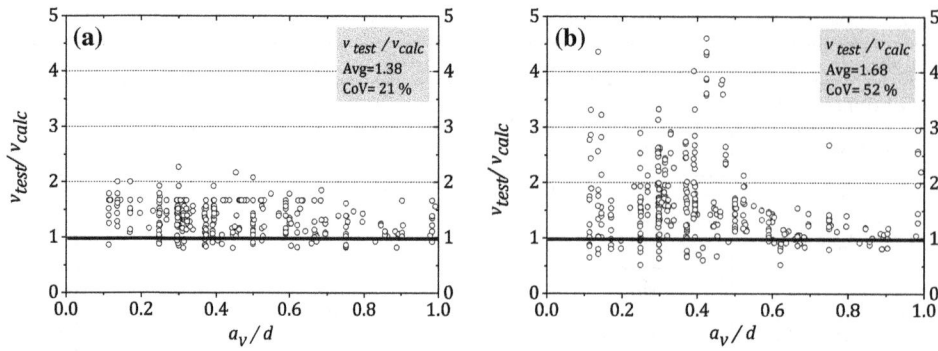

Fig. 7 Effect of shear span-to-depth ratio, a_v/d, on shear strength predictions by means of: **a** proposed strut-and-tie model approach; and **b** the shear-friction model used by the ACI 318-11 for 357 reinforced concrete corbels.

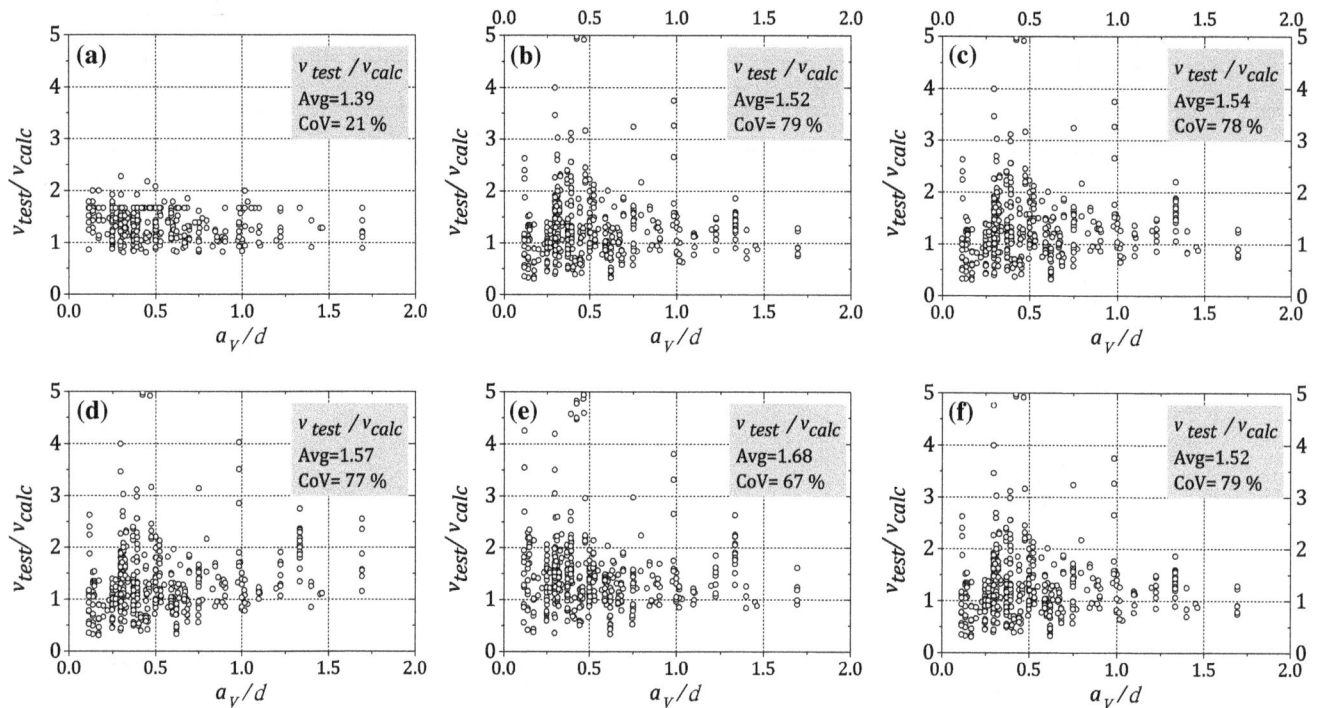

Fig. 8 Variation of ratio of measured-to-calculated strength by means of: **a** proposed strut-and-tie model approach; **b** ACI 318-11; **c** AS 3600; **d** CSA A23.3-04; **e** Eurocode 2; and NZS 3101-1 with shear span-to-depth ratio a_v/d.

yield force of main longitudinal reinforcement ratio ranged from 0 to 1.56. The corbel thickness ranged from 51 to 600 mm and overall thickness varied between 140 and 1140 mm.

11.2 Code Provisions and Analytical Models

Although several methods to compute the strength of RC corbels are adopted in design codes around the world, little is known about the accuracy and conservativeness of design procedures based on different rationales. American (ACI Committee 318 2011), Australian (AS 3600), Canadian (CSA A23.3-04), European (Eurocode 2), and New Zealand (NZS 3101-1) code recommendations include special provisions for corbels design. The main aim of the recommendations is to give practical design rules to avoid brittle shear failure ensuring the development of a well-defined

strength mechanism that generally occurs in the formation of a strut-and-tie resistant mechanism. Several other methods are available to estimate the shear capacity of corbels, including empirical equations (Fattuhi 1994), shear-friction approach (Hermansen and Cowan 1974, Mattock 1976), and strut-and-tie models(Solanki and Sabnis 1987; Siao 1994; Hwang et al. 2000b; Russo et al. 2006) including plastic truss models(Campione et al. 2007). The load-carrying capacity of the 455 corbels was calculated using the proposed analysis method, the shear-friction based approach provided by the ACI 318-11(American Concrete Institute 2011) and the strut-and-tie model proposed by different code provisions [ACI Committee 318; Australian code AS 3600; Canadian code (CSA A23.3-04); Eurocode 2 and New Zealand code (NZS 3101-1)].

The shear-friction based approach provided by the ACI 318-11 is valid for corbels made from both normal and high

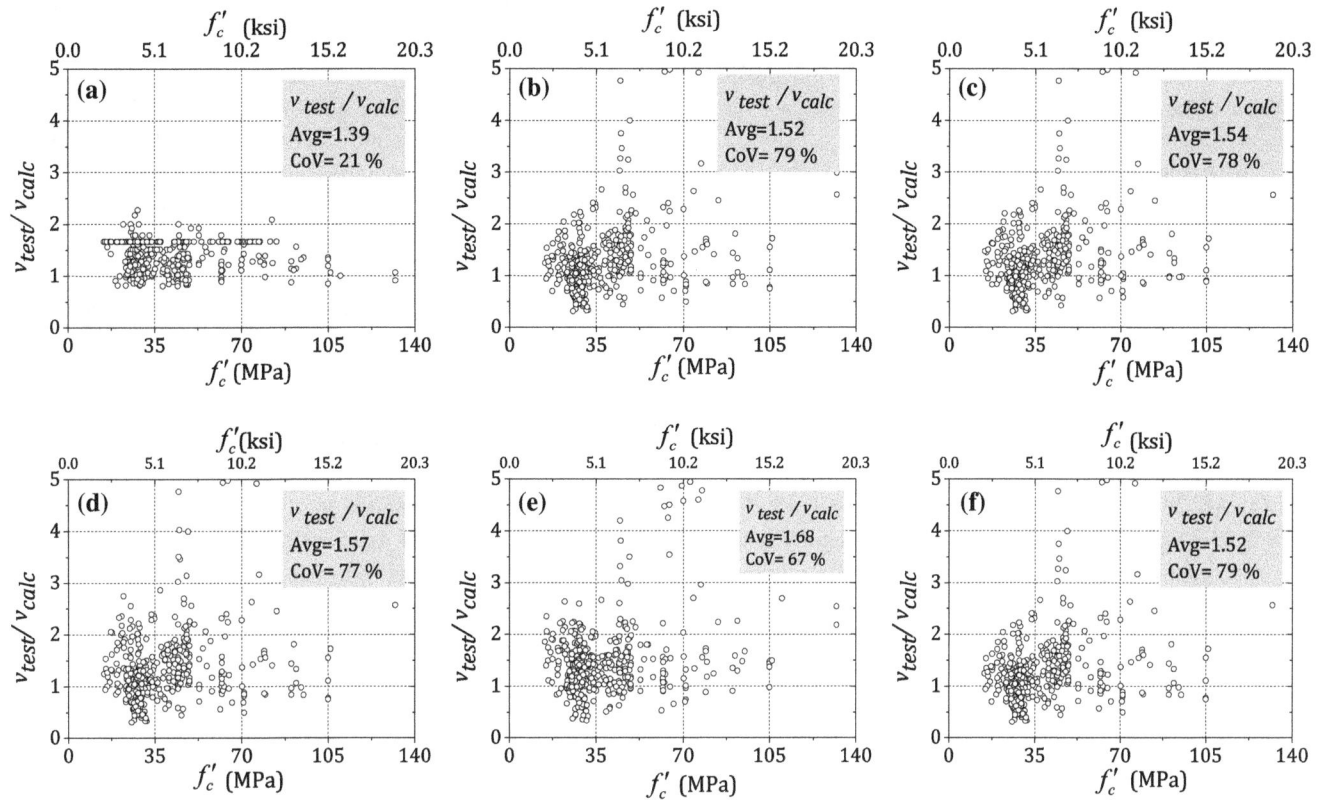

Fig. 9 Variation of ratio of measured-to-calculated strength by means of: **a** proposed strut-and-tie model approach; **b** ACI 318-11; **c** AS 3600; **d** CSA A23.3-04; **e** Eurocode 2; and NZS 3101-1 with concrete strength, f_c.

strength concrete with span-to-depth ratio less than unity. This procedure refers to two typical modes of failure: the first is the failure mode due to shear constraint occurring at the interface between column and corbel, and which occurs with very small shear-span ratios and reduced percentages of reinforcement. For a shear failure, the shear strength of a corbel is given by:

$$V_{cv} = \rho_{vf} f_{ys} \mu b d \tag{29}$$

where $\rho_{vf} = (\rho_s + \rho_h)$ is the frictional reinforcement ratio; f_{ys} is the yield strength of the friction reinforcement; ρ_s and ρ_h is the principal reinforcement ratio and horizontal web reinforcement ratio, respectively; μ is the coefficient of friction (taken as 1.4 for monolithic construction); and b is the corbel width. The second mode of failure is due to flexural yielding of the principal longitudinal reinforcement, and the carrying capacity can be estimated as:

$$V_{cv} = \frac{\rho_s f_y j d}{[a_v(1 + \alpha(h - d))]} bd \tag{30}$$

where α is the horizontal-to-vertical loads ratio; and jd is the lever arm calculated by $jd = d - \frac{(A_s f_{ys} - Nu)}{0.88 f_c' b}$. The corbel strength is taken as the minimum value of Eqs. (29) and (30). Moreover, the code imposes an upper limit on the load-carrying capacity with a maximum value of V_{cv} shall not exceed the smallest of $0.2 f_c' bd$, $(3.3 + 0.08 f_c')$ bd and $11 bd$.

Several code recommendations (CSA Committee A23.3 2004, NZS 3101 2006) specify the strut-and-tie models for the design of corbels, while only for corbels having shear span-to-depth ratio greater than 1.0, the ACI 318-11 recommends the use of a strut-and-tie model described in ACI 318-11, Appendix A. However, it does not provide detailed guidance on strut-and-tie models for different cases. It is well known that in using the strut-and-tie model, the designer is free to select the form and dimensions of the load-resisting truss to transfer the applied forces to the supports. More than one strut-and-tie model is usually feasible and thus there is no unique design solution as there typically is with the use of the conventional sectional design procedures. The safety of the strut-and tie model approach is highly dependent on the suitability of the assumption in lower-bound plasticity theory that the structure is adequately ductile to allow the load to be supported in the way chosen by the designer.

The experimental results are compared with predictions made on the basis of a simplified strut-and-tie model (STM) accounting for the main tie steel only, a refined strut-and-tie model accounting for the secondary crack-control reinforcement (Reineck 2003). These two models were reported to be very conservative and assume corbels failure are due to yielding of the tie and/or horizontal web reinforcement, thus prevent the assessment of codes provisions (Yang et al. 2012). Instead, two strut-and-tie models for a double-sided corbel and a single corbel projected from a column shown in Fig. 5, are proposed and it is assumed corbel failure is due to either crushing of the horizontal and diagonal concrete strut, crushing of the compression zone, or yielding of principal tensile reinforcement, similar to that assumed in the proposed strut-and-tie based method.

11.3 Comparison of Load-Carrying Capacity

Very few studies found in the literature on the validity of load-carrying capacity models of RC corbels in code provisions including strut-and-tie models. Table 1 summarizes the average, Avg, standard deviation, SD, and coefficient of variation, CoV, of the ratio between measured and calculated capacities, v_{test}/v_{calc}, of RC corbels considered, based on the proposed strut-and-tie based method, the shear-friction based approach provided by the ACI 318-11, the strut-and-tie model proposed by five codes of practice examined (ACI 318-11; (c) AS 3600; (d) CSA A23.3-04; (e) Eurocode 2; and NZS 3101-1). The distribution of average strength ratios for the specimens in the database against the concrete strength, f_c', and shear span-to-depth ratio, a_v/d, is shown in Figs. 6, 7, 8 9, where Avg and CoV values are also reported.

For the comparison with the shear-friction based approach provided by the ACI 318-11, only 357 corbels with shear span-to-depth ratio less than unity have been taken into account. Careful examination of the results shows that the shear strength ratios, v_{test}/v_{calc}, using the shear-friction based approach provides highly conservative and scattered estimates of the strength of corbels over a wide range of concrete strength and shear span-to-depth ratio. The coefficient of variation is quite high, with a value of 52 %, thus a low 5 % fractile value is to be expected. Altogether, 51 tests exhibit unconservative estimations, which is remarkably more than the 5 % fractile of 18 tests. Therefore, the results from this new database, which is much larger and more comprehensive than that used to calibrate the shear-friction based approach of ACI 318-11 in the 1980s, are clearly unsafe. In particular, the shear-friction based approach is unconservative in the prediction of the load-carrying capacity for corbels with concrete strength less than 50 MPa, (see Fig. 6b). By contrast, the calculated capacities by the proposed strut-and tie based method are both accurate and conservative with low scatter or trends for RC corbels with shear-to-span depth ratios ranging from 0 to 1 (see Fig. 7a).

The selected strut-and-tie models shown in Fig. 5 produce results that are quite similar to each other, refer to Table 1. Figures 7 and 8 present the effect of shear span-to-depth ratio, a_v/d, and concrete strength, f_c' on the load-carrying capacity predictions of the strut-and-tie based method and the five codes of practice examined for the double-sided corbel model only, respectively. On the whole, the predictions of the proposed method are very consistent for a broad range of concrete strengths and shear span-to-depth ratio, indicating that Avg, SD and COV are 1.39, 0.29, and 21 %, respectively. On the other hand, the overall performances of the five codes of practice examined are very similar, highly conservative and scattered. This conservatism may be attributed to the conservatism in the effective depth of the diagonal strut (Park and Kuchma 2007). They consistently underestimate the load-carrying capacity of corbels with

concrete strength greater than 35 MPa and shear span-to-depth ratio greater than 0.3. The largest average of the shear strength ratios, (v_{test}/v_{calc}) of all STM models appear as specified in Eurocode 2. The size of this test database and the use of these five code provisions are enough to obtain valuable insight into the behaviour of RC corbels from a strut-and-tie perspective. Considering the width of the compiled database, the obtained results are considered to be adequately fair to suggest that the proposed strut-and-tie based method provides a reliable and safe means of predicting the load-carrying capacity of reinforced concrete corbels.

12. Summary and Conclusions

A strut-and-tie based method intended for calculating the load-carrying capacity of reinforced concrete corbels has been presented. In addition to the normal strut-and-tie force equilibrium requirements, the proposed model accounts for strain compatibility and stress–strain relationship of cracked reinforced concrete, and uses a secant stiffness formulation. Based on the available test results in the literature and their comparison with the proposed model and the shear-friction based approach provided by the ACI 318 formulas as well as the strut-and-tie provisions in the American, Canadian, New Zealand, Eurocode and Australian codes., the following conclusions may be drawn:

1. The calculated load-carrying capacities by the proposed method were both accurate and conservative with limited scatter or trends for reinforced concrete corbels with shear span-to-depth ratios ranging from 0.1 to 2 and made from normal or fibrous concrete.
2. The shear-friction based approach provided by the American code is highly conservative and scattered estimates of the strength of corbels over a wide range. By contrast, the calculated capacities by the proposed strut-and tie based method are both accurate and conservative with low scatter or trends for RC corbels with shear-to-span depth ratios ranging from 0 to 1.
3. The predictions by the proposed strut-and-tie based method are adequately conservative and accurate to conclude that it provides a safe and reliable means of calculating the load-carrying capacity of concrete corbels.
4. Based on the conclusions drawn from this research, the proposed strut-and-tie should be adopted in future adjustments to code provisions and in the development of design guidelines for all types of D-regions in structural concrete. Furthermore, both experimental and mathematical studies are still needed to investigate the applicability and limitations of the proposed strut-and-tie method when applied to a wide range of D-regions.

Acknowledgment

The Author would like to thank Prof. Keun-Hyeok Yang, Kyonggi University, South Korea for providing some information on tests of corbels and assistance in populating the corbels database.

References

Abdul-Wahab, H. M. (1989). Strength of reinforced concrete corbels with fibers. *ACI Structural Journal, 86*(1), 60–66.

Alameer, M. (2004). Effects of fibres and headed bars on the response of concrete corbels. M SC thesis, Department of Civil Engineering and Applied Mechanics, McGill University, Montreal, Canada.

Ali, M., & White, R. (2001). Consideration of compression stress bulging and strut degradation in truss modeling of ductile and brittle corbels. *Engineering Structures, 23*(3), 240–249.

American Concrete Institute. (2011). *Building Code Requirements for Structural Concrete (ACI 318-11) and Commentary (ACI 318R-11)*. Farmington Hills, MI: ACI.

Australian code AS 3600. (2009). *Australian Standard for Concrete Structures* (p. 213). North Sydney, Australia: Standards Australia.

Bourget, M., Delmas, Y., & Toutlememonde, F. (2001). Experimental study of the behaviour of reinforced high-strength concrete short corbels. *Materials and Structures, 34*(3), 155–162.

British Standards Institution. (2004). *Eurocode 2: Design of concrete structures—Part 1–1: General rules and rules for buildings*. London, UK: British Standards Institution.

Campione, G., La Mendola, L., & Mangiavillano, M. L. (2007). Steel fiber-reinforced concrete corbels: Experimental behavior and shear strength prediction. *ACI Structural Journal, 104*(5), 570–579.

Chakrabarti, P. R., Farahi, D. J., & Kashou, S. I. (1989). Reinforced and precompressed concrete corbels-an experimental study. *ACI Structural Journal, 86*(4), 132–142.

Clottey, C. (1977). Performance of lightweight concrete corbels subjected to static and repeated loads. PhD thesis, Oklahoma State University, Ann Arbor, MI, pp. 127–127.

CSA Committee A23.3. (2004). *Design of concrete structures*. Mississauga, Canada: Canadian Standard Association 232.

Fattuhi, N. (1987). SFRC corbel tests. *ACI Structural Journal, 84*(2), 119–123.

Fattuhi, N. (1990). Strength of SFRC corbels subjected to vertical load. *Journal of Structural Engineering, 116*(3), 701–718.

Fattuhi, N. (1994). Reinforced corbels made with plain and fibrous concretes. *ACI Structural Journal, 91*(5), 530–536.

Foster, S. J., Powell, R. E., & Selim, H. S. (1996). Performance of high-strength concrete corbels. *ACI Structural Journal, 93*(5), 555–563.

He, Z.-Q., Liu, Z., & Ma, Z. J. (2012). Investigation of load-transfer mechanisms in deep beams and corbels. *ACI Structural Journal, 109*(4), 467–476.

Hermansen, B. R., & Cowan, J. (1974). Modified shear-friction theory for bracket design. *ACI Journal Proceedings, 71*(2), 55–60.

Hwang, S.-J., Fang, W.-H., Lee, H.-J., & Yu, H.-W. (2001). Analytical model for predicting shear strength of squat walls. *Journal of Structutral Engineering, ASCE, 127*(1), 43–50.

Hwang, S.-J., & Lee, H.-J. (1999). Analytical model for predicting shear strengths of exterior reinforced concrete beam-column joints for seismic resistance. *ACI Structural Journal, 96*(5), 846–857.

Hwang, S.-J., & Lee, H.-J. (2000). Analytical model for predicting shear strengths of interior reinforced concrete beam-column joints for seismic resistance. *ACI Structural Journal, 97*(1), 35–44.

Hwang, S.-J., & Lee, H.-J. (2002). Strength prediction for discontinuity regions by softened strut-and-tie model. *Journal of Structural Engineering, 128*(12), 1519–1526.

Hwang, S.-J., Lu, W.-Y., & Lee, H.-J. (2000a). Shear strength prediction for deep beams. *ACI Structural Journal, 97*(3), 367–376.

Hwang, S.-J., Lu, W.-Y., & Lee, H.-J. (2000b). Shear strength prediction for reinforced concrete corbels. *ACI Structural Journal, 97*(4), 543–552.

Kriz, L. B., & Raths, C. H. (1965). Connections in precast concrete structures—Strength of corbels. *PCI Journal, 10*(1), 16–61.

Lu, W.-Y., Lin, I.-J., & Hwang, S.-J. (2009). Shear strength of reinforced concrete corbels. *Magazine of Concrete Research, 61*(10), 807–813.

Lu, W. Y., Lin, I. J., Hwang, S. J., & Lin, Y. H. (2003). Shear strength of high-strength concrete dapped-end beams. *Journal of the Chinese Institute of Engineers, 26*(5), 671–680.

MacGregor, J., & Wight, J. (2009). *Reinforced concrete: Mechanics and design*. Singapore: Prentice Hall and Pearson Education South Asia.

Mattock, A. H. (1976). Design proposals for reinforced concrete corbels. *PCI Journal, 21*(3), 18–42.

NZS 3101. (2006). *Part 1: Code of practice for the design of concrete structures and Part 2: Commentary on the design of concrete structures*. Wellington, New Zealand: Standards Association of New Zealand.

Park, J., & Kuchma, D. (2007). Strut-and-tie model analysis for strength prediction of deep beams. *ACI Structural Journal, 104*(6), 657–666.

Paulay, T., & Priestley, M. (1992). *Seismic design of reinforced concrete and masonry buildings*. New York, NY: Wiley.

Reineck, K. (2003). Examples for the design of structural concrete with strut-and-tie models. *ACI International, SP-208*, 128–141.

Russo, G., Venir, R., Pauletta, M., & Somma, G. (2006). Reinforced concrete corbels-shear strength model and design formula. *ACI Materials Journal, 103*(1), 3–10.

Schlaich, J., Schäfer, K., & Jennewein, M. (1987). Toward a consistent design of structural concrete. *PCI Journal, 32*(3), 74–150.

Siao, W. B. (1994). Shear strength of short reinforced concrete walls, corbels, and deep beams. *ACI Structural Journal, 91*(2), 123–132.

Solanki, H., & Sabnis, G. M. (1987). Reinforced concrete corbels-simplified. *ACI Structural Journal, 84*(5), 428–432.

Vecchio, F. J. (1989). Nonlinear finite element analysis of reinforced concrete membranes. *ACI Structural Journal, 86*(1), 26–35.

Vecchio, F. J., & Collins, M. P. (1993). Compression response of cracked reinforced concrete. *Journal of Structural Engineering, 119*(12), 3590–3610.

Yang, K.-H., & Ashour, A. F. (2012). Shear capacity of reinforced concrete corbels using mechanism analysis. *Proceedings of the ICE-Structures and Buildings, 165*(3), 111–125.

Yang, J., Lee, J., Yoon, Y., Cook, W., & Mitchell, D. (2012). Influence of steel fibers and headed bars on the serviceability of high-strength concrete corbels. *Journal of Structural Engineering, 138*(1), 123–129.

Yong, Y., & Balaguru, P. (1994). Behavior of reinforced high-strength-concrete corbels. *Journal of Structural Engineering, 120*(4), 1182–1201.

Yong, Y., McCloskey, D. H., & Nawy, E. G. (1985). *Reinforced corbels of high-strength concrete.* London, UK: ACI Special Publication. **87**.

15

Effect of Autoclave Curing on the Microstructure of Blended Cement Mixture Incorporating Ground Dune Sand and Ground Granulated Blast Furnace Slag

Omer Abdalla Alawad[1,2],*, Abdulrahman Alhozaimy[1], Mohd Saleh Jaafar[2], Farah Nora Abdul Aziz[2], and Abdulaziz Al-Negheimish[1]

Abstract: Investigating the microstructure of hardened cement mixtures with the aid of advanced technology will help the concrete industry to develop appropriate binders for durable building materials. In this paper, morphological, mineralogical and thermogravimetric analyses of autoclave-cured mixtures incorporating ground dune sand and ground granulated blast furnace slag as partial cementing materials were investigated. The microstructure analyses of hydrated products were conducted using scanning electron microscopy (SEM), energy dispersive X-ray spectroscopy (EDX), differential thermal analysis (DTA), thermo-graphic analysis (TGA) and X-ray diffraction (XRD). The SEM and EDX results demonstrated the formation of thin plate-like calcium silicate hydrate plates and a compacted microstructure. The DTA and TGA analyses revealed that the calcium hydroxide generated from the hydration binder materials was consumed during the secondary pozzolanic reaction. Residual crystalline silica was observed from the XRD analysis of all of the blended mixtures, indicating the presence of excess silica. A good correlation was observed between the compressive strength of the blended mixtures and the CaO/SiO_2 ratio of the binder materials.

Keywords: ground dune sand, slag, autoclave curing, SEM, EDX, DTA, TGA, XRD.

1. Introduction

Concrete is a composite material consisting of aggregate, water and Portland cement (PC). When PC is placed in contact with water, several chemical reactions occur (Hewlett 2003; Taylor 1997). These reactions produce many phases, such as calcium silicate hydrates (C–S–H gel), calcium hydroxide (CH or portlandite), ettringite (AFt), and monasulfomonite (AFm) (Englehardt and Peng 1995; Hewlett 2003). To ensure the continuity of cement hydration, the cast mixture should be cured under appropriate conditions (Mehta and Monteiro 2006; Topçu and Uygunoğlu 2007). The typical curing conditions used in concrete technology include normal curing, low pressure steacuring, high pressure steam curing and membrane curing. The so-called normal curing is conducted under a moist and ambient temperature. Under these conditions, the hydration process and strength development rate of cement paste are slow (Liu

et al. 2005). Consequently, concrete takes several days, or sometimes several months, to reach its ultimate strength (Gutteridge and Dalziel 1990; Topçu and Uygunoğlu 2007). As a result, in precast concrete plants and other applications where higher early strength is of great concern, normal curing conditions are not preferred (Erdoğdu and Kurbetci 2005; Hanson 1963; Kołakowski et al. 1994).

In precast concrete technology, accelerated curing conditions, such as high-pressure steam curing (autoclave curing), are adopted (Mindess et al. 1981). The advantages of autoclave curing include that compressive strength equivalent to that of 28 days of normal curing can be achieved within 24 h under autoclave curing, less drying shrinkage, elimination of efflorescence, and better resistance to sulfate attacks (Kołakowski et al. 1994; Menzel 1934; Mindess et al. 1981; Neville 1973). Another advantage of autoclave curing is that many types of siliceous materials can be used as supplementary cementing materials even though some siliceous materials may not be feasible to react under normal curing conditions (Kalousek 1954).

The microstructure of cement paste cured under autoclave conditions is different than that produced at ambient temperature (Menzel 1934). The primary formed C–S–H gel converts to crystalline alpha dicalcium silicate hydrate (α-C_2SH) or $C_3SH_{1.5}$ (Jupe et al. 2008; Mindess et al. 1981; Taylor 1997). These phases are porous and weak, which leads to a deterioration in the compressive strength and concrete durability (Eilers et al. 1983; Grabowski and Gillott 1989). To avoid the formation of undesired phases (α–C_2SH

[1]Civil Engineering Department and Center of Excellence for Concrete Research and Testing, College of Engineering, King Saud University, Riyadh, Saudi Arabia.
*Corresponding Author; E-mail: oalawad@ksu.edu.sa
[2]Civil Engineering Department, Faculty of Engineering, Universiti Putra Malaysia, Seri Kembangan, Malaysia.

and $C_3SH_{1.5}$), fine siliceous material should be added at 30–40 % by weight of the cement. The added siliceous materials react with the CH generated from the PC hydration to produce new C–S–H phases, such as tobermorite and xonotlite (Berardi et al. 1975; Eilers et al. 1983; Mindess et al. 1981; Taylor 1997). The hydration product features of autoclaved mixtures depend on many parameters, including the curing conditions, calcium to silica ratio (Ca/Si), and type of added silica (Bresson et al. 2002; Eilers et al. 1983; Hope 1981; Kołakowski et al. 1994). It has been stated that, under autoclave conditions, the use of crystalline silica produces high strength compared to the use of amorphous silica (Assarsson and Rydberg1956; Grabowski and Gillott 1989; Jupe et al. 2008; Luke 2004; Sanders and Smothers 1957; Yazici 2007). Additionally, it has been demonstrated that, at each autoclaving temperature, there is an optimum period of curing that results in good mechanical and physical properties (Hanson 1963; Menzel 1934; Neville 1973; Shi and Hu 2003; Yazıcı et al. 2008).

Recently, many tools have been used to examine the concrete structure not only at microscale but also at the nano-level (Chae et al. 2013; Kar et al. 2014). Scanning electron microscopy (SEM), energy dispersive X-ray spectroscopy (EDX), X-ray diffraction (XRD) and thermo-graphic analysis (DTA and TGA) are being used to study the microscale changes that occur in the cement paste and concrete structure (Murmu and Singh 2014; Singh et al. 2015). Understanding the hydration mechanism and microstructure properties with the aid of advanced technology will help the concrete industry develop appropriate binders for durable building materials (Lange et al. 1997).

Previous studies by the authors have reported that ground dune sand and ground granulated blast furnace can be used as high volume cement replacement materials in autoclave concrete production (Alawad et al. 2015). However, the effect of autoclave conditions on microstructure of blended cement mixtures containing ground dune sand and slag have

not been studied in depth. This study aimed to investigate the microstructures of autoclave-cured products of blended cement paste mixtures that incorporate ground dune sand and ground granulated blast furnace slag as cement replacement materials. Microstructure analyses were performed using SEM, EDX, XRD, DTA, and TGA.

2. Experimental Work

2.1 Raw Material
PC complying with ASTM C 150 and ground granulated blast furnace slag of grade 100 (ASTM C 989) were used in this study. The chemical composition and physical properties of the PC and slag are presented in Table 1. The dune sand used in this study was obtained from Riyadh, Saudi Arabia. The particle size analysis of natural dune sand is shown in Fig. 1. The natural dune sand was mechanically ground until 95 % passed a 45-μm sieve. The chemical composition and specific surface area of the ground dune sand (GDS) are presented in Table 1. The XRD patterns and SEM images of the GDS and slag are shown in Figs. 2 and 3, respectively.

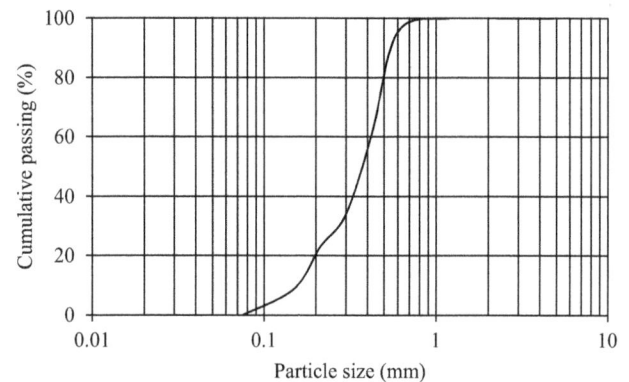

Fig. 1 Particle size distribution of natural dune sand.

Table 1 Chemical and physical properties of PC, GDS and Slag.

Chemical oxide	Chemical composition (%)		
	PC	GDS	Slag
SiO_2	22.62	93.4	33.33
Al_2O_3	6.11	0.19	13.74
Fe_2O_3	3.69	0.32	0.44
CaO	57.96	0.45	43.1
MgO	2.16	–	5.76
K_2O	0.98	0.04	0.41
Na_2O	0.17	–	0.2
SO_3	2.99	–	1.89
LOI	3.02	1.15	1.34
Specific surface area, BET cm^2/g	3012	2574	5314
Specific gravity	3.15	2.64	2.94

(a)

(b)

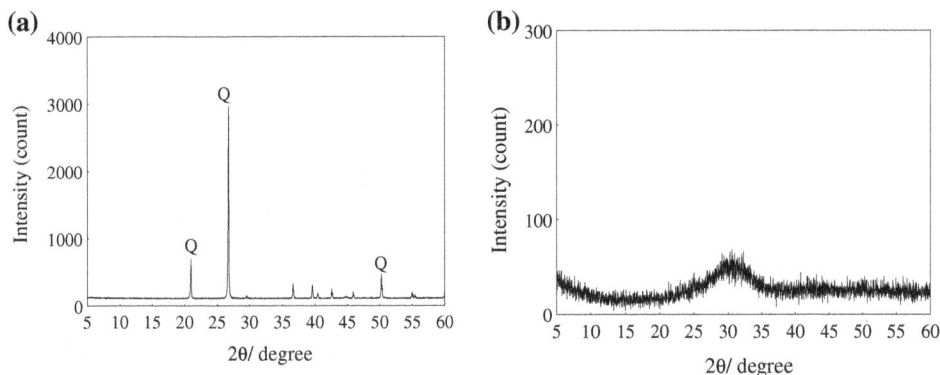

Fig. 2 XRD pattern **a** GDS and **b** slag.

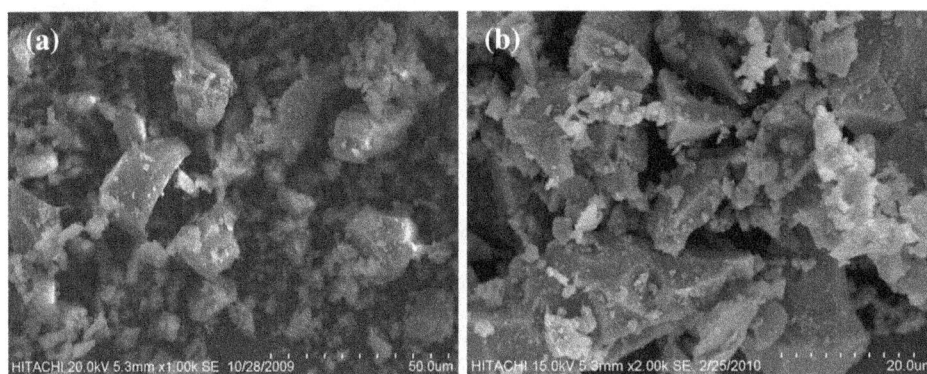

Fig. 3 SEM images **a** GDS and **b** slag.

As shown in Fig. 2a, a very sharp quartz peak at 26.5° (2θ) reflecting the crystallinity of silica in the GDS was observed. However, no stand-out peak of the slag was detected, representing the non-crystallinity of slag (Fig. 2b) (Divsholi et al. 2014). The SEM of GDS and slag presented that both materials have an angular particle texture (Fig. 3). Mining sand with a maximum particle size of 2.36 mm was used as a fine aggregate to fabricate mortar mixtures. The specific gravity and fineness modulus of the fine aggregate were 2.5 and 2.49, respectively.

2.2 Sample Preparation

Five mixtures of paste, a control mixture and four blended mixtures, were prepared. The control mixture (M1) consisted of 100 % PC as the binder material. For the blended mixtures, GDS and slag were incorporated as PC replacement materials. The replacement level of GDS was held constant

at 40 %, while the concentrations of slag were 0, 15, 30, and 45 %. The blended mixtures were denoted as M2, M3, M4, and M5, and their binder composition details are summarized in Table 2. To examine the performance of the GDS and slag with regard to compressive strength, the control and the four blended mortar mixtures were formulated with a binder composition similar to that used for paste mixtures. The mortar had a binder:fine aggregate:water ratio of 1:3:0.3.

The paste and mortar mixtures were prepared according to ASTM C 305 and cast in 50 mm cubic steel molds. The cast samples were covered with plastic sheets and kept under laboratory conditions (23 ± 3 °C and 50 ± 5 relative humidity) for 24 h. After being demolded at the age of 24 h, the samples were immersed in water at 23 ± 3 °C (normal curing conditions) for 16 h to develop initial strength and then placed in the autoclave chamber. The chamber

Table 2 Binder proportions for the ternary blended cement mixtures.

Mixture ID	Ingredient %			Water/binder ratio	Molar CaO/SiO$_2$
	PC	GDS	Slag		
M1	100	–	–	0.5	2.75
M2	60	40	–	0.5	0.74
M3	45	40	15	0.5	0.67
M4	30	40	30	0.5	0.61
M5	15	40	45	0.5	0.56

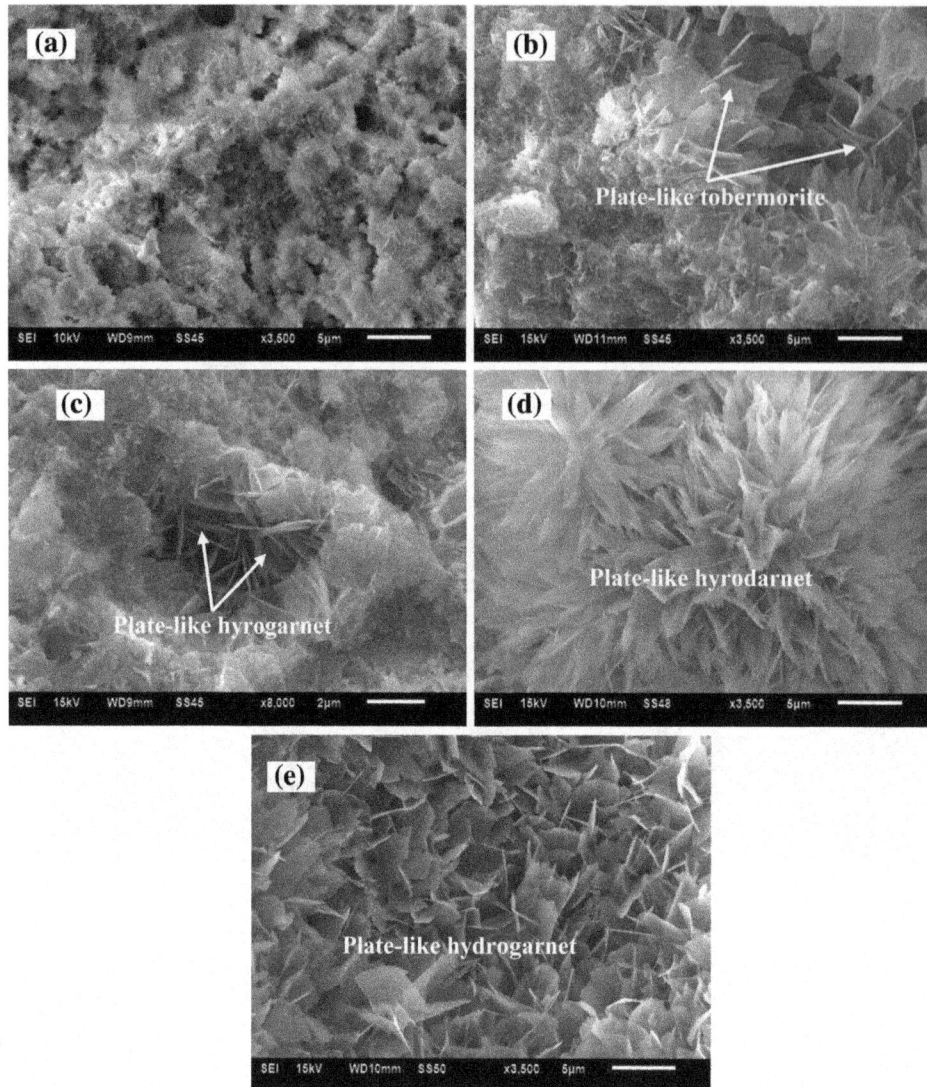

Fig. 4 SEM images of autoclave cured mixtures a M1, b M2, c M3, d M4 and e M5.

temperature was increased from room temperature to 182 ± 3 °C within 1 h. Consequently, the pressure was increased from atmospheric pressure to 1.0 MPa. The adopted autoclave conditions (temperature and pressure) are similar to those used by Yang et al. (2000). The temperature and pressure were kept constant at 182 ± 3 °C and 1.0 MPa for 5 h, then the autoclave heater was turned off and the chamber was allowed to cool naturally. Room temperature was reached in approximately 1.5 h.

2.3 Sample Characterization

Microstructural and microanalytical characterizations of the hydrated pastes were conducted using a Jeol JSM 6610LV coupled with EDX. The samples for SEM analysis were prepared by taking fractured surface specimens of the cured pastes. The specimens were glued to carbon stubs with carbon paint prior to the SEM analysis. The SEM analysis was carried out using accelerating voltages of 10 and 15 kV and magnifications of ×3500 and ×8000. The chemical composition analysis of selected spots (fields of view) was carried out using EDX.

Thermogravimetric analysis (DTA and TGA) was carried out using a TA instrument (model SDT Q600). DTA and

TGA analyses were conducted to monitor the phase changes and to evaluate the amount of CH consumed in the cured samples. A predefined amount of the selected sample in powder form was weighed, placed in a platinum sample pan and then heated from room temperature to 1000 °C at a heating rate of 10 °C min^{-1} in a nitrogen gas flow. XRD analysis was conducted using a Shimadzu XRD-6000 diffractometer with a scanning rate of 2°/min from 10° to 60° (2θ) to obtain the mineralogical information for each sample. The samples for XRD analysis were prepared by grinding pieces of hydrated pastes into a powder form that could pass a 75-μm sieve. The compressive strength test of the cast mortars was carried out in accordance with ASTM C109.

3. Results and Discussion

3.1 Morphological Study

The SEM images and EDX analyses of the hardened pastes of the control and blended mixtures cured under autoclave conditions are shown in Figs. 4 and 5, respectively. Fibrous-like hydrated products and pores of a dark color were

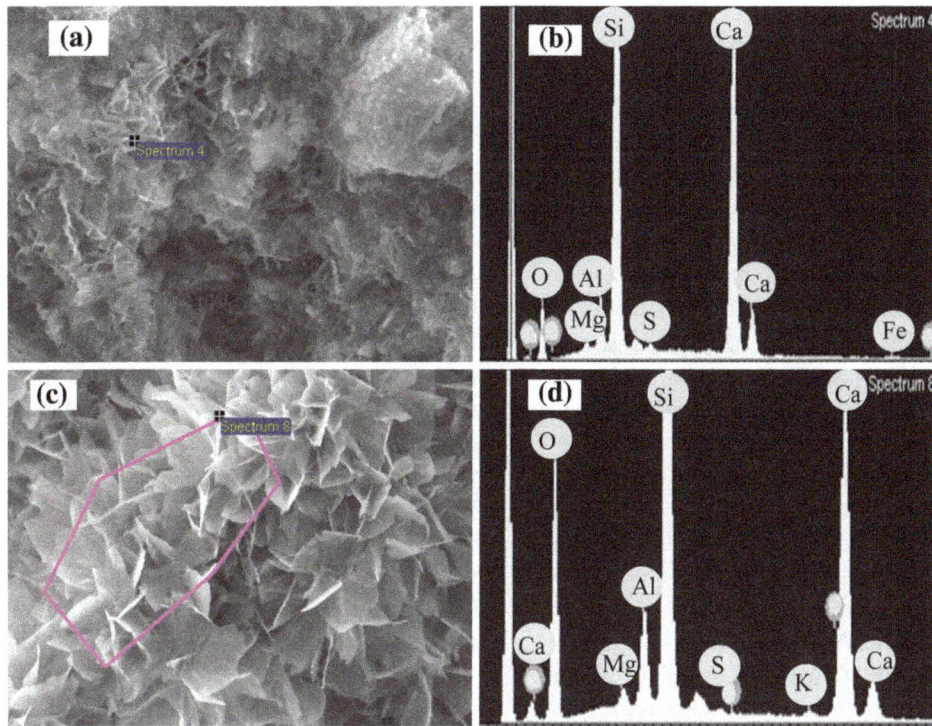

Fig. 5 EDX analysis of M2 and M4 mixtures.

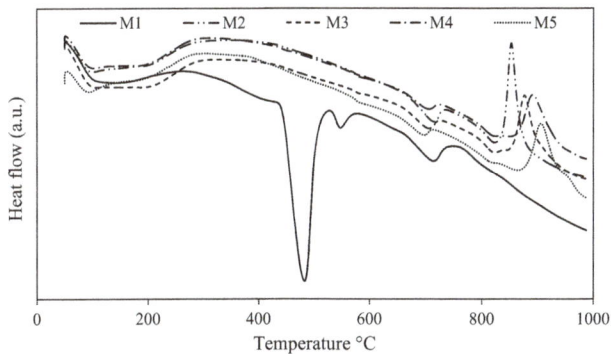

Fig. 6 DTA curves of autoclave cured mixtures.

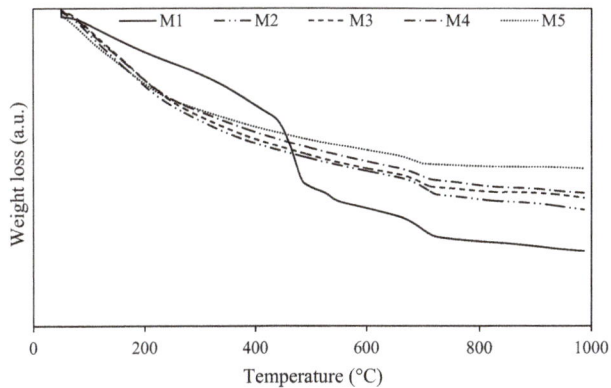

Fig. 7 TGA curves of autoclave cured mixtures.

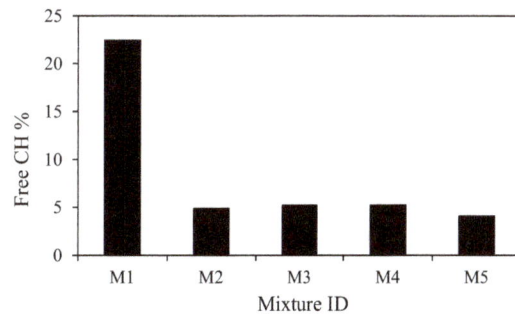

Fig. 8 Free CH of autoclave cured mixtures.

Kjellsen et al. 1991). For the M2 mixture, plate-like structures of tobermorite ($C_5S_6H_5$) with dense and closed network structures were observed, as shown in Fig. 4b. For the blended mixtures containing slag (M3, M4 and M5), plate-like structures of tobermorite ($C_5S_6H_5$) and hydrogarnet (C_4ASH_4) were observed, as shown in Fig. 4c–e (Klimesch and Ray 1998; Kyritsis et al. 2009). The hydrogarnet phases were dominant when the content of slag was increased. In particular, thin plate-like structures with broken edges were observed in the M4 mixture. In general, the blended mixtures revealed dense, compacted and closed network structures. This structure justified the enhancement effect in the compressive strength of the blended mixtures cured under autoclave conditions (Kondo et al. 1975; Wongkeo et al. 2012).

Figure 5 presents the EDX analysis of the M2-AC and M4-AC mixtures. The image and EDX analysis of the M2-AC mixture are shown in Fig. 5a and b, respectively. High peak intensities of Ca and Si with a Ca/Si ratio approximately equal to unity were observed from the EDX analysis (Fig. 5b). This analysis indicated that the crystalline

observed in the control mixture (Fig. 4a). This image also revealed that the hydrated products are associated with loose structures. These features explained the negative effect of the performance of the plain cement mixture cured under autoclave conditions (Alawad et al. 2014; Bakharev et al. 1999;

structure phases are newly formed CSH (i.e., tobermorite). The formation of CSH phases with a Ca/Si ratio close to unity is a favored result for the concrete strength and physical properties (Eilers et al. 1983; Yazıcı et al. 2008). The image and EDX analysis of the M4-AC mixture are shown in Fig. 5c and d, respectively. The presence of slag leads to the formation of thin crystalline structure phases of newly formed aluminum bearing CSAH (i.e., C_4SAH_4) (Klimesch and Ray 1998; Kyritsis et al. 2009; Mostafa et al. 2009). The reason for the formation of C_4SAH_4 phases is attributed to the presence of the element Al in the system as the slag contains a significant amount of Al_2O_3 (Table 1). The EDX analysis indicated that the presence of GDS and slag not only prevented the formation of weak and permeable phases (α-C_2SH) but also introduced new CSH phases, such as tobermorite ($C_5S_6H_5$) and hydrogarnet (C_4SAH_4), which are associated with high strength and low permeability (Yazıcı et al. 2008).

3.2 Thermogravimetric Analysis

Figures 6 and 7 show the DTA and TGA curves of the control and blended mixtures cured under autoclave conditions. The DTA curve of the M1 mixture revealed the existence of a clear endothermic peak located between 450 and 550 °C (Fig. 6). This endothermic peak was attributed to the de-hydroxylation of CH generated from the PC hydration (Alarcon-Ruiz et al. 2005; Oner and Akyuz 2007). For the blended mixture containing 40 % GDS (M2), no endothermic peak in the CH zone was observed, indicating the consumption of CH in the secondary pozzolanic reaction between the SiO_2 of GDS and the CH generated from the PC. Additionally, a significant exothermic peak at 850 °C, due to the crystallization of CSH and belite, was observed in the M2 mixtures (Klimesch et al. 1996). For ternary blended mixtures (M3–M5), no endothermic peak in the CH zone was observed, and the clear formation of an exothermic peak at 850 °C, due to the crystallization of CSH, was instead observed. However, the intensity of the exothermic peaks at 850 °C of the slag mixtures were less than that of the M2 mixture (PC-GDS). It was also observed that all the tested mixtures exhibit an endothermic peak at 750 °C (Fig. 6). This endothermic peak was ascribed to carbonation that may take place during sample preparation (Saikia et al. 2006).

Figure 7 shows the TGA curves for the control and blended mixtures cured under autoclave conditions. With respect to the CH change, which occurred between 450 and

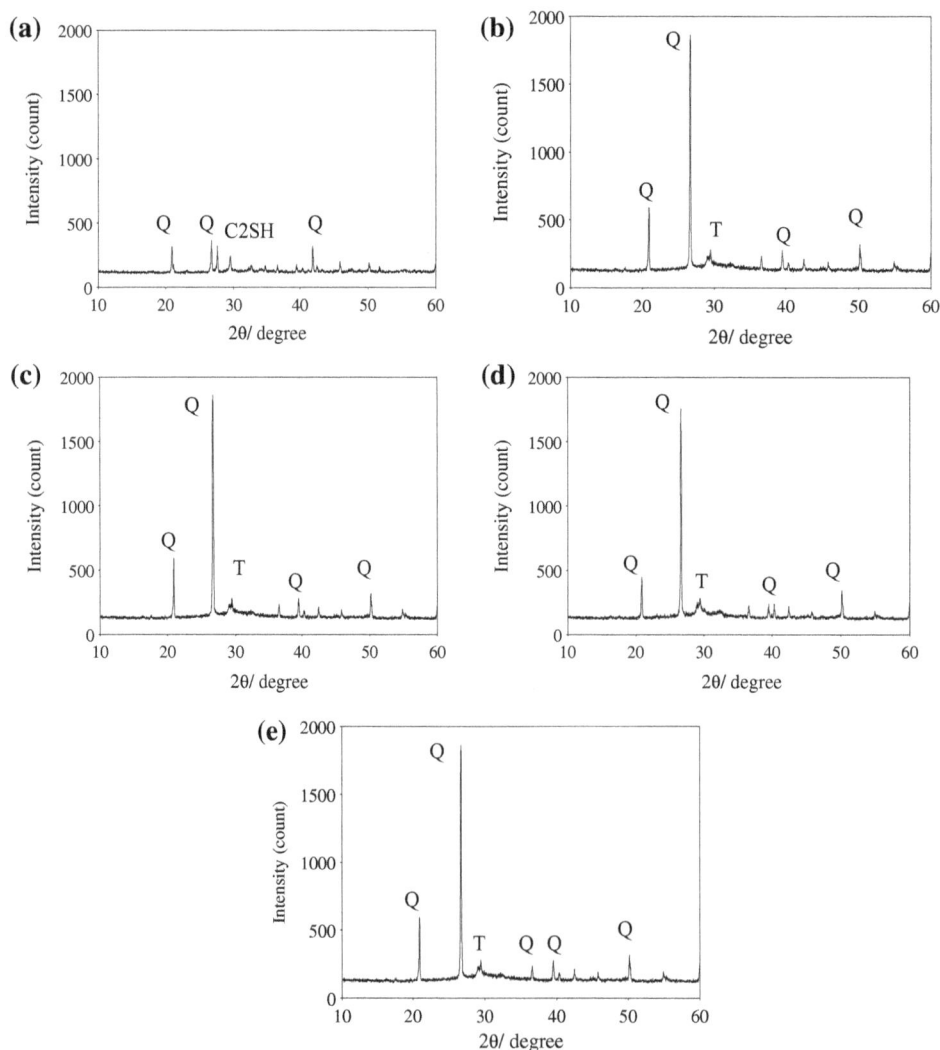

Fig. 9 XRD patterns of autoclave cured mixtures **a** M1, **b** M2, **c** M3, **d** M4 and **e** M5.

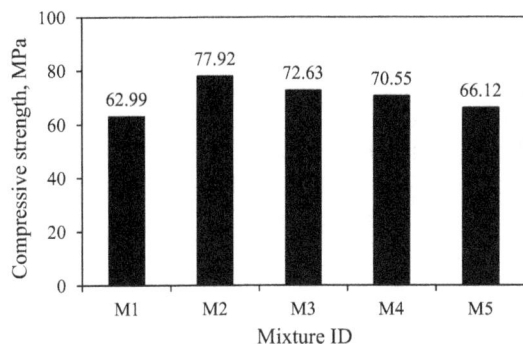

Fig. 10 Compressive strengths of autoclave cured mixtures.

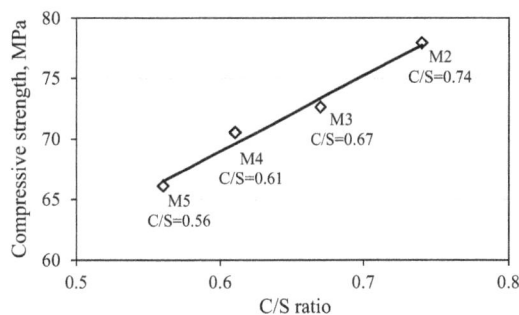

Fig. 11 Correlation of compressive strength and molar CaO/SiO$_2$ ratio (C/S).

550 °C, an obvious mass loss in M1 was observed, while all the blended samples showed less mass change in this zone. The calculated value of free CH of M1 was found to be 22.4 %, whereas, for the blended mixtures (M2–M5), it was found to be approximately 5 %, irrespective of the slag content, as shown in Fig. 8. The low mass change in the blended mixtures is evidence of the consumption of CH in the newly formed CSH phases, such as tobermorite.

3.3 Mineralogical Analysis

The XRD patterns of the control and blended mixtures cured under autoclave conditions are shown in Fig. 9. For the M1 mixture, a crystalline peak at 29.5°–31° (2θ) corresponding to the α-dicalcium silicate hydrate (α-C$_2$SH) was observed, as shown in Fig. 9a. This peak might overlap with the calcite peak that took place during the sample curing and preparation. Although the thermogravimetric analysis indicated the presence of CH (portlandite), no portlandite peaks were observed from the XRD of M1. The observed non-consistency in thermogravimetric and XRD results could be attributed to the phenomenon of preferential orientation exhibited by portlandite crystals during XRD (Wee et al. 2000). For M2, residual quartz peaks at 2θ° = 21° and 26.5° were detected, but no portlandite peaks were observed, as shown in Fig. 9b. These residual quartz peaks (1890 count) indicated the presence of excess crystalline silica in the M2 mixture. For the M3, M4 and M5 mixtures, the incorporation of slag as a ternary blended element did not consume the remaining crystalline silica, as shown in Fig. 9c–e. The quartz peak intensities for the M3, M4 and M5 mixtures at 26.5° were 1900, 1880 and 1910 counts, respectively. The results indicated that the reactivity of slag is greater than that of GDS (Kondo et al. 1975). Consequently, the CaO provided from the slag reacted rapidly with the SiO$_2$ and Al$_2$O$_3$ generated from the slag itself, resulting in there being no free CaO to combine with the un-reacted crystalline silica of GDS. Therefore, to utilize the excess crystalline silica, materials rich in CaO should be added.

3.4 Compressive Strength

The compressive strengths of the control and blended mixtures (M1–M5) cured under autoclave conditions are shown in Fig. 10. The compressive strength results presented are an average of three samples. The average of the strength

values of the three samples is within the limits of 5 %. As can be observed from Fig. 10, all the blended mixtures yielded compressive strengths higher than that of the control (M1). The low compressive strength of M1 is attributed to the formation of lime-rich alpha dicalcium silicate hydrate (α-C$_2$SH) (Bakharev et al. 1999; Kjellsen et al. 1991). The microstructure analysis of M1 showed the formation of fibrous-like hydrated products and loose structures. These features explained the negative effect of autoclave curing on the compressive strength performance of the control mixture. The inclusion of 40 % GDS enhanced the compressive strength of M2 by approximately 16 % compared to that of M1. The enhancement in the compressive strength could be ascribed to the fact that, under autoclave curing, the SiO$_2$ of GDS reacted with the calcium hydroxide (CH) liberated from the hydration of the cement to form additional CSH phases, such as tobermorite (C$_5$S$_6$H$_5$). The formation of tobermorite filled the pores and enhanced the compactness of the hydrated mixture. It has been reported that tobermorite is associated with high strength and lower permeability features (Eilers et al. 1983, Jupe et al. 2008). The inclusion of slag as a further PC replacement (M3, M4 and M5) maintained the compressive strength between 66 (M1) and 72 MPa (M2). The slight decrease in the compressive strength in comparison with M2 could be attributed to the formation of hydrogarnet (C$_4$ASH$_4$) phases instead of tobermorite (C$_5$S$_6$H$_5$). It has been stated that the performance of the mechanical properties of hydrogarnet is lower than that of tobermorite (Kondo et al. 1975). In general, the slag-bended mixtures revealed a dense, compacted and closed network of thin plate-like structures. Thin plate-like structures were formed due to the presence of the element aluminum in the system. The compressive strength result indicated that a combination of 40 % GDS and 45 % slag could be used to reduce the total weight of PC by up to 85 % and still achieve compressive strength performance comparable to that of the control mixture (M1). It should be mentioned that a mixture containing 40 % GDS and 60 % slag (no cement) was not cast in this study. The literature review showed that the use of cement-less mixtures had resulted in cracking during autoclave curing (Shi and Hu 2003). The cracking was due to the lack of strength development, and the specimens were not strong enough to withstand the high temperature and pressure during the autoclaving process.

Figure 11 presents the relationship between the compressive strength and molar CaO/SiO$_2$ ratio of the raw binder materials. The molar CaO/SiO$_2$ ratios of the control and blended mixtures are presented in Table 2. It was found that there is a good correlation between the compressive strength and molar CaO/SiO$_2$ ratio. The compressive strength decreases almost linearly as the CaO/SiO$_2$ ratio decreases ($R^2 = 0.99$). The literature review demonstrated that the optimum CaO/SiO$_2$ ratio for autoclave curing to achieve concrete strength is around unity (Alhozaimy et al. 2012; Eilers et al. 1983; Taylor 1997). In this study, the highest compressive strength achieved was 72 MPa (M2-AC) with a CaO/SiO$_2$ ratio of 0.74. Clearly, this value for the CaO/SiO$_2$ ratio is lower than that found in the literature. The incorporation of slag as a further PC replacement caused a further reduction in the CaO/SiO$_2$ ratio due to the low content of CaO in the slag compared to that of PC (Table 1). Therefore, to increase the CaO/SiO$_2$ ratio, materials rich in CaO content should be added.

4. Conclusions

This paper studied the effect of autoclave curing on the microstructure properties of mixtures incorporating GDS and ground granulated blast furnace slag as a PC replacement material. The following conclusions can be drawn:

1. The SEM and EDX results indicated that the incorporation of GDS and slag produces denser and more homogeneous microstructures. The SEM showed the clear formation of crystalline plate-like tobermorite and hydrogarnet in the blended mixture cured under autoclave conditions. In addition, the EDX indicated that the CSH products formed had C/S approximately equal to unity.
2. The DTA and TGA confirmed the consumption of CH and clear formation of an exothermic peak at 850 °C. The intensity of exothermic peaks at 850 °C of the PC-GDS mixture (M2) was higher than those of the PC-GDS-slag mixtures (M3, M4, M5).
3. The XRD indicated the presence of residual crystalline silica of the PC-GDS mixture (M2), whereas the CH peaks disappeared. The incorporation of slag did not affect the peak intensities of residual crystalline silica.
4. The incorporation of GDS significantly enhanced the compressive strength of the autoclave-cured mixture, whereas the inclusion of slag maintained the compressive strength to be higher or comparable to that of the control mixture.
5. A good correlation was found between the compressive strength and the CaO/SiO$_2$ ratio. The compressive strength decreased linearly as the CaO/SiO$_2$ ratio decreased. The authors recommend using materials rich in CaO to increase the CaO/SiO$_2$ ratio and to utilize the excess SiO$_2$ of GDS in the pozzolanic reaction.

Acknowledgments

This study is part of a joint research project between King Saud University, Saudi Arabia and Universiti Putra Malaysia titled "Development of local sand as a cementitious material for high-performance concrete". The funding of this work by King Saud University is gratefully acknowledged.

References

Alarcon-Ruiz, L., Platret, G., Massieu, E., & Ehrlacher, A. (2005). The use of thermal analysis in assessing the effect of temperature on a cement paste. *Cement and Concrete Research, 35*(3), 609–613.

Alawad, O., Alhozaimy, A., Jaafar, M., Al-Negheimish, A., & Aziz, F. (2014). Microstructure analyses of autoclaved ground dune sand–Portland cement paste. *Construction and Building Materials, 65*, 14–19.

Alawad, O. A., Alhoziamy, A., Jaafar, M. S., Aziz, A., Noor, F., & Al-Negheimish, A. (2015). Blended cement containing high volume ground dune sand and ground granulated blast furnace slag for autoclave concrete industry. *Applied Mechanics and Materials, 754–755*(1), 395–399.

Alhozaimy, A., Al-Negheimish, A., Alawad, O., Jaafar, M., & Noorzaei, J. (2012). Binary and ternary effects of ground dune sand and blast furnace slag on the compressive strength of mortar. *Cement & Concrete Composites, 34*(6), 734–738.

Assarsson, G. O., & Rydberg, E. (1956). Hydrothermal reactions between calcium hydroxide and amorphous silica. *The Journal of Physical Chemistry, 60*(4), 397–404.

Bakharev, T., Sanjayan, J., & Cheng, Y.-B. (1999). Effect of elevated temperature curing on properties of alkali-activated slag concrete. *Cement and Concrete Research, 29*(10), 1619–1625.

Berardi, M. C., Chiocchio, G., & Collepardi, M. (1975). The influence of precuring on the autoclave hydration of quartz-tricalcium silicate mixtures. *Cement and Concrete Research, 5*(5), 481–487.

Bresson, B., Meducin, F., Zanni, H., & Noik, C. (2002). Hydration of tricalcium silicate (C3S) at high temperature and high pressure. *Journal of materials science, 37*(24), 5355–5365.

Chae, S. R., Moon, J., Yoon, S., Bae, S., Levitz, P., Winarski, R., & Monteiro, P. J. (2013). Advanced nanoscale characterization of cement based materials using X-ray synchrotron radiation: a review. *International Journal of Concrete Structures and Materials, 7*(2), 95–110.

Divsholi, B. S., Lim, T. Y. D., & Teng, S. (2014). Durability properties and microstructure of ground granulated blast furnace slag cement concrete. *International Journal of Concrete Structures and Materials, 8*(2), 157–164.

Eilers, L. H., Nelson, E. B., & Moran, L. K. (1983). High-temperature cement compositions-pectolite, scawtite, truscottite, or xonotlite: Which do you want? *Journal of Petroleum Technology, 35*(7), 1373–1377.

Englehardt, J. D., & Peng, C. (1995). Pozzolanic filtration/solidification of radionuclides in nuclear reactor cooling water. *Waste Management, 15*(8), 585–592.

Erdoğdu, Ş., & Kurbetci, Ş. (2005). Influence of cement composition on the early age flexural strength of heat-treated mortar prisms. *Cement & Concrete Composites, 27*(7), 818–822.

Grabowski, E., & Gillott, J. (1989). Effect of replacement of silica flour with silica fume on engineering properties of oilwell cements at normal and elevated temperatures and pressures. *Cement and Concrete Research, 19*(3), 333–344.

Gutteridge, W. A., & Dalziel, J. A. (1990). Filler cement: the effect of the secondary component on the hydration of Portland cement: part I. A fine non-hydraulic filler. *Cement and Concrete Research, 20*(5), 778–782.

Hanson, J. (1963). Optimum steam curing procedure in precasting plants. *ACI Journal Proceedings, 60*(1), 75–100.

Hewlett, P. (2003). *Lea's chemistry of cement and concrete.* Oxford, UK: Butterworth-Heinemann.

Hope, B. B. (1981). Autoclaved concrete containing flyash. *Cement and Concrete Research, 11*(2), 227–233.

Jupe, A. C., Wilkinson, A. P., Luke, K., & Funkhouser, G. P. (2008). Class H cement hydration at 180 °C and high pressure in the presence of added silica. *Cement and Concrete Research, 38*(5), 660–666.

Kalousek, G. L. (1954). Studies on the cementious phases of autoclaved concrete products made of different raw materials. *ACI Journal Proceedings, 50*(1), 365–378.

Kar, A., Ray, I., Halabe, U. B., Unnikrishnan, A., & Dawson-Andoh, B. (2014). Characterizations and quantitative estimation of alkali-activated binder paste from microstructures. *International Journal of Concrete Structures and Materials, 8*(3), 213–228.

Kjellsen, K. O., Detwiler, R. J., & Gjørv, O. E. (1991). Development of microstructures in plain cement pastes hydrated at different temperatures. *Cement and Concrete Research, 21*(1), 179–189.

Klimesch, D. S., & Ray, A. (1998). Hydrogarnet formation during autoclaving at 180 °C in unstirred metakaolin-lime-quartz slurries. *Cement and Concrete Research, 28*(8), 1109–1117.

Klimesch, D. S., Ray, A., & Sloane, B. (1996). Autoclaved cement-quartz pastes: the effects on chemical and physical properties when using ground quartz with different surface areas part I: quartz of wide particle size distribution. *Cement and Concrete Research, 26*(9), 1399–1408.

Kołakowski, K., De Preter, W., Van Gemert, D., Lamberts, L., & Van Rickstal, F. (1994). Low shrinkage cement based building components. *Cement and Concrete Research, 24*(4), 765–775.

Kondo, R., Abo-El-Enein, S. A., & Daimon, M. (1975). Kinetics and mechanisms of hydrothermal reaction of granulated blast furnace slag. *Bulletin of the Chemical Society of Japan, 48*(1), 222–226.

Kyritsis, K., Meller, N., & Hall, C. (2009). Chemistry and morphology of hydrogarnets formed in cement based CASH hydroceramics cured at 200 °C to 350 °C. *Journal of the American Ceramic Society, 92*(5), 1105–1111.

Lange, F., Mörtel, H., & Rudert, V. (1997). Dense packing of cement pastes and resulting consequences on mortar properties. *Cement and Concrete Research, 27*(10), 1481–1488.

Liu, B., Xie, Y., & Li, J. (2005). Influence of steam curing on the compressive strength of concrete containing supplementary cementing materials. *Cement and Concrete Research, 35*(5), 994–998.

Luke, K. (2004). Phase studies of pozzolanic stabilized calcium silicate hydrates at 180 °C. *Cement and Concrete Research, 34*(9), 1725–1732.

Mehta, P. K., & Monteiro, P. J. (2006). *Concrete: microstructure, properties, and materials.* New York, NY: The McGraw-Hill Companies Inc.

Menzel, C. A. (1934). Strength and volume change of steam-cured portland cement mortar and concrete. *ACI Journal Proceedings, 31*(11), 125–148.

Mindess, S., Young, J. F., & Darwin, D. (1981). *Concrete.* Englewood Cliffs: Prentice-Hall.

Mostafa, N. Y., Shaltout, A. A., Omar, H., & Abo-El-Enein, S. A. (2009). Hydrothermal synthesis and characterization of aluminium and sulfate substituted 1.1 nm tobermorites. *Journal of Alloys and Compounds, 467*(1), 332–337.

Murmu, M., & Singh, S. P. (2014). Hydration products, morphology and microstructure of activated slag cement. *International Journal of Concrete Structures and Materials, 8*(1), 61–68.

Neville, A. M. (1973). *Properties of concrete.* London, UK: Pitman.

Oner, A., & Akyuz, S. (2007). An experimental study on optimum usage of GGBS for the compressive strength of concrete. *Cement & Concrete Composites, 29*(6), 505–514.

Saikia, N., Kato, S., & Kojima, T. (2006). Thermogravimetric investigation on the chloride binding behaviour of MK–lime paste. *Thermochimica Acta, 444*(1), 16–25.

Sanders, L. D., & Smothers, W. J. (1957). Effect of tobermorite on the mechanical strength of autoclaved portland cement-silica mixtures*. *ACI Journal Proceedings, 54*(8), 127–139.

Shi, C., & Hu, S. (2003). Cementitious properties of ladle slag fines under autoclave curing conditions. *Cement and Concrete Research, 33*(11), 1851–1856.

Singh, L. P., Goel, A., Bhattachharyya, S. K., Ahalawat, S., Sharma, U., & Mishra, G. (2015). Effect of morphology and dispersibility of silica nanoparticles on the mechanical behaviour of cement mortar. *International Journal of Concrete Structures and Materials, 9*(2), 207–217.

Taylor, H. F. W. (1997). *Cement chemistry*. London, UK: Telford Services Ltd.

Topçu, İ. B., & Uygunoğlu, T. (2007). Properties of autoclaved lightweight aggregate concrete. *Building and Environment, 42*(12), 4108–4116.

Wee, T. H., Suryavanshi, A. K., & Tin, S. S. (2000). Evaluation of rapid chloride permeability test (RCPT) results for concrete containing mineral admixtures. *ACI Materials Journal, 97*(2), 221–232.

Wongkeo, W., Thongsanitgarn, P., & Chaipanich, A. (2012). Compressive strength and drying shrinkage of fly ash-bottom ash-silica fume multi-blended cement mortars. *Materials and Design, 36*, 655–662.

Yang, Q., Zhang, S., Huang, S., & He, Y. (2000). Effect of ground quartz sand on properties of high-strength concrete in the steam-autoclaved curing. *Cement and Concrete Research, 30*(12), 1993–1998.

Yazici, H. (2007). The effect of curing conditions on compressive strength of ultra high strength concrete with high volume mineral admixtures. *Building and Environment, 42*(5), 2083–2089.

Yazıcı, H., Yiğiter, H., Karabulut, A. Ş., & Baradan, B. (2008). Utilization of fly ash and ground granulated blast furnace slag as an alternative silica source in reactive powder concrete. *Fuel, 87*(12), 2401–2407.

Permissions

All chapters in this book were first published in IJCSM, by Springer; hereby published with permission under the Creative Commons Attribution License or equivalent. Every chapter published in this book has been scrutinized by our experts. Their significance has been extensively debated. The topics covered herein carry significant findings which will fuel the growth of the discipline. They may even be implemented as practical applications or may be referred to as a beginning point for another development.

The contributors of this book come from diverse backgrounds, making this book a truly international effort. This book will bring forth new frontiers with its revolutionizing research information and detailed analysis of the nascent developments around the world.

We would like to thank all the contributing authors for lending their expertise to make the book truly unique. They have played a crucial role in the development of this book. Without their invaluable contributions this book wouldn't have been possible. They have made vital efforts to compile up to date information on the varied aspects of this subject to make this book a valuable addition to the collection of many professionals and students.

This book was conceptualized with the vision of imparting up-to-date information and advanced data in this field. To ensure the same, a matchless editorial board was set up. Every individual on the board went through rigorous rounds of assessment to prove their worth. After which they invested a large part of their time researching and compiling the most relevant data for our readers.

The editorial board has been involved in producing this book since its inception. They have spent rigorous hours researching and exploring the diverse topics which have resulted in the successful publishing of this book. They have passed on their knowledge of decades through this book. To expedite this challenging task, the publisher supported the team at every step. A small team of assistant editors was also appointed to further simplify the editing procedure and attain best results for the readers.

Apart from the editorial board, the designing team has also invested a significant amount of their time in understanding the subject and creating the most relevant covers. They scrutinized every image to scout for the most suitable representation of the subject and create an appropriate cover for the book.

The publishing team has been an ardent support to the editorial, designing and production team. Their endless efforts to recruit the best for this project, has resulted in the accomplishment of this book. They are a veteran in the field of academics and their pool of knowledge is as vast as their experience in printing. Their expertise and guidance has proved useful at every step. Their uncompromising quality standards have made this book an exceptional effort. Their encouragement from time to time has been an inspiration for everyone.

The publisher and the editorial board hope that this book will prove to be a valuable piece of knowledge for researchers, students, practitioners and scholars across the globe.

List of Contributors

Keyvan Ramin
Structural & Mechanical Department, Advance Researches & Innovations, Aisan Disman Consulting Engineers Inc., 6718783559 Kermanshah, Iran

Mitra Fereidoonfar
Department of Structural Engineering, Asians Disman Consulting Engineers, Kermanshah, Iran

S. Y. Seo and S. M. Jeon
Department of Architectural Engineering, Korea National University of Transportation, Chungju, Korea

K. T. Kim
B&K Construction Technology, Chungju, Korea

M. Kuroki and K. Kikuchi
Department of Architecture and Mechatronics, Oita University, Oita, Japan

Pratanu Ghosh and Quang Tran
Department of Civil and Environmental Engineering, California State University, Fullerton, Fullerton, USA

Wei Ren
Key Laboratory of Bridge Inspection and Reinforcement Technology of China Ministry of Communications, Chang'an University, Xi'an 710064, Shaanxi, China

Lesley H. Sneed, Yang Yang and Ruili He
Department of Civil, Architectural & Environmental Engineering, Missouri University of Science and Technology, Rolla, MO 65409, USA

Lok Pratap Singh, Sriman Kumar Bhattachharyya, Saurabh Ahalawat, Usha Sharma and Geetika Mishra
CSIR-Central Building Research Institute, Roorkee 247667, India

Anjali Goel
Gurukul Kangri University, Haridwar 249404, India

Ashhad Imam
Department of Civil Engineering, King Fahd University of Petroleum & Minerals, Dhahran 31261, Saudi Arabia

Fatai Anifowose
Center for Petroleum and Minerals, Research Institute, King Fahd University of Petroleum & Minerals, Dhahran 31261, Saudi Arabia

Abul Kalam Azad
Department of Civil & Environmental Engineering, College of Engineering Sciences, King Fahd University of Petroleum & Minerals, Dhahran 31261, Saudi Arabia

Abdurrahmaan Lotfy
Lafarge Canada Inc., Toronto, ON, Canada

Khandaker M. A. Hossain and Mohamed Lachemi
Department of Civil Engineering, Ryerson University, Toronto, ON, Canada

M. Morga
Mobility Department – Transportation Infrastructures Technologies, AIT Austrian Institute of Technology, Vienna, Austria

G. C. Marano
Department of Civil Engineering and Architecture, Technical University of Bari, Bari, Italy

Aref Mohamad al-Swaidani
Faculty of Architectural Engineering, Arab International University (AIU), Damascus, Syria

Samira Dib Aliyan
Syrian Arab Organization for Standardization and Metrology (SASMO), Damascus, Syria

Wei Ren
Key Laboratory of Bridge Detection Reinforcement Technology Ministry of Communications, Chang'an University, Xi'an 710075, Shaanxi, China

Lesley H. Sneed and Xin Kang
Department of Civil, Architectural & Environmental Engineering, Missouri University of Science and Technology, Rolla, MO 65409, USA

Yiting Gai
CCCC First Highway Consultants CO., LTD, Xi'an 710075, Shaanxi, China

Dali Bondar
Department of Civil, Architectural, and Building, Coventry University, Coventry CV1 5BF, UK

C. J. Lynsdale
Department of Civil and Structural Engineering, University of Sheffield, Sheffield S1 3JD, UK

N. B. Milestone
Milestone and Associates Ltd, Lower Hutt 5010, New Zealand

N. Hassani
Research Centre of Natural Disasters in Industry, P.W.U.T., Tehran, Iran

Tae-Seok Seo
Hyundai Engineering and Construction Co., Ltd., Yongin 446-716, Korea

Moon-Sung Lee
Division of Architecture and Architectural Engineering, Hanyang University, Ansan 426-791, Korea

A. A. Akindahunsi and H. C. Uzoegbo
University of the Witwatersrand (WITS), Johannesburg 2050, Gauteng, South Africa

Wael Kassem
Engineering at Al-Qunfudah, Umm Al-Qura University, P.O. BOX 288, Al-Qunfudah 21912, Saudi Arabia

Omer Abdalla Alawad
Civil Engineering Department and Center of Excellence for Concrete Research and Testing, College of Engineering, King Saud University, Riyadh, Saudi Arabia
Civil Engineering Department, Faculty of Engineering, Universiti Putra Malaysia, Seri Kembangan, Malaysia

Abdulrahman Alhozaimy and Abdulaziz Al-Negheimish
Civil Engineering Department and Center of Excellence for Concrete Research and Testing, College of Engineering, King Saud University, Riyadh, Saudi Arabia

Mohd Saleh Jaafar and Farah Nora Abdul Aziz
Civil Engineering Department, Faculty of Engineering, Universiti Putra Malaysia, Seri Kembangan, Malaysia